P9-DWG-592
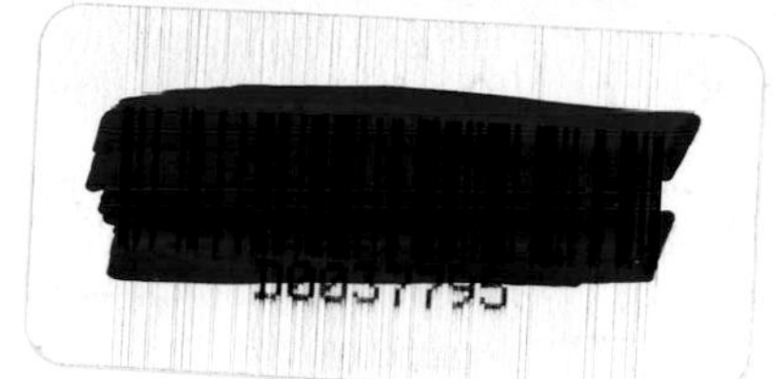

Significance of Medical Microbiology in the Care of Patients

Significance of Medical Microbiology in the Care of Patients

VICTOR LORIAN, M.D., Editor

Chairman, Division of Microbiology and Epidemiology,
Department of Pathology,
The Bronx-Lebanon Hospital Center, Bronx, New York
Professor of Laboratory Medicine
Albert Einstein College of Medicine
Bronx, New York

The Williams & Wilkins Co.
Baltimore

Made in the United States of America

Library of Congress Cataloging in Publication Data

Main entry under title:

Significance of medical microbiology in the care of patients.

Bibliography: p.
Includes index.
1. Medical microbiology. I. Lorian, Victor. [DNLM: 1. Microbiology. QW4 S578]
QR46.S55 616.01 76-21297

683-05164-4

Composed and printed at the
Waverly Press, Inc.
Mt. Royal and Guilford Aves.
Baltimore, Md. 21202, U.S.A.

Preface

When we step into a jet we feel assured that the flying machine is in perfect order, that the pilot is a man of excellence, that the crew will take care of us during the flight, and that our lives will be safe. Patients who step into the hospital to be cured of their illnesses and who put their bodies and lives into the hands of the hospital's staff should feel the same kind of confidence. The microbiology laboratory certainly contributes to the security in health care that most of our hospitals provide. The microbiologist who meets the needs of the clinician with a prompt contribution toward diagnosis and treatment deserves recognition as an indispensable and fundamental element in the complex organization of patient care.

During recent years bacterial taxonomy had to be revised repeatedly because advancing technology has enabled the microbiologist to distinguish between closely related species which were previously lumped together. A multitude of antibacterial agents, together with techniques developed for determining antibiotic susceptibility, have become available. Microbiologists' varying concepts about what efforts will gain maximal information continues to generate a great deal of discussion on exactly where to set the limits of taxonomic investigation and antibiotic sensitivity, how detailed, accurate and fast a method need be, and where to draw the line between research interest and patient care. Indeed, technological progress will soon make rapid yet accurate results a reality and learned controversy will be directed toward other interests.

This volume attempts to furnish guidelines, cut-offs, and detailed descriptions about how the needs of patients and physicians should best be filled in the hospital microbiology laboratory. Most of the chapters deal with what, rather than how, things should be done. Of utmost interest in many chapters is the question of identification of species and their susceptibility to antibacterial agents—the main endeavors of a clinical microbiology laboratory. In many instances definite, strong statements are made, some of which contrast sharply with traditional practices. There are wide differences in opinion about both the various areas of activity and interpretation in light of limited budgets. There are, therefore, no simple solutions; only intelligent choices. We have to assess our immediate as well as our long range priorities.

I have brought together experts in various areas of microbiology and infectious diseases who wrote their chapters out of long personal experience and encyclopedic knowledge of their specialized fields. Part of the material in this book was presented at my 8th and 9th annual courses in "Recent Advances in Medical Microbiology" held in New York City in 1975 and 1976.

I wish to extend my gratitude to the contributors for taking time from their many activities and for forging their experience into conclusions and recommendations. Their cooperation made the task of editing a real pleasure. I am sure their efforts will prove to be invaluable to all those concerned with the practice and teaching of microbiology.

Special recognition must go to Dr. Paul Ellner and to Mr. Arthur Press who enthusiastically urged me to undertake the publication of this volume. My gratitude also goes to my Chief and friend, Dr. Leopold Reiner, and to Dr. Seymour Cohen for their cooperation in organizing my courses and for creating conditions that allowed me to concentrate on the many endeavors required for the editing of this volume.

I would also like to thank Sara A. Finnegan whose expertise and kindness made the process of publishing at The Williams & Wilkins Co. an enjoyable experience.

VICTOR LORIAN

New York, 1977

Contributors

Donald Armstrong, M.D., Chief, Infectious Disease Service, Director, Microbiology Laboratory, Memorial Sloan-Kettering Cancer Center, Professor of Medicine, Cornell University Medical College, New York, New York

Barbara Atkinson, B.S. MT (ASCP), The Bronx-Lebanon Hospital Center, Bronx, New York

Raymond C. Bartlett, M.D., Director, Division of Microbiology, Department of Pathology, Hartford Hospital, Hartford, Connecticut

Charlotte C. Campbell, B.S., Sc.D., Professor and Chairman, Department of Medical Sciences, Southern Illinois University School of Medicine, Springfield, Illinois

R. Gordon Douglas, Jr., M.D., Professor of Medicine and Microbiology Head, Infectious Disease Unit, University of Rochester, School of Medicine, Rochester, New York

Paul D. Ellner, Ph.D., Professor of Microbiology, Columbia University College of Physicians and Surgeons, New York, New York, Director, Clinical Microbiology Service, Presbyterian Hospital, New York, New York

G. L. Gilardi, Ph.D., Director, Microbiology Division, Department of Laboratories, Hospital for Joint Diseases and Medical Center, New York, New York

Caroline Breese Hall, M.D., Assistant Professor of Pediatrics and Medicine, in Infectious Disease, University of Rochester School of Medicine, Rochester, New York

Henry D. Isenberg, Ph.D., Attending Microbiologist, Long Island Jewish-Hillside Medical Center, New Hyde Park, New York, Professor of Clinical Pathology, State University of New York at Stony Brook, Stony Brook, New York

Lawrence J. Kunz, Ph.D., Associate Professor, Department of Microbiology and Molecular Genetics, Harvard Medical School, Director, Francis Blake Bacteriology Laboratories, Massachusetts General Hospital, Boston, Massachusetts

Bennett Lorber, M.D., Assistant Professor of Medicine, Microbiology and Immunology, Temple University Health Sciences Center, Philadelphia, Pennsylvania

Victor Lorian, M.D., Chairman, Division of Microbiology and Epidemiology, The Bronx-Lebanon Hospital Center, New York, New York, Professor of Laboratory Medicine, Albert Einstein College of Medicine, New York, New York

Donald B. Louria, M.D., Professor and Chairman, Department of Preventive Medicine and Community Health, New Jersey College of Medicine and Dentistry, Newark, New Jersey

Sarabelle Madoff, Head, Mycoplasma Laboratory, Departments of Medicine and Bacteriology, Massachusetts General Hospital, Boston, Massachusetts

George J. Moore, President, Bac-Data, Medical Information Systems, Inc., Clifton, New Jersey

William R. McCabe, M.D., Professor of Medicine and Microbiology, Boston University School of Medicine, Director, Microbiology Laboratory, and Division of Infectious Diseases, University Hospital, Boston University School of Medicine, Boston, Massachusetts

Erwin Neter, M.D., Professor of Microbiology, State University of New York, Director of Bacteriology, Children's Hospital, Buffalo, New York

Willy N. Pachas, M.D., Clinical Associate in Medicine, Department of Medicine, Massachusetts General Hospital, Boston, Massachusetts

L. D. Sabath, M.D., Professor of Medicine, Head, Section of Infectious Disease, University of Minnesota, Minneapolis, Minnesota

Christine C. Sanders, Ph.D., Assistant Professor, Department of Medical Microbiology, Creighton University School of Medicine, Omaha, Nebraska

W. Eugene Sanders, Jr., M.D., Professor and Chairman, Department of Medical Microbiology, Professor of Medicine, Creigton University School of Medicine, Omaha, Nebraska

J. C. Sherris, M.D., Professor and Chairman, Department of Microbiology and Immunology, University of Washington, Seattle, Washington

Robert M. Swenson, M.D., Associate Professor of Medicine, Microbiology and Immunology, Head, Section of Infectious Diseases, Temple University Health Sciences Center, Philadelphia, Pennsylvania

John A. Washington II, M.D., Head, Section of Clinical Microbiology, Mayo Clinic and Mayo Foundation, Rochester, Minnesota

Emanuel Wolinsky, M.D., Professor of Medicine, Case Western Reserve University, School of Medicine, Director of Clinical Microbiology, Metropolitan General Hospital, Cleveland, Ohio

Contents

1

Conventional Methods, Speciation and the Computer

LAWRENCE J. KUNZ

At the outset it may be worthwhile to redefine the elements of the title of this presentation in order to set the limits of the discussion and to prevent unnecessary confusion or misunderstanding.

To begin with, the term "conventional methods" is intended to include the multi-test systems introduced just a few years ago for the identification of bacteria. After extensive evaluation of these so-called diagnostic kits by several groups of workers, one or another of the systems has been adopted by so many laboratories that they now ought to be considered as conventional methods. They include the API System (Analytab Products Inc.), Enterotube (Roche Diagnostics), Inolex (Wilson Pharmaceutical and Chemical Corp.), Minitek (Bioquest), Pathotec (General Diagnostics), and the R/B System (Corning Medical Diagnostics). In addition to aiding in identification of bacteria, three of these systems – API, Enterotube, and R/B – provide numbers, each consisting of 7 or fewer digits, for designating biotypes of the identified species. In general, these kits are useful in identifying Gram-negative rods, particularly the Enterobacteriaceae.

Secondly, for the purposes of discussing in this presentation the taxonomic level to which identification of certain microorganisms ought to be pursued, the term "speciation" will be inadequate. Indeed, a rationale will be presented to support the policy of extending the identification of certain organisms to subspecies level and, in certain instances, to infrasubspecific subdivisions, such as serotype, morphotype, phagetype, biotype, and other forms designated by similar descriptive terms. The use of infrasubspecific forms is described in the International Code of Nomenclature of Bacteria (1).

Finally, the term "computer" in this presentation refers to the computer system of the bacteriology and chemistry laboratories of the Massachusetts General Hospital (7). This system is an on-line system, dedicated to bacteriology and chemistry laboratory applications. The transmission of results of laboratory determinations to patient locations is accomplished through the computer system which in the normal course of operations accumulates and stores for future retrieval and analysis enormous amounts of laboratory data. Some applications of the computer system and of its data bank to the promotion of the objectives of diagnostic bacteriology will be described subsequently.

In order to provide some dimensional context for the data that will be presented, it might be worthwhile to refer to certain statistical information in Table 1.1 concerning the Massachusetts General Hospital and its Bacteriology Laboratory. The total number of beds, between 1000 and 1100, has not changed substantially in the last decade or so, although there has been an increase in facilities providing for the intensive care of critically ill patients. The average number of specimens per month has surprisingly and inexplicably remained rather constant during the years 1973–1975, with unaccountable wide fluctuations from month to month.

Table 1.2 shows the frequency of occurrence of various groups of microorganisms isolated during a single month. These groups are numerically ranked in essentially the same order at the present time. Noteworthy is the fact that the Enterobacteriaceae account for a third of the total number of organisms identified. Understandably, this is the group of bacteria for which the diagnostic kits were designed.

CONVENTIONAL METHODS OF IDENTIFICATION

The same consideration of their frequency of occurrence made for many bacteriologists the resolution of the uncertain status of the family Enterobacteriaceae most welcome. The clarification of the taxonomy of the family by Ewing and Edwards in 1960 (2) not only provoked lively interest but also

TABLE 1.1
Bacteriology Laboratory: The Massachusetts General Hospital

	July 1968	October 1974
Number of beds	1,065	1,098
Number of specimens	9,521	15,618
Microorganisms identified	7,577	8,369
Number of designations used	75	69

TABLE 1.2
Massachusetts General Hospital—July 1968—Type and frequency of isolates

Isolate	Frequency
Enterobacteriaceae	2,468
Staphylococci	1,551
Streptococci	1,191
Non-fermentative G-rods	667
Yeasts and other fungi	596
Other "pyogens"	503
"Minor" G+ bacteria	473
Anaerobic bacteria	112
Mycobacteria	16
Total	7,577

stimulated the design of new methods for identification. Figure 1.1 outlines the method developed at the Massachusetts General Hospital. It is based on the use of Kligler's and Lysine Iron Agars and depends on several selected secondary series of biochemical tests for identification of certain members of the family not readily identified by primary differential media, and for speciation of others (5). Table 1.3 indicates the level to which identifications were made in July 1963 when the schema was first adopted. Note the relative paucity of binomials, denoting failure to identify organisms to species level in many instances.

In order to encounter more occasions of successful species identification it was obvious that more biochemical reactions were required. Accordingly, about 2 years ago we adopted the API System for identification of Gram-negative rods. It consists of a strip of 20 plastic test modules containing substrates for 22 biochemical reactions.

Three factors entered into the decision to change methods:

1. Better identification of bacteria through use of more biochemical tests, as already noted.
2. Elimination of rehandling of subcultures of those organisms that required secondary series of biochemical tests.
3. Capability of generating biotype numbers for use in epidemiological studies and for other purposes.

Figure 1.1

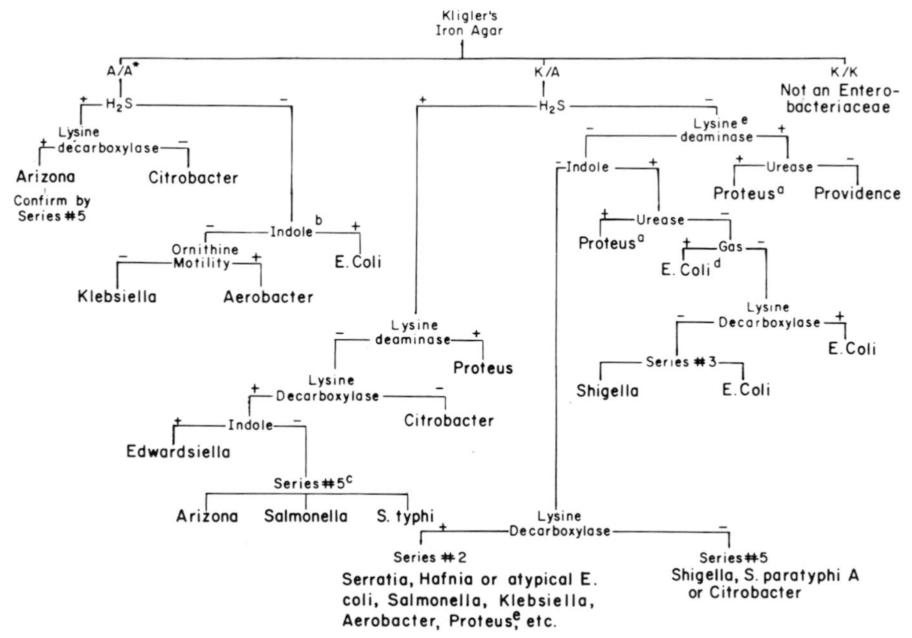

Schema for differentiation and identification of Enterobacteriaceae. *A = acid; K = alkaline; letter over oblique line = reaction on slant; letter under line = reaction in butt. Superscripts denote steps in procedure where additional biochemical tests are required for better differentiation and identification. (Adapted from J. G. Johnson, L. J. Kunz, W. Barron, and W. H. Ewing: *Applied Microbiology*, *14*:212–217, 1966, courtesy American Society for Microbiology, Washington, D. C.)

TABLE 1.3
Frequency of Isolation of Enterobacteriaceae from urine specimens—July 1963

Isolate	Frequency
Shigella	0
Escherichia coli	307
Salmonella	0
Arizona group	0
Citrobacter	14
Klebsiella	157
Enterobacter	24
Serratia	6
Proteus vulgaris	1
Proteus mirabilis	119
Proteus morganii	21
Proteus rettgeri	40
Providence group	14

It was apparent that the average cost of media required for identification of each isolate was somewhat greater for the 20-test system than for that then employed for incomplete identification. It was also obvious that use of the 20-test system effected savings in personnel costs through elimination of additional handling of subcultures in the case of organisms requiring secondary series of biochemical tests by our existing method.

Consideration of the respective expenses for media and for personnel led to the conclusion that it was possible to achieve excellent identification of a large proportion of bacteria encountered in routine diagnostic work at little or no additional cost over that being expended for incomplete, tentative, or otherwise unsatisfactory identifications. Moreover, the added benefits of biotyping could be achieved at no additional cost during routine identification procedures and without the need for special techniques, expertise, or libraries of reagents such as are required for phagetyping, serotyping, and other specialized infrasubspecific subdividing of organisms.

In recent years there has been considerable discussion about the relative importance of the actual cost of identifying microorganisms cultured from specimens submitted for bacteriologic examination. Two aspects of the subject have attracted particular attention and have been given special emphasis in conferences and discussions in the literature and elsewhere. The first concerns the needless expending of resources of the laboratory on specimens of doubtful value either because they are replicates of previously submitted specimens or because they are of poor quality for one reason or another. There is no argument with the futility of such testing and this matter will not be dealt with in this presentation. The second concerns questioning of the clinical relevance of the various taxonomic levels to which identification of microorganisms may be pursued and of the accuracy and proficiency with which the identification should be made. This aspect of the matter will be discussed in the next portion of the presentation.

VALUE OF SPECIATION AND INFRASUBSPECIFIC IDENTIFICATION

As a preamble, it may be useful to reflect on the fact that a great many episodes of infection are cured either spontaneously or by specific or nonspecific therapy prescribed by a physician either knowingly or intuitively, but without proved identification of the etiologic agent. On the other hand, it would be reckless to ignore the fact that there are infectious diseases, the understanding and successful therapy of which require precise and accurate recognition of the offending agent.

Whereas illnesses amenable to therapy or needless of it usually tolerate deliberate, discriminitive neglect of diagnostic microbiology on the part of attending physicians, such discretion cannot be permitted to the laboratorian who, usually uninformed clinically, is in danger of erroneously extending neglect to the seriously ill patient. Because critically ill patients with infectious diseases are very often hospitalized, hospital bacteriologists are particularly vulnerable to judgemental errors because the policy of benign, discriminative neglect frequently requires interpretation and implementation.

While considerations such as this might be sufficient motivation for providing excellent yet practical bacteriologic service especially in a hospital-based laboratory, there are other compelling reasons for doing so. Following are some of the more obviously important reasons why speciation or infrasubspecific differentiation in clinical microbiology may be important. They will be illustrated whenever it appears necessary, interesting, or appropriate.

Usefulness of Specific and Infrasubspecific Identification

In Providing Actual Diagnosis. There are many obvious examples of diseases in which identification of the etiologic agent provides the diagnosis of the disease. Examples of species of bacteria associated with particular diseases are *Salmonella typhi* and typhoid fever, *Francisella tularensis* and tularemia, *Bacillus anthracis* and anthrax. Certain specific infectious diseases, such as brucellosis and leptospirosis, may be diagnosed through isolation and identification merely of the genus of the etiologic agent. It is arguable whether the physician makes a specific diagnosis which suggests the identity of the etiologic agent more frequently than the bacteriologist identifies the etiologic agent and thereby suggests to the physician the disease entity of the particular patient.

In the case of infections of certain body sites, recognition of the association of specific organisms with the disease process provides the physician not only with a clinical diagnosis but also with the identity and nature of the etiologic agent, basis for effective treatment, insight into epidemiologic and public health problems that might be involved, and with measures to be taken to anticipate or prevent sequellae. Those diseases already mentioned in which these factors are important are infrequently encountered in urban settings in our society in this decade. With similar implications, however, and much more frequent are streptococcal pharyngitis, gonorrhoeae, tuberculosis, shigellosis, and meningococcal menigitis in which the isolation and identification of the etiologic agent signifies the presence of a bacterial agent (rather than a virus, fungus, or protozoa) that can be effectively treated, and for which measures can be taken to anticipate and prevent unnecessary complications. The dangers of the imprecision or inadequacy of inaccurate or incomplete speciation in each of the cases cited can be appreciated for one or another of the factors (epidemiology, choice of therapy, etc.) of importance in the total treatment of the illness.

Microbiologic examination of specimens derived from body sites normally populated with saprophytic or commensal microorganisms requires particular attention of both the microbiologist and the patient's physician. The isolation, identification and enumeration by the laboratory of the organisms present, for instance, in a urine specimen can provide the physician with information on which to base judgments or decisions as to:

1. Whether the organisms present in the specimen represent normal flora or potential pathogens,
2. Whether or not to initiate antimicrobial therapy,
3. Whether or not to await results of antimicrobial susceptibility tests,
4. Whether to treat for primary, recurrent or relapsing infection.

Some of these factors will be dealt with in detail in subsequent sections of this chapter.

In Indicating Severity of Illness. Several examples may be given: Staphylococcal endocarditis in which *Staphylococcus aureus* usually causes acute endocarditis whereas *Staphylococcus epidermidis* is associated with subacute infection; salmonellosis in which *Salmonella choleraesuis* suggests the probability of hematogenous spread and localization; typhoid fever in which perforation of the bowel, localization, and protracted carrier state should be anticipated; and finally shigellosis due to *Shigella dysenteriae* in which neurotoxic effects may add to the severity of the disease.

In Suggesting Effective Therapy. This is exemplified very well in group D streptococci in which the killing of enterococcal species is known to require use of synergistic antimicrobial agents; *Streptococcus bovis*, the nonenterococcal group D human pathogen, is readily killed by penicillin alone. Many examples of the organism-antibacterial relationship are presented in other chapters. Their inclusion here would be redundant. As already noted, it is true that many infectious processes can be effectively treated without knowledge of the etiologic agent, either blindly or intuitively or after testing the antimicrobial susceptibility of the organism. Whenever such success has been achieved, further knowledge of the organism is very often superfluous. Until we can anticipate in which cases specific identification of the offending agent would be superfluous, we are probably committed, at least in the case of hospitalized patients, in continuing to accumulate stores of information on which to base future predictions and decisions.

In Analyzing the Nature of Chronic Infections. The use of infrasubspecific forms for this purpose was neatly demonstrated by Turck et al. (13) in studies of antibacterial therapy of a series of patients with chronic bacteriuria. Using species identification as well as serotyping of *Escherichia coli* in comparing strains of bacteria sequentially isolated from their subject patients (Fig. 1.2), Turck and co-workers were able to distinguish between relapse (infection with the same organism) and reinfection (with another organism). These studies enabled them to modify antibacterial therapy, based on the differentiation. The value of sero-, phage-, and pyocine typing and their general unavailability were emphasized by this group in 1966:

> "Unfortunately, it is not feasible at present for most routine diagnostic laboratories to incorporate serologic identification of *Esch. coli* into the usual evaluation of a patient with chronic urinary-tract infection. Furthermore, other technics, such as phage and pyocine typing of pseudomonas and serologic typing of Klebsiella-aerobacter organisms, will make the identification of Gram-negative pathogens even more sophisticated. Because the data presented here emphasize the importance of differentiating relapse from reinfection in patients with chronic bacteriuria, and thereby provide a rational basis for duration of therapy, a means whereby these technics are made more generally available to the practicing physician should be found" (13).

Information similar to that derived in this study through serotyping has now become more widely available to physicians who utilize laboratories in which biotyping is performed and reported. However, in order to make comparisons of biotypes of organisms derived from cultures in follow up studies with those of previous cultures, either original cultures may be saved for later comparison or biotyping must be done on a routine basis.

In Epidemiology, Particularly in Nosocomial Infections. The use of infrasubspecific identification (serotyping of salmonella, phagetyping of staphylococci, pyocine typing of *Pseudomonas aeruginosa*, etc.) in epide-

Figure 1.2

		RELAPSE		
Before Rx *Escherichia coli* 06: H1/12	→	During Rx No growth	→	After Rx *E. coli* 06: H1/12
		REINFECTION		
Before Rx *E. coli* 06: H1/12	→	During Rx No growth	→	After Rx *E. coli* 04: H5
		OR		OR
	→	*E. coli* 04: H5	→	*Proteus mirabilis*
		OR		
	→	*P. mirabilis*		

Differentiation of two patterns of recurrent infections of the urinary tract. (Adapted from M. Turck, K. N. Anderson, and R. G. Petersdorf: *New England Journal of Medicine*, *275*:70–73, 1966.)

miological studies has a long history and hardly needs documentation. In our own experience, serotyping of Salmonellae has been exceedingly helpful in elucidating the source and spread of three different types of salmonellosis in the hospital: (i) through tube feedings containing brewer's yeast contaminated with three serotypes of *Salmonella enteritidis* (6); (ii) a foodborne outbreak due to serotype Heidelberg (12); (iii) a protracted epidemic due to a diagnostic medium, carmine dye, contaminated with serotype Cubana (9).* A multistate epidemic due to several serotypes involving a contaminated food supplement containing brewer's yeast and other dietary ingredients also exemplified the usefulness of serotyping of Salmonella for epidemiologic purposes (10). There is no evidence yet that biotyping of Salmonella will prove to be as useful in epidemiologic studies as serotyping has been. In the case of other Enterobacteriaceae however, the application of biotyping would appear to provide a wider range of usefulness to microbiological and epidemiological investigators in many institutions.

In Preventive Medicine and Public Health. Recognition of particular genera, species or strains of bacteria as agents of particular disease states can be documented with numerous examples. Thus, *Mycobacterium marinum* is recognized as the agent of "swimming pool granulomata," certain serotypes of *Streptococcus pyogenes* as incitants of acute glomerulonephritis, *Leptospira canicola* as a human pathogen of canine origin, etc. Several examples already cited for other purposes are also applicable here. Public health measures, appropriate sanitary precautions and timely public warnings can be introduced when the presence of the disease and the corresponding etiologic agent are recognized and appreciated.

In Global Biology. The protracted controversy over the potential risk of creating through recombinant DNA a highly virulent microbe resistant to all available antimicrobial agents has stimulated many and varied discussions intended to assess the odds of occurrence, magnitude of pathogenicity, and extent of resistance of the putative monster. Evaluation is continuing on methods for monitoring, containment, treatment, and other aspects of the possible generation and spread of such an organism. Examination in our laboratory of the biochemical characteristics of the proposed DNA propagating factory for such experimentation, i.e., *Escherichia coli* strain K-12, has demonstrated that it is a biotype very infrequently seen in human clinical material examined bacteriologically in our laboratory. This rarely occurring API biotype—*E. coli* 5044510—could easily be monitored and its presence detected in studies relating to colonization, and consequent persistence, commensalism, symbiosis, saprophytism, parasitism, or lethalism, provided pertinent biochemical phenotypic expression were not altered.

In Study of Antibiotic Resistance. An increase in number of organisms resistant to a particular antibiotic or antibiotics may occur in a variety of ways regardless of the genetic basis or mechanisms of resistance entailed.

* Serotyping performed in the laboratory of Dr. Joseph Winters at N.Y. Salmonella Center.

In the case of a newly recognized resistant organism in a restricted environment such as a hospital, it is probably important to determine: (i) whether a single resistant strain of the organism has spread throughout a given location or from one location to others, (ii) whether many resistant strains of a given species have been selected, (iii) whether most or all species of a given genus, or (iv) whether several genera or groups of species have become resistant and consequently been selected by pressure of antibiotics in the environment. Such knowledge can be acquired by proper species identification or, in the case of the involvement of a single species, infrasubspecific typing; the latter may be necessary to determine whether a single resistant strain has been disseminated or whether many strains or types of the same species have emerged in the resistant state.

There are other circumstances and problems in medicine in which infrasubspecific subdivisions are useful in aiding the understanding of relationships of microorganisms in situations, events or phenomena. Much of our understanding of infectious disease has been based on recognition, differentiation, and identification of microorganisms in the context of particular problems. It is unlikely that, in the future, careful, precise, specific identification of microorganisms will be less rewarding than approximate, gross taxonomic designations.

THE COMPUTER

The laboratory computer system at the Massachusetts General Hospital effectively performs its primary function in expediting the transmission of test results from the respective laboratories to the patients' charts. The programs, designed and maintained by the hospital's Laboratory of Computer Science, permit the efficient entry of masses of data into the computer by means of a language of mnemonic code words. The storage system and programs for retrieval of the data permit the analysis of results of testing of 150,000–200,000 specimens processed in the Bacteriology Laboratory per year (7). Some of the secondary benefits derived from manipulation of this data base aid in the accuracy and precision with which bacterial identifications are made in the laboratory. Some uses of the computer for this purpose will be discussed here.

For reasons that are detailed elsewhere (8), we have chosen not to use the laboratory computer directly for identification of bacteria, although several groups of workers have employed the computer for this purpose (for example see (3), (4), (11)). However, the computer system does include programs that extend the usefulness of speciation and infrasubspecific subdivision of isolated bacteria for patient care, epidemiology, quality control of laboratory work, and research capability. This is accomplished through certain functions, tabulations, listings, and analysis of data that are printed either automatically daily or are retrieved upon special callout either daily, monthly, or at other periodic intervals. They include the following.

1. Daily listings of patients whose tests or cultures signify the presence of communicable diseases. Originally intended to assist in the

compliance with requirements for official reporting of diseases to health departments, these listings are useful in control of intrahospital infections.

2. Daily listings, by organism, of patients with positive blood cultures (Fig. 1.3). For each organism represented on the list, data are presented showing the number of patients with blood cultures positive for the same organism during the previous quarter of the year, during the present quarter to date, and during the present month to date. The usefulness of these types of data to detect actual or incipient epidemics of bacteremia is obvious.
3. Monthly lists of numbers of organisms isolated, according to type of specimen from which they were isolated. This listing provides suggestion of trends in frequency of various species and a kind of gross test of quality of performance of identifications.
4. Monthly lists of biotypable organisms according to several types of associated information. These include number and percentage of the various biotypes of each bacterial species; number and percentage of strains with each of the antibiotic susceptibility patterns associated with each biotype as well as the reciprocal of this information, i.e., a listing of the numbers of biotypes associated with each susceptibility pattern for each biotypable organism.
5. Histograms of sizes of zones of inhibition with numbers of strains for each principal organism for each antibiotic (Fig. 1.4). These histograms serve as guides to quality control of the test process and as indices of changes in susceptibility of the various species to different antibiotics. They are printed every month and for 6- and 12-month periods.
6. Monthly tabulation of percentages of susceptible strains of each major organism for all antibiotics tested. These data, derived from computer interfacing with the electronic zone analyzer, are compared with the previous half year's experience and are broken down by specimen sources, by certain hospital locations, and according to whether the cultures were derived from inpatients or outpatients.
7. Ad hoc lists of special searches for unusual organisms, unusual

Figure 1.3

```
STAPH AUREUS
( 001 THIS MONTH, 016 THIS QUARTER, 035 LAST QUARTER)
                                BM5    3584   08/23/74

STAPH EPIDERMIDIS
( 002 THIS MONTH, 041 THIS QUARTER, 093 LAST QUARTER)
                                MREC   4016   08/23/74
                    I           WB12   6832   08/29/74

STREPTOCOCCI PROBABLY ENTEROCOCCI
( 001 THIS MONTH, 010 THIS QUARTER, 014 LAST QUARTER)
                                PH6    7366   08/30/74
```

Portion of report of positive blood cultures printed 9/2/74 for epidemiological analysis. Patients' unit numbers and names have been partially obliterated to maintain confidentiality. (From J. E. Prier, J. T. Bartola, and H. Friedman (Editors): *Modern Methods in Medical Microbiology: Systems and Trends*, University Park Press, Baltimore, 1976.)

Figure 1.4

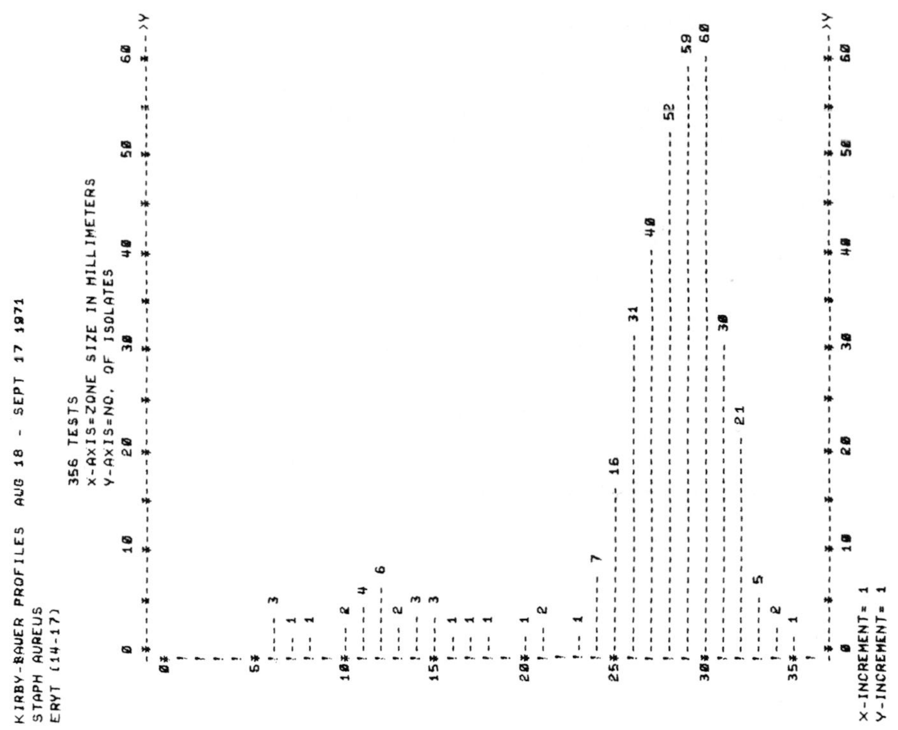

Computer-generated histogram of zone sizes occurring in tests of susceptibility of 356 strains of *Staphylococcus aureus* to erythromycin. (From J. E. Prier, J. T. Bartola, and H. Friedman (Editors): *Modern Methods in Medical Microbiology: Systems and Trends*, University Park Press, Baltimore, 1976.)

biotypes, and organisms or biotypes of special interest because of in-house, local, regional, or national interest, etc. Searches such as these can be instituted within a few hours and implemented on the same day as, for example, when there is suspicion that a particular organism or biotype may be involved in nosocomial infection or other episodes of epidemiologic interest. The search of computer files continues nightly and the results are printed out each morning for as long as need or interest indicates.

8. Printouts of instances of possible discrepancies between the species identification of certain organisms and the expected, acceptable patterns of susceptibility of the species to antibacterial agents. These printouts provide not only line-listings of real or potential discrepancies between the susceptibility pattern and respective identification but also the steps to be taken to test and, if necessary, confirm or change the apparently unsatisfactory, contradictory data (Fig. 1.5).

These programs provide insight into the relationships of etiology, antimicrobial susceptibility, epidemiology, and the quality of performance involved in the study of the complex biological environment that is constituted by the patients, personnel, and microorganisms present in a given

Figure 1.5

```
                        7451-24    MISC-   (ENTEROCOCCI--METH)

PROBLEM:  METHICILLIN SENSITIVE ENTEROCOCCI

   1. REPEAT SENSI                    STILL 14 OR MORE?    YES....NO....
        ZONE SIZE....

   2. IF 'YES' SET UP THE FOLLOWING:
        ARGININE                                POS....NEG....
        SUCROSE AGAR (GUMDROP COLONIES)         POS....NEG....
        STARCH HYDROLYSIS                       POS....NEG....

   3. IS ORGANISM S. BOVIS?                     YES....NO....

   4. IF NOT BOVIS REFER FOR GROUPING           GROUP.........
```

Example of computer-generated report of one type of apparent discrepancy between antibiotic susceptibility and identification of an organism. Procedure to be followed for resolving the apparent conflict is included in this printout. Patient's name and unit number have been partially obliterated to maintain confidentiality. (From J. E. Prier, J. T. Bartola, and H. Friedman (Editors): *Modern Methods in Medical Microbiology: Systems and Trends*, University Park Press, Baltimore, 1976.)

hospital. Because it can present enormous amounts of accumulated data in concise tables and figures for inspection and analysis (Fig. 1.6), the computer has the capability of assisting in our understanding of the relationships that influence the control or spread of infectious diseases among our

Figure 1.6

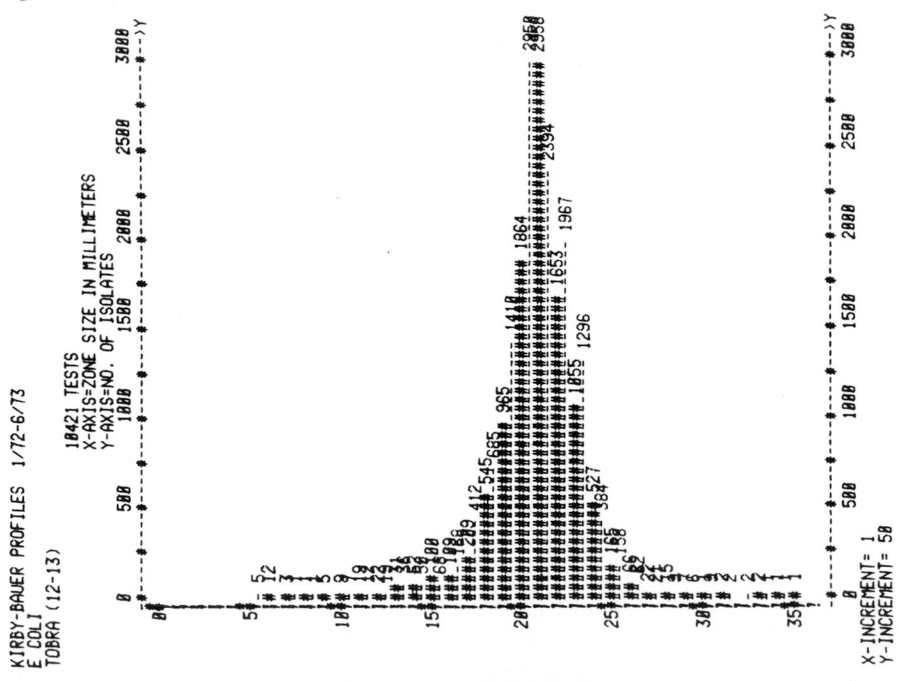

Comparison of distribution of zone sizes for gentamicin (vertical dashes) and for tobramycin (cross-hatched figures) against 10,421 strains of *Escherichia coli*. Computer-generated plots for organisms tested during an 18-month period are superimposed, slightly out of register. (From J. E. Prier, J. T. Bartola, and H. Friedman (Editors): *Modern Methods in Medical Microbiology: Systems and Trends*, University Park Press, Baltimore, 1976.)

patients. The decisions suggested by analysis of the data will depend to a large extent on the accuracy and precision of the data—in the case of microorgsnisms, on the accuracy of their identification and the confidence with which the identifications are made. Use of the computer emphasises the wisdom buried in the aphorism: "The value of the conclusion is proportional to the value of the premises."

SUMMARY AND CONCLUSIONS

1. Speciation or infrasubspecific identification is of immediate importance and in some instances may actually be necessary for precise diagnosis of certain diseases.
2. Epidemiological studies often depend on infrasubspecific identification for the demonstration of relatedness between microorganisms isolated from individual patients and sources of infection.
3. Analysis of detailed identification of organisms sequentially isolated during protracted infections may provide the basis for differentiating between relapse and recurrent infection and for prescribing therapy appropriate to the condition defined.
4. Collection, tabulation and analysis of accumulated data is prerequisite to the study and eventual understanding of certain types of infection, method of spread, treatment, or prevention.
5. Specific and infrasubspecific identification must constantly be performed, at least in certain settings, to provide for the continued advancement of knowledge of the various aspects of infectious disease; this is particularly applicable to the understanding of opportunistic infections of the human host with compromised defenses. The extent to which the laboratory of any particular institution should pursue the study and identification of bacteria will be influenced not only by the needs of the patients but also by the characteristics of the laboratory: the intellectual, physical, and financial resources available for performance of diagnostic bacteriology.
6. Finally, let a warning be raised of possibly the most serious consequences of the practice of discriminative neglect of diagnostic bacteriology. These are: the gradual erosion of professional pride and the slowing of the pursuit of excellence—intellectual and spiritual qualities that have been traditionally associated with microbiologists. Their loss would be profoundly felt.

LITERATURE CITED

1. Editorial Board of the Judicial Commission of the International Committee on Nomenclature of Bacteria (Ed.). International Code of Nomenclature of Bacteria. *Int J Syst Bacteriol 16:*459–490, 1966.
2. Ewing, W. H., and Edwards, P. R. The principal divisions and groups of Enterobacteriaceae and their differentiation. *Int Bull Bacteriol Nomencl Taxon 10:*1–12, 1960.
3. Friedman, R. B., Bruce, D., MacLowry, J., and Brenner, V. Computer-assisted identification of bacteria. *Am J Clin Pathol 60:*395–403, 1973.
4. Gavan, T. L., and Ma, C. A. Computer-assisted identification of Enterobacteriaceae. Abstract. Annual Meeting of the American Society for Microbiology, 1972.
5. Johnson, J. G., Kunz, L. J., Barron, W., and Ewing, W. H. Biochemical differentiation of the Enterobacteriaceae with the aid of Lysine-Iron-Agar. *Appl Microbiol 14:*212–217, 1966.

6. Kunz, L. J., and Ouchterlony, O. T. G. Salmonellosis originating in a hospital: A newly recognized source of infection. *N Engl J Med 253:*761–763, 1955.
7. Kunz, L. J., Poitras, J. W., Kissling, J., Mercier, B. A., Cameron, M., Lazarus, C., Moellering, R. C., Jr., and Barnett, G. O. The Role of the Computer in Microbiology, p. 181–193. In J. E. Prier, J. T. Bartola, and H. Friedman (Eds.). *Modern Methods in Medical Microbiology: Systems and Trends*. Baltimore: University Park Press, 1976.
8. Kunz, L. J. Symposium on advances in medical microbiology – Part I. Computerization in microbiology. *Hum Pathol 7:*169–175, 1976.
9. Lang, D. J., Kunz, L. J., Martin, A. R., Schroeder, S. A., and Thomson, L. A. Carmine as a source of nosocomial salmonellosis. *N Engl J Med 276:*829–832, 1967.
10. McCall, C. E., Collins, R. N., Jones, D. B., Kaufmann, A. F., and Brachman, P. S. An interstate outbreak of salmonellosis traced to a contaminated food supplement. *Am J Epidemiol 84:*32–39, 1966.
11. Rypka, E. W., Clapper, W. E., Bowen, I. G., and Babb, R. A model for identification of bacteria. *J Gen Microbiol 46:*407–424, 1967.
12. Smith, L. H. (Ed.). Cases from the medical grand rounds of the Massachusetts General Hospital. *Am Pract Digest Treat 8:*118–122, 1957.
13. Turck, M., Anderson, K. N., and Petersdorf, R. G. Relapse and reinfection in chronic bacteriuria. *N Engl J Med 275:*70–73, 1966.

2

Medical Microbiology: How Fast to Go—How Far to Go

RAYMOND C. BARTLETT

COST-BENEFIT AND PRIORITIES

How fast we should go, or how far we should go, in the reporting of microbiologic information from clinical material are questions that cannot be answered on the basis of patient care benefit alone. Usually added benefits incur additional cost. The highest priorities for more rapid reporting, or production of more refined information, must be assigned to those measures which provide the most benefit for the least cost. Unfortunately objective measurement of patient care benefit is difficult, time consuming, and inaccurate as is comparative cost accounting of laboratory methods. Ultimately experienced administrators must apply their judgment to establish priorities for allocation of resources to the multitude of new services which are proposed not only by clinical laboratories but other departments throughout the entire health care facility on the basis of what little objective data that may be available.

For decades physicians and laboratory workers, including microbiologists, have viewed the introduction of almost any available new technique as essential to the improvement of patient care, a priori. This attitude contributed to a tripling of laboratory costs while hospital costs doubled over a recent 5-year period (7). It is becoming increasingly apparent that both benefit and cost will have to be more objectively evaluated in the future. Otherwise we would always go as fast as we could go and as far as we could go. In fact, if we do not begin to look for ways to objectively establish our own priorities for the most cost effective utilization of our resources, continued escalation in costs will bring about further public pressure for government agencies to set these priorities for us.

Unique policies have been introduced in our laboratory which we believe expedite reporting of information and concentrate use of our resources on the production of the most useful information for the care of patients (1–3).

HOW FAST TO GO

Collection and Transmission of Specimens

Proper containers and transport media are required to prevent deterioration or contamination. At the Hartford Hospital the time of collection must be stated on all specimens and no more than a 2-hr delay is acceptable except for blood cultures collected in broth and throat swabs in transport medium. Over 97% of specimens are currently received within these time limits. When improper containers are used, or transit time exceeds 2 hr, or time of collection is not given, specimens are held for 5 days and both oral telephone and written reports are rendered requesting repeat collection or a consultation by the physician with laboratory personnel regarding processing.

Figure 2.1 (*A*, below; *B*, facing page)

(A)

Doe, John N8R
12343210
Dr. JONES 72

Date	Time	By	Specimen and Request	Report (ANTIMICROBIALS See Reverse Note N)	Amp	Ceph	Clin	Ery	Gent	Kana	Meth	Nit	Pen	Tmx	Tobra
1 Collected 2 Received 3 Reported															
1 3-30-76	9 am		Left Leg	Piease Consult Microbiology for Processing: See Reverse Note A											
2 3-30-76	9^{30} am	LA	Wound	( ) Superficial material (squamous cells)											
3 3-30-76		LA		(✓) No evidence of inflamation (neutrophils)											
1 3-30-76	9^{15} am		Leukens Sputum	OROPHARYNGEAL CONTAMINATION: PLEASE REPEAT											
2 3-30-76	9^{45} am	JB		See Reverse A											
3 3-30-76		JE													
1 3-30-76	11 am		Leukens	**Gram Stain:** Preliminary Report. Cells and Bacteria Suggested. Neut (3) Round ( ) No Bacteria ( ) Pneumo (✓) Yeast ( ) Bacteroides Q (2) Mucus ( ) RBC (0) Mixed Morph ( ) Staph ( ) Enteric (✓) ( ) Squam (2) No Cells ( ) No Pred. Type ( ) Strep ( ) Pseudom ( ) Hemophilus											
2 3-30-76	11^{30} am	JO													
3 3-31-76		JE													
1				**Culture:** Mixed Culture of S. pneumo (✓) E. coli ( ) Bacteroid ( ) Coryn ( ) Staph aur ( ) Kleb/Ent (✓) lac-Gm-rd (✓) Clostrid ( ) Staph ep ( ) Proteus ( ) Yeast ( ) See Reverse Note B Bacillus ( ) __Strep ( ) Pseudom ( ) ____ ( )											
2															
3															
1 3-30-76	7^{30} am		Chest Drainage	**Gram Stain:** Preliminary Report. Cells and Bacteria Suggested. Q (2) Round ( ) No Bacteria ( ) Strep ( ) Enteric (✓) Bacteroides Neut (1) RBC (3) Coryn ( ) Yeast ( ) Pseudom ( ) ( ) Squam (0) No Cells ( ) Staph (✓) Clost ( ) Neisser ( ) Hemophilus											
2 3-30-76	8am	JK													
3 4-1-76		CS													
1				Staph aureus 3+			S	S			S		S		
2				E. coli 2+	S	S			S	S					S
3				(Pseudomonas aeruginosa 1+ - See Reverse Note B)											
1 3-30-76	1pm		trach Asp.	**Gram Stain:** Preliminary Report. Cells and Bacteria Suggested. Neut (3) Round ( ) No Bacteria ( ) Pneumo ( ) Yeast ( ) Bacteroides Q (3) Mucus ( ) RBC (3) Mixed Morph ( ) Staph (✓) Enteric (✓) ( ) Squam (0) No Cells ( ) No Pred. Type ( ) Strep ( ) Pseudom (✓) Hemophilus											
2 3-30-76	1^{49} pm	NTK													
3 3-31-76		JE													
1				**Culture:** Mixed Culture of S. pneumo ( ) E. coli ( ) Bacteroid ( ) Coryn ( ) Staph aur (✓) Kleb/Ent (✓) lac-Gm-rd ( ) Clostrid ( ) Staph ep ( ) Proteus (✓) Yeast ( ) See Reverse Note C Bacillus ( ) __Strep ( ) Pseudom (✓) ____ ( )											
2															
3															
1 3-30-76	10^{15} am		Clean Catch	**Culture:** Mixed Culture of S. pneumo ( ) E. coli (10^4) Bacteroid ( ) Coryn ( ) Staph aur ( ) Kleb/Ent ( ) lac-Gm-rd ( ) (✓) See Reverse Note D Clostrid ( ) Staph ep (10^3) Proteus ( ) Yeast ( ) Bacillus ( ) Strep ( ) Pseudom ( ) ( ) ( ) See Reverse Note L											
2 3-30-76	10^{45} am	BM	Urine												
3 3-31-76		JE													

HH 6574 DISCARD ANY PREVIOUS COPIES HAVING THIS PAGE NUMBER **MICRO** 1

Specimens are accessioned on card kept in laboratory (*A*). Copies are made and sent to be included in the patient's record each time results are recorded. Previous copies of the page are discarded. The system allows laboratory personnel to follow the results of sequential specimens.

Results of examination of direct smears or of "gross" reports of mixed cultures are recorded on tape which is affixed to the card. Stamps are also used to convey information. Reference is made to explanatory notes (*B*) which appear on the reverse of the report. These explain reason for disposition of specimen and indicate that the specimen or culture will be held to provide the physician with an opportunity to request further evaluation if clinically indicated.

Evaluation of Specimen Quality

Smears of respiratory secretions, wound exudates, body fluids (excluding urine), and tissue are routinely Gram stained, examined, and reported on the same day. Figure 2.1 depicts the reporting format which is used. Master cards are kept in the laboratory and specimens are accessioned onto these as they are received. As information is recorded on these reports, copies are prepared and sent for inclusion in the patient's record. Systems applied in our laboratory for the evaluation of the quality of specimens

(B)

PROCESSING AND REPORTING OF MICROBIOLOGIC SPECIMENS

* = new, corrected or updated information not previously reported.

1. Cells in direct smears: 0 = none; 1 = few; 2 = moderate; 3 = many
2. Antimicrobials: Amp = Ampicillin; Ceph = Cephalosporin; Clin = Clindamycin; Ery = Erythromycin; Gent = Gentamicin; Kana = Kanamycin; Meth = beta lactamase resistant Penicillins; Nit = Nitrofurantoin; Pen = Penicillin; TMX = Trimethoprim - sulfamethoxasole; Tob = Tobramycin
3. Roladex file on each nursing unit describes availability of services and proper mode for submission of specimens.
4. Notes A-N refer to incompletely processed specimens which are held in the laboratory for five days. During this time physicians may consult with Microbiology for more complete processing. In many instances repeat collection is suggested.

Note A: Abundant squamous cells and few or no neutrophils suggest that this material is superficial and may contain contaminating or colonizing bacteria unrelated to infection. See footnotes 1 and 2.

Note B: Correlation with gram stain suggests that this(these) isolate(s) may not relate to infection and may represent colonization or contamination. See footnotes 1 and 2.

Note C: Gram stain demonstrates neutrophils and few or no squamous cells. Four or more potential pathogens were isolated and correlation with the gram stained direct smear does not help identify any as more probably pathogens than others. Complete speciation and antibiograms have not been performed because of the doubtful value of such information for direction of therapy against mixed infections containing this many potential pathogens. See footnote 2.

Note D: Bacteria isolated probably represent urethral or perineal flora and are insufficient in number to suggest urinary tract infection. See footnote 2.

Note E: This isolate may represent an atypical strain of E. coli or Enterobacter, Serratia, Proteus, Providencia, Salmonella, Citrobacter or a non-fermentative gram negative rod. See footnote 2.

Note F: Additional clinical information is required to assure proper processing and production of useful information from this specimen. See footnote 2.

Note G: If complete speciation, typing or grouping of this isolate is clinically indicated, please consult Microbiology.

Note H: This specimen was received in a condition inappropriate for optimum production of clinically useful information. Collection of another specimen is suggested. See footnote 2.

Note J: See Roladex under General Information "Laboratory night Emergency Service Card 4" and "Bacti Note 1, card 1" regarding specimens submitted after 9:30 PM. Collection of another specimen is suggested. See footnote 2.

Note K: This specimen represents one of two or more submitted from the same apparent body site by the same method of collection on the same day. The one of best apparent quality has been selected for processing. See footnote 2.

Note L: This is a mixed culture. Although one or more species exceeds 10^5/ml, the probability of contamination is high. See footnotes 1 and 2.

Note M: Complete speciation is performed on alternate days and antibiograms are repeated on the fifth day when isolates show the same gross colony morphology in cultures of consecutively submitted daily specimens. See footnote 2.

Note N: Susceptibility when placed in brackets (S) is highly predictable based on clinical data, previous testing of this species and was not established by in vitro testing of this isolate.

Footnotes:

1. Collection of another specimen is suggested avoiding superficial sources of contamination.
2. Please consult Microbiology if clinical considerations warrant further processing of this specimen. (Specimen will be held five days).

R. C. Bartlett, M.D.
Division of Microbiology
Department of Pathology
ext. 2206

based on the Gram stain are described elsewhere (3). This report confirms through cost accounting that the time expended on evaluation of specimen quality by examination of Gram-stained direct smears is more than compensated for by elimination of complete identification and susceptibility testing of all bacteriologic isolates observed in poor quality specimens. Empirical criteria are used to establish a "Q" score which reflects the numbers of leukocytes and squamous cells present. When an abundance of squamous cells and few leukocytes are seen, specimens are considered of insufficient quality to be cultured. When specimens yield mixed cultures and demonstrate a mixture of leukocytes and squamous cells on direct smear suggesting that indigenous and contaminating flora are present, an attempt is made to correlate isolates with the morphologic types observed in the Gram-stained direct smear. Correlating isolates are identified and tested for antimicrobial susceptibility. Isolates which do not correlate are reported on the basis of gross colony morphology without antibiograms. The report indicates to the clinician that correlation with the Gram-stained direct smear suggests that these isolates do not relate to infection. When specimens are not cultured or when isolates are reported on the basis of gross colony morphology without antibiograms, they are held for 5 days as in the case of delayed transmission and a consultation by the physician with laboratory personnel or collection of another specimen is suggested. The medical value of complete identification will be reviewed in more detail later in this section.

Reporting of Bacteria in Gram-Stained Direct Smears

When specimens qualify for culture, bacteria and yeasts observed in the Gram-stained direct smear are reported. These are divided into the following genus and family categories based on morphologic types; staphylococci, streptococci, *Streptococcus pneumoniae*, corynebacteria, clostridia, yeast, "enterics," *Pseudomonas, Neisseria,* and *Bacteroides-Haemophilus*. Workers are urged to check more than one category on the report if the morphology is not uniquely suggestive of any single type. This provides more useful information to the physician than such terms as "Gram-negative pleomorphic coccobacilli," "Gram-positive cocci," etc. Most microbiologists have been reluctant to be this specific in written reports. Justification for this policy was based on a correlation of the reports of direct smears and cultures of 277 specimens. Specimens were excluded in which two or more morphologically related types were seen in the direct smear. For example, specimens suggesting both Pseudomonas and "enterics" in the direct smear were excluded but those showing "enterics" and "Staphylococci" were included. Table 2.1 demonstrates a 40–80% correlation between presumptive identification by Gram stain and culture. Different, but morphologically related species, were isolated in a maximum of 31% of the specimens. An unexpected finding was the predictive value of identifying *Pseudomonas aeruginosa* in Gram-stained direct smears. Pseudomonads were suspected when uniform rods 0.5–0.8 μ in width and 1.5–3 μ in length were seen, often arranged end to end, and staining intensely with

TABLE 2.1
Gram-stained direct smears containing mixtures of morphologically unrelated Gram-positive and Gram-negative bacteria were correlated with cultures; 44–83% of specimens showed agreement between presumptive identification and culture. Morphologically similar species were isolated in 0–31% of specimens

Gram Stain	Total	Culture								
		Staphylococcus	Streptococcus	Pneumococcus	Enterics	Pseudomonas	Bacteroides-Haemophilus	Yeast	Clostridium	Sterile
Staphylococcus	34	(83)	15	0	24	3	6	3	0	0
Streptococcus	35	11	(54)	17	26	0	17	9	11	9
Pneumococcus	26	4	0	(50)	31	0	0	4	0	27
Enterics	73	8	10	5	(63)	21	3	5	3	5
Pseudomonas	15	0	0	0	20	(80)	7	0	0	0
Bacteroides-Haemophilus	51	12	14	14	31	6	(44)	4	2	16
Yeast	33	3	3	3	39	3	6	(45)	0	27
Clostridium	10	10	20	0	40	0	40	10	(40)	10
	277									

safranin without bipolarity or vacuoles. Enteric rods are usually more pleomorphic, larger, often show bipolar staining and vacuoles, and stain less intensely with safranin. *Pseudomonas* was isolated from 80% of the specimens in which its presence was suggested by the Gram-stained direct smear. Conversely, when enteric bacteria were suspected in the Gram-stained direct smear, *P. aeruginosa* was isolated in only 21%. Similarly, *Haemophilus* or *Bacteroides* were isolated in 44% of the specimens in which their presence was suggested by the Gram-stained direct smear but in a maximum of 31% when they were not suggested by the direct smear. In each category the Gram stain appeared to be of predictive value in anticipating the outcome of cultures (Table 2.1).

Expediting Reports

Reporting of specimens which are unsuitable for culture or of isolates in mixed culture on the basis of gross colony morphology provides information 24–48 hr earlier than would occur if complete identification and susceptibility testing had been performed. This speeds the collection of better quality specimens. Any laboratory could submit preliminary reports containing this type of information on all specimens but this generally proves to be too much of a clerical burden. We allow no specimen to spend more than 48 hr in our laboratory without rendering some kind of report.

Other very simple means of expediting reporting of information are frequently neglected. The mere fact that a specimen collected from a normally sterile body site shows growth or no growth is of considerable clinical value and often may be reported days before a final report is rendered.

The cost of both oral and written reporting on poor quality specimens as well as the examination and reporting of Gram-stained direct smears is offset in our laboratory by the reduction in time and expense which would have been applied to identification and susceptibility testing of isolates obtained from specimens of less than optimum quality (3).

Methods for rapid speciation and antimicrobial susceptibility testing are available. It is not yet clear whether these can be performed at a cost comparable to that of conventional methods. If this could be established it would eliminate the need to demonstrate any added patient care benefit. If there will be an added cost, carefully controlled clinical studies must be conducted to establish the effect on diagnosis, treatment, morbidity, mortality, and duration of hospitalization.

Simple techniques that will speed the reporting of useful information include the reading of Bauer-Kirby plates after 6–7 hr of incubation (4), examination of blood cultures twice a day, or the routine subculturing of blood cultures after 12–14 hr of incubation (8) and the examination of culture plates with a colony microscope or hand lens. The latter has proved especially useful for examination of cultures for Mycobacteria (18). Others include direct methods of examination employing fluorescent antibodies (5) and detection of microbial antigen by counterimmunoelectrophoresis (14) and gas chromatography (10). The β-lactamase test speeds recognition of *Haemophilus influenzae* which are ampicillin resistant (17).

HOW FAR TO GO

If there were no cost limitation on the amount of information which could be derived and reported from clinical specimens, there might appear to be no disadvantage in reporting this to physicians. We must consider the time which is required for evaluation of complex microbiologic reports and the risk of misinterpretation which may cause errors in diagnosis and treatment. Generally, production of additional information results in additional cost and we are faced again with establishing priorities which relate to patient care benefit. There are a number of procedures which are of little, if any, clinical value that are routinely performed in many microbiology laboratories. Recommendations to guide laboratories in production of more clinically useful information and application of policies to control and assess specimen quality have been developed by an ad hoc committee of the Connecticut State Department of Health (12). Similar recommendations are currently being developed by the College of American Pathologists.

Throat cultures should be examined routinely only for group A β-hemolytic streptococci. Predominating growth of *H. influenzae* type b or *S. pneumoniae* in children under the age of five should be reported, although this often represents only harmless colonization. Cervical-vaginal specimens should be examined routinely only for Neisseria. Smear and culture for *Corynebacterium vaginale* and Candida should be conducted only on special request or when "vaginitis" is stated on the request slip. Staphylococci, *Streptococcus* sp., and enteric organisms should be sought only from cervical-vaginal specimens associated with surgical wounds or when "endometritis" is stated on the request slip.

Anaerobic microbiology should be applied selectively. Sputum, tracheal secretions, bronchial washings, urine, or vaginal secretions should not be cultured for anaerobes. If requests are received on other specimens, their suitability for anaerobic culture may be assessed by examining the Gram-stained direct smear. If squamous cells are seen without neutrophils the specimen should not be cultured anaerobically. Instead, a request should be made for a repeat collection avoiding contamination from the skin or mucous membranes. Direct anaerobic cultures should be performed routinely on blood, or exudate collected from a source which is unlikely to be contaminated with indigenous flora. If Gram stains of direct smears of any exudate show neutrophils and an absence of squamous cells, direct anaerobic culture should be performed. Gassed out tubes are optimal for collection but Amies' medium may be used for swab collected specimens (16).

Selective Media

Some limits are imposed on the number of selective media that are employed. We find anaerobic colistin-nalidixic acid (CNA) medium, anaerobic blood agar plates (BAP), and modified Thayer-Martin medium suitable for anaerobes. The latter is a substitute for vancomycin-kanamycin medium. It has the virtue of being used also for isolation of Neisseria and has the one disadvantage of failing to support Fusobacterium.

We prefer modified Thayer-Martin medium for primary isolation of *Neisseria gonorrhoeae* and depend upon Amies' modification of Cary-Blair medium as a transport system. Many workers are concerned with the extent to which unusual enteric pathogens should be sought in feces from patients with enterocolitis. We examine feces routinely only for Salmonella and Shigella. The former practice of looking for *Staphylococcus aureus* has been discontinued because of the low frequency of staphylococcal enterocolitis. Current confusion over the relationship betweeen enteropathogenic *Escherichia coli* and enterotoxogenic *E. coli* suggests that routine serotyping is no longer useful. When stools are negative for Salmonella and Shigella, we suggest in writing that the physician consult with laboratory staff if symptoms persist. At that time we would inform the physician of the availability of techniques for the isolation of Vibrio, Yersinia and more obscure causes of enterocolitis.

Microbial Identification

Perhaps nothing arouses more violent differences of opinion among microbiologists than a comparison of criteria which are applied to microbial identification between one laboratory and another. We must recognize that it does not necessarily follow that taxonomic differences in bacteria always signify differences in infectivity, response to treatment and prognosis. It is not really the name of the organisms that counts. Perhaps someday a numerical coding system will be developed which will instantaneously convey to the clinician what he really needs to know. Is the organism normally indigenous to the site of collection? What is the probability that this organism is colonizing this site without producing infection? What is the relative invasiveness of this organism compared to others which are relatively opportunistic? What is the relative prognosis between this organism and other potentially invasive species from this site? Is therapy of this organism reliably directed by in vitro susceptibility testing? If not, what is the recommended therapy? What is the likelihood that this isolate represents reinfection with the same strain previously isolated from this site? Is this isolate epidemiologically related to similar isolates being found in other patients?

While some large laboratories in referral centers should conduct more costly and complete identification on all isolates to further knowledge of the distribution of indigenous flora, frequency of colonization and emergence of new pathogens, it is clear that the majority of the nation's laboratories cannot afford this practice nor can the clinicians whom they serve provide better patient care as a result of it. Indeed physicians who deal intermittently with infectious disease are more likely to misinterpret and abuse this information. New commercially available systems make it possible to apply a large number of biochemical tests to provide more complete and accurate identification of isolates at a cost equal to or below that which results from application of a more limited number of conventional tests. This is a great step forward but uncontrolled application to all isolates, especially those in mixed culture from specimens of less than optimal quality, may cause a further increase in cost because of the convenience of their use. We believe that many isolates should not be "completely identified."

Mixed Cultures

We conduct complete identification of isolates in pure culture or in mixed culture when up to three isolates are found in the presence of neutrophils and in the absence of squamous cells. When additional isolates are found in such specimens, it is our conviction that complete identification and antibiograms do not contribute more to diagnosis and treatment than is provided by a general indication of the species present based on gross colonial morphology sometimes supplemented by a few rapid tests (coagulase, spot bile solubility, oxidase, etc.). When mixed cultures are observed in specimens which display both squamous cells and neutrophils in the direct smear, complete identification is applied only to those isolates which correlate with morphologic types observed in the direct smear. Those which do not correlate are incompletely identified.

Incomplete identification is also applicable to isolates present in numbers less than 10^5/ml in clean catch urine specimens or when two or more isolates are obtained from clean catch specimens or specimens collected from closed drainage systems.

Obviously, incomplete identification based on gross colonial morphology carries with it some risk of error in addition to lumping otherwise distinguishable species and genera. The terms "coliform" or "Klebsiella/Enterobacter" may include *Escherichia coli, Citrobacter* sp., *Klebsiella* sp., *Enterobacter* sp., some strains of *Aeromonas hydrophila* and rare *Salmonella* and *Arizona*.

Strains of *Pseudomonas aeruginosa* which produce obvious pyocyanin are oxidase positive, provide typical antibiograms, and are easily and confidently speciated by these criteria alone. The swarming of lactose-negative colonies of *Proteus mirabilis* and *Proteus vulgaris* permits these to be reported as *Proteus* species. With these exceptions "lactose-negative Gram-negative rod colonies" must be reported as such. This large group may include nonpigmented strains of *Pseudomonas aeruginosa* as well as other *Pseudomonas* species, *Edwardsiella*, late lactose fermenting variants of the genera *Escherichia, Enterobacter, Klebsiella, Serratia, Providencia, Yersinia, Erwinia, Vibrio, Aeromonas, Flavobacterium, Pasteurella, Moraxella, Acinetobacter,* and other more obscure nonfermentative Gram-negative rods. It could include *Salmonella* and *Shigella* if these were isolated in mixed culture from contaminated material from body sites other than feces with no specific request for their isolation.

While incomplete identification resulting in "gross" reports which lump many genera and species may seem an abomination to many microbiologists, it seems questionable that the added value of more complete identification of such isolates which are obtained from mixed cultures and which fail to correlate with Gram stain morphology are worth the added effort.

Other Limitations on "Complete Identification"

Even when "complete identification" is indicated there are limits to which this should be carried out. Many subspecies-specific procedures such as serotyping and biotyping are available. These procedures speed identification and often provide useful clinical information when applied to *Haemophilus influenzae, Salmonella, Shigella,* and *Streptococcus pyogenes*. They may prove useful in the rapid identification of *Clostridium* species, *Yersinia pestis, Pseudomonas pseudomallei,* and *Francisella tularensis* but many of these are conducted only in referral centers. The direct examination of spinal fluid for meningococci, pneumococci, and *Haemophilus* with fluorescent antibody conjugates has proven to be practical in some laboratories. Otherwise subspecies-specific classification methods for *Pseudomonas aeruginosa, Escherichia coli, Salmonella enteritidis, Klebsiella pneumoniae, Serratia marcescens, Flavobacterium, Neisseria,* and *Listeria* are not of practical value and should be applied only on the basis of a specific epidemiologic investigation which is directed and controlled by the infection control committee of the institution or in specific instances when reinfection must be distinguished from treatment failure, i.e. *E. coli* urinary tract infection.

In many instances species of various genera provide no more useful information than the genus itself. This includes *Citrobacter* and *Entero-*

bacter (*Enterobacter agglomerans* should be identified and reported), *Serratia*, *Providencia*, and *Aeromonas*. It is difficult to establish compelling criteria for the separation of indole-positive *Proteus* species. *Bacteroides* should be divided into *Bacteroides fragilis* not subspecies *fragilis*, *Bacteroides fragilis* subspecies *fragilis*, *Bacteroides melaninogenicus*, *Bacteroides* species (not conforming to the others), and *Fusobacterium* sp. *Eikenella corrodens* is sufficiently unique both microbiologically and clinically to be worth recognizing and reporting.

Minimum Criteria for Identification

Once having determined when complete identification will be conducted, one then has to establish the number of criteria which will be applied to provide a statistically adequate level of confidence for that identification. No one has established what level of statistical confidence is medically necessary for bacterial identification. It may be that these levels will vary depending upon the body site, the species, and the medical significance of potential errors in identification. We established a set of minimum criteria for microbial identification of common isolates which would assure that the most probable identification was at least 20 times greater than the next most likely identification (>95%) using a computer. The data base consisted of several recent sources of biochemical reactions for Gram-negative bacteria (6, 9, 19) and antibiogram profiles established in our laboratory. Bayes' theorem was applied to establish the probability of correct identification when any given combination of results was provided. Table 2.2 reveals that the probability of correct identification of *Escherichia coli* is 93% more probable than the next most likely species if only the presence of acid colonies on MacConkey and the expected antibiograms are observed. Addition of both oxidase and indole tests would increase the probability to 99%.

Swarming *Proteus* species are separated into *P. vulgaris* and *P. mirabilis* by a single test, ornithine decarboxylase. Isolates which appear to belong to the *Klebsiella-Enterobacter* group are identified with at least 95% confidence by six tests performed in five tubes (citrate, sulfide-indole-motility, ornithine, and lysine and decarboxylase base). Most lactose nonfermenting isolates are characterized to a confidence level of 95% by the same group with the addition of triple sugar iron agar and the oxidase test. Additional tests and serogrouping are sometimes required. Some uncommon combinations of reactions require up to 18 additional tests to establish an identification which exceeds 95% probability. The results of testing such isolates are repeatedly entered into the computer program by technologists until sufficient criteria have been satisfied to yield this level of confidence.

Antibiograms obtained on these isolates are entered into a computer program along with the presumptive species identification. The program uses Bayes' theorem to compute the probability that the antibiogram is consistent with those of recent proven isolates of the same species. About 10% are rejected. These are respeciated using the five-tube scheme outlined above for lactose nonfermenters. Two-thirds turn out to be correctly and antibiogram results observed in author's laboratory. Reactions critical to identification at confidence levels listed must agree with those in the table. Where results of tests are not listed, these are not crucial to identification and may be ignored. When results do not fit this table, additional tests are performed**

I. MacConkey Positive Acid Colonies															
								Antimicrobials (+ = Sensitive)							
	Oxidase	Indole	Citrate	Motility	H_2S	Ornithine	Lysine	Ampicillin	Cephalosporin	Gentamicin	Kanamycin	Polymyxin	Probable Identification	Identities of Less Probability	RP‡
Resembling *Escherichia coli*								+	+	+	+	+	*E. coli*	*Serratia rubidaea* *Aeromonas* *Vibrio* *Actinobacillus* *Arizona*	93 94 94 94
	0							+	+	+	+	+	*E. coli*	*Serratia rubidaea* *Arizona*	93 95
	0	+						+	+	+	+	+	*E. coli*	*Citrobacter diversus*	99
		0	+					0	+/0	+	+	+	*C. freundii*	—	100
Resembling *Klebsiella-Enterobacter*		0	+	0		0	+	0	+/I	+	+	+	*Klebsiella pneumoniae*	*Klebsiella ozaenae*	99
		+	+	0	0	0	+						*K. pneumoniae*	*K. ozaenae* *S. rubidaea*	84 98
		+	+				0	0	0/+	+	+	+	*C. diversus*	*Citrobacter freundii* *Enterobacter agglomerans* *K. pneumoniae*	97 98 99
		0	+	+	0	+	+	0	0	+	+	+	*Enterobacter aerogenes*	*Enterobacter cloacae*	99
		0	+	+	0	+	0	0	0	+	+	+	*E. cloacae*	*E. aerogenes* *C. freundii*	97 99
		0	+	+		+	0	0	0	+	0	+	*E. cloacae*	*C. freundii* *E. aerogenes* *Streptococcus liquefaciens*	96 97 98

Table 2.2—Continued

II. MacConkey Positive Non-Acid Colonies

Pigment	Oxidase	Indole	Citrate	Motility	H_2S	Ornithine	Lysine	Pad	Urea	Arabinose	Ampicillin	Cephalosporin	Gentamicin	Kanamycin	Polymixin	OF Oxid	Growth Cetrimide	Probable Identification	Identities of Less Probability	RP‡
0				S							+	+	+	+	0			*Proteus mirabilis*	*Vibrio alginolyticus*	99
0				S		0					0	0	+	+	0			*Proteus vulgaris*	*Proteus rettgeri*	99
0	0	+	0	0	0	+	0	0	+									*Yersinia enterocolitica*	*Proteus morganii*	88
	0	+		S	+	+		+	+									*Proteus mirabilis*	*P. vulgaris*	76
0	0	+	0		0			0	0		+	+	+	+	+			*E. coli*	*Shigella*	97
	0	+	0		+			0	0									*Edwardsiella*	*E. coli*	86
0		+		+		+		+	+		0	0	+	+	0			*P. morganii*	*P. rettgeri* *Pasteurella pneumotropica*	99 99
0		+				0		+	+		0	0	+	+	0			*P. rettgeri*	*P. vulgaris* *P. morganii* *Providencia*	95 95 99
0		+				0		+	0		0	0	+	+	0			Providencia sp.	*P. rettgeri*	98
0	0	0	+		+	0	+	0	0									Salmonella¶	*Arizona* *S. rubidaea* *K. ozaenae*	89 93 97
0	0	0	+	0	0	0		0	0		0	0	+	+	+	+		*Acinetobacter calcoaceticus*		100
0	0	0				+				0	0	0	+	+	0/+			*Serratia marcescens*	*P. morganii*	99
0	0	0	0	0		0	0	0	+									*Yersinia pseudotuberculosis*	*K. ozaenae* *Y. enterocolitica* *Yersinia pestis*	88 94 95
0	0		0	0			0	0	0									*Shigella*§	*Y. pestis* *Klebsiella rhinoscleromatis* *K. ozaenae* *E. coli* HB-5	51 63 78 93 94

Table 2.2—Continued

II. MacConkey Positive Non-Acid Colonies

Pigment	Oxidase	Indole	Citrate	Motility	H_2S	Ornithine	Lysine	Pad	Urea	Arabinose	Ampicillin	Cephalosporin	Gentamicin	Kanamycin	Polymixin	OF Oxid	Growth Cetrimide	Probable Identification	Identities of Less Probability	RP‡
0	0			+	0	0	0			+								*E. agglomerans*	*Y. pseudotuberculosis* *S. rubidaea* *Chromobacterium violaceum* *C. freundii*	66 72 78 85
Green yellow											0	0	+	0	+			*Pseudomonas aeruginosa*	*Pseudomonas maltophilia*	99
0	+			+							0	0	+	0	+	+	+	*P. aeruginosa*		100
0	0	+	0	0		+	0	0	+									*Y. enterocolitica*	*P. morganii*	88

* Bayes theorem:

Absolute probability (*AP*) = percent of strains of species "X" that demonstrate any given combination of observed reactions (OR^a, OR^b, OR^c, etc.)

AP = (percent sp. "X" = OR^a) × (percent sp. "X" = OR^b) × (percent sp. "X" = OR^c) etc.

Probability of identification (*PI*) of species *A* for a given combination of reactions for a file of species extending from *A* to *Z* =

$$\frac{\text{Frequency sp. } A \times AP \text{ sp. } A}{(\text{Frequency sp. } A \times AP \text{ sp. } A) + (\text{frequency sp. } B \times AP \text{ sp. } B) + (\text{frequency sp. } C \times AP \text{ sp. } C) + \cdots (\text{frequency sp. } Z \times AP \text{ sp. } Z)}$$

‡ Relative probability (*RP*) = probability that most probable identification (ID^1) is correct with respect to next most probable species (ID^2)

$$RP = 100(PI^{ID^1}/PI^{ID^1} + PI^{ID^2})$$

speciated but display unusual antibiograms. The remainder represents mixed cultures, and errors in speciation or performance and interpretation of the susceptibility test.

Cost of Microbial Identification

An examination was conducted of the cost of applying this system of identification to the approximately 6500 Gram-negative isolates which are examined each year in the Hartford Hospital laboratory. Table 2.3 displays the costs for operation of the Microbiology Division. Workload accounting was conducted using the Workload Recording Method of the College of American Pathologists (20). Production of 2,800,000 units at a cost of $869,300.00 less $85,019 cost for media production yielded a cost of $0.28 per unit of work performed in the laboratory. Table 2.4 represents a breakdown of costs for media preparation.

Gram-negative rod isolates are identified in our laboratory by the criteria shown in Table 2.2 and described previously on page 24. The average cost rises as the number of tests increase to an average cost of $15.84 for those isolates which are subject to a battery of 18 tests (Table 2.5).

Ewing and Martin (6) and Hugh and Gilardi (9) have provided tables which suggest that a minimum of 18 tests be applied to the identification of all gram-negative rod isolates. Application of these criteria by the system in use in our laboratory using conventional media would increase the total cost for Gram-negative rod speciation from $22,553 per year to $102,263 (6500 × $15.84). If we purchased prepared media the cost per tube would be doubled. This would increase total cost from $0.88 a tube to $1.06 ($0.88 + $0.18), a 20% increase. This would increase the total cost for the system in use to over $27,000 per year and the cost for routine application of 18 tests to each isolate to over $123,000 per year. The result would be a 450% increase in material and labor costs to increase confidence levels from 90–95% to 99%. The cost-benefit of this has not been established and seems doubtful.

Cost of Commercial Identification Systems

A comparison was conducted of the relative cost of the system in use in our laboratory employing minimum criteria and conventional biochemical tests (hereafter referred to as "H.H. system") and three commercially available identification systems, Enterotube (Roche Diagnostics, Nutley, N. J.) and API 10S and API 20E (Analytab Products Incorporated, Plainview, N. J.). Several publications have established the accuracy of these systems when correlated with other commercial identification systems and conventional biochemical tests (11, 15).

Fifty consecutive isolates requiring 5–7 tubes for identification were selected from our daily workload and transferred to MacConkey plates. These were identified by an experienced technologist using the H.H. system. On the same day, isolates were picked from MacConkey plates, inoculated into Enterotubes and API 10S and API 20E strips under our

TABLE 2.3
Cost of operation of Microbiology Laboratory includes indirect cost allocations and a portion of general laboratory expenses.

	Dollars		Dollars
CHARGES	823308	*STATISTICAL DATA:*	
		Fulltime equivalent employees	31.8 (15% total)
GENERAL FUND EXPENSE:			
Salaries	351601	Square feet	5500
Supplies	85964	Pounds laundry	5609*
Depreciation		CSS requisitions	521†
Moveable equipment	8470	Pharmacy requisitions	845
Fixed equipment, buildings	4662	Personnel living in	0
Subtotal	450697	*GENERAL LABORATORY EXPENSES:* Transfer to microbiology	
Employee benefits	53187		
Total	503884		%
COST ALLOCATIONS:		Salaries, benefits, staff, other divisions	0
Administration	20335		
Accounting	3286	Salaries—benefits, staff Division of microbiology	100
Billing and collection	36146		
Communications	5841	Salaries—benefits, director	15†
Personnel department	5390	Salaries—benefits, personnel	15†
Public relations	739		
Purchasing	546	Materials	15†
Housekeeping	4683	Cost allocations (17% laboratory total)	15†
Laundry	1254		
Resident personnel	0		
Plant operation	15142		
Repairs and maintenance	9197		
Medical care	15405		
Nurse education	23188		
House staff education	32400		
Central supply	771		
Pharmacy	2106		
Cafeteria	13971		
Subtotal	190400		
Subtotal	694284		
General laboratory expense	175087		
Total	869371		
Balance Gain or (loss)	(46063)		

* Based on 25% of total laboratory laundry use.
† Based on personnel; microbiology = 15% total laboratory.

observation by technologists experienced with each system. The total time required to reach a complete identification by each system was recorded (Table 2.6). The actual time observed in this study for processing a single tube of conventional medium was 239.3/486.5 = 0.49 min (see Table 2.6). Although one CAP work unit was originally intended to signify 1 min of work, it is apparent that about one-fifth of the time is required for the processing of a tube of medium than is implied by the allocation of 2.5 units. Work units were calculated (see Table 2.6) for the Enterotube and API 10S and API 20E systems (8.5, 11.0, and 12.2, respectively) which are compatible with the CAP workload recording method by establishing a ratio with minutes of work and CAP units for the H.H. system using

TABLE 2.4
Cost for preparation of media. Allocation of depreciation, maintenance, cost allocations, and general laboratory expense is based on ratio of man-hours applied to media production to total microbiology division man-hours

Allocation	Cost	
	Amount	Basis
Materials	$44,276	Actual
Wages	$16,017	Actual (75 hrs/week)
Benefits	$ 2,402	0.15 wages
Depreciation, maintenance, cost allocations general expense	$22,324	Ratio man-hours: media = 75, division = 1272
Total	$85,019	
Pieces of media	482,652	
Cost/piece	$ 0.18	

TABLE 2.5
Cost of each biochemical test is established by adding materials cost (Table 2.4) to use cost. Use cost is based on allocation of 2.5 units to each biochemical test according to Workload Recording Method of the College of American Pathologists. The unit cost was established by dividing total cost of operation (Table 2.3) less the cost of media production (Table 2.4) by the total number of units of work less media production units ($869,371 − 85,019/2,800,000 = $0.28). Total cost per tube = materials cost ($0.18) + cost/unit × units/test ($0.28 × 2.5) = $0.88.

	Tests	Tubes	Number/Year	Cost	Cost/Isolate
a. *Escherichia coli*	0	0	× 2518 × $0.88 =	$0	$0
b. Swarming *Proteus*	1	1	× 581 × $0.88 =	$511	$0.88
c. *Klebsiella/Enterobacter*	6	5	× 710 × $0.88 =	$3124	$4.40
d. Lactose nonfermenters (Plus 10% a, b, and c (381) antibiogram rejects, see text)	8	7	× 2467 × $0.88 =	$15196	$6.16
e. Problem isolates (0.07 total = 235)		18	× 235 × $0.88 =	$3722	$15.84
f. Total			6511	$22553	$3.46

Escherichia coli isolates which fit acceptable antibiogram profiles are identified without biochemical tests (Table 2.2). Swarming *Proteus* species require only one test, *Klebsiella/Enterobacter* species six tests performed in five tubes, and lactose nonfermenters eight tests in seven tubes. Problem isolates group includes most strains which are nonfermentative, oxidative, and nonoxidative.

The average cost for this system appears lower than the cost for other systems shown in table 2.7 because it includes substantial numbers of isolates for which few or no biochemical tests were performed.

conventional media (H.H. system units per isolate/H.H. system minutes per isolate:commercial system units per isolate/commercial system minutes per isolate). The average number of minutes required to process the Enterotube and API systems are about one-fifth the number of workload recording units which this computation suggests. Maintenance of this proportionality in units is essential if laboratories are to assess comparative workloads using the CAP workload accounting system. These units were multiplied by the cost per CAP unit for our laboratory. The materials cost (list) and the tube of saline (considered equal to a tube of culture medium for cost accounting) was added to obtain the final cost per isolate (Table 2.7).

TABLE 2.6
Time required for inoculation and reading of tests performed on 50 isolates was recorded for the H.H. system, Enterotube, and API 10S, and API 20E systems. Isolates were consecutive strains which required six to eight tests for identification by H.H. system

A. Comparative technical time for H.H. minimum criteria identification system and Enterotube, API 20E applied to 50 isolates

H.H. system	Isolates	Time (Min)	Tubes	Min/Tube
7 tubes	33	150	231	0.47
5 tubes	17		85	
Secondary tests (× 1.5)	21	15	31.5	0.48
2nd day tests (× 5)	17	35	85	0.41
Computer time		15		
Problem isolates	3	24.3	54	0.45
	50	239.3	486.5	
Enterotube	50	84		
API 10S	50	108		
API 20E*	50	120		

* Includes computer time:

B. Workload recording units per isolate

$$\text{H.H. system units/isolate} = \frac{2.5 \times \text{total tubes (486.5)}}{\text{total isolates (50)}} = 24.3$$

$$\frac{\text{H.H. system units/isolate}}{\text{H.H. system minutes/isolate}} = \frac{\text{Commercial system units/isolate}}{\text{Commercial system minutes/isolate}}$$

$$\text{Enterotube} = \frac{24.3}{239/50} = \frac{X}{84/50} = 8.5$$

$$\text{API 10S} = \frac{24.3}{239/50} = \frac{X}{108/50} = 11.0$$

$$\text{API 20E} = \frac{24.3}{239/50} = \frac{X}{120/50} = 12.2$$

It is necessary to establish an appropriate number of workload units for the Enterotube and API systems. For laboratories using the CAP Workload Recording system this is accomplished by establishing a ratio between the average number of units per isolate (applying 2.5 CAP units to each tube of conventional media) divided by the average minutes per isolate using the H.H. system; and the units to be assigned per isolate divided by the average minutes required per isolate using the commercial system.

More than 50 typical consecutive isolates would have been required to compare the accuracy of these systems. With the H.H. system, we were unable to speciate, or established the incorrect species, for three isolates among the genera *Proteus*, *Citrobacter*, and *Enterobacter*. In two instances cultures became contaminated. With Enterotube, we were unable to identify an indole-positive isolate of *Klebsiella pneumoniae*. Both API 10S and API 20E provided complete and accurate identification of all isolates.

The cost of use and length of time required to provide complete identification for Enterotube and API systems compared to the H.H. system is displayed in Tables 2.8 and 2.9. The API 10S system may be supplemented with API 20E or other secondary tests to provide an accuracy of 99%. Robertson and MacLowry reported that use of the API 10S system by a method of "best judgment" produced 95% accuracy (13). When two or more identifications are suggested by the API 10S strip results alone, the isolate having the highest frequency of occurrence is reported. Use of Enterotube

TABLE 2.7
Cost per isolate for commercial systems is based on materials cost (list), relative CAP units expended and ancillary costs (saline, oxidase tests, computer use). Accuracy of API 10S is improved by performing API 20E on ambiguous identifications.

	Materials cost		Units		Cost/Unit		Other costs	Total
Enterotube	$1.95	+	8.5	×	$0.28	+	$0.07[a]	$4.40
API 10S	$1.50	+	11	×	$0.28	+	$0.18[b]	$4.76
API 10S	$1.50	+	11	×	$0.28	+	$0.18[b]	
						+	$1.13[c]	$5.90
API 20E	$2.17	+	12.2	×	$0.28	+	$0.18[b]	
						+	$0.05[d]	$5.64

[a] Oxidase test on 8% of isolates.
[b] Tube of saline.
[c] Cost of repeating ambiguous identifications (about 20%) by API 20E.
[d] Annual cost of API computer service (Analytical Profile Index Series B)/6500.

TABLE 2.8
Speed and accuracy of identification.
API 20E system identified isolates most rapidly. H.H. system was unable to speciate, or incorrectly speciated, 3 isolates and required repeat testing in two instances because cultures became mixed with contaminants. Published evaluations indicate accuracy >95% for API 20E and Enterotube (11, 15).

	Cumulative Percent Identified to Species Level			
	H.H.	API 10S	API 20E	Enterotube
Identified				
First day				
Primary tests	58	80	100	84
Secondary tests	66			86
Second day	90	96		94
Unidentified				
Insufficient data	6	0	0	2
Mixed culture	4	0	0	0

appeared more economical than that of API 10S. API 20E provides more rapid identification at somewhat greater cost. Microbiologists must individually assess the patient care benefit of these alternative systems, but each would appear to provide more speed and accuracy per dollar spent than conventional methods employing individual tubes of media. Combinations of selected conventional tests, antibiograms, and commercial systems offer the maximum in accuracy and economy (see Table 2.9).

CONCLUSIONS

How fast or how far we can go in clinical microbiology is constrained both by economics and technology. Even with progressing semiautomation we cannot assume that developments which offer the production of more information or more accurate information or produce it more rapidly are always desirable.

TABLE 2.9
Summary of increase or decrease in cost (±100%) of different approaches to identification of Gram-negative rods in Division of Microbiology, Department of Pathology, Hartford Hospital. Cost of $3722 for identifcation of problem isolates is based on observed frequency of 7% with the H. H. System (Table 2.5). The increased speed and accuracy of identification of the other systems which are displayed in Table 2.9 may result in reduction in the percentage of problem isolates and the resulting cost of their identification.

	Escherichia coli	Swarming Proteus	Klebsiella- Enterobacter	LNF	Problem Isolates
H	$ 0	$ 511	$ 3124	$15195	
H + ox + ind	$ 4432				
Enterotube	$11079	$2556	$ 3124	$10855	
A10BJ	$11986	$2766	$ 3380	$11743	
A20	$14202	$3277	$ 4004	$13914	
A10	$14856	$3428	$ 4189	$14555	
Ewing and Martin	$39885	$9203	$11246	$39077	$3722

E. coli	Swarm Proteus	K/E	LNF	Problems	Total	Change in Cost
H	H	E	E	EM	$ 18212	81%
H	H	A10BJ	A10BJ	EM	$ 19356	86%
H	H	A20	A20	EM	$ 22151	98%
H	H	H	H	EM	$ 22553	100%
H	H	A10	A10	EM	$ 22978	102%
H + ox + ind	H	A20	A20	EM	$ 26583	118%
H + ox + ind	H	A10	A10	EM	$ 27410	121%
E	E	E	E	EM	$ 31336	139%
A20	A20	A20	A20	EM	$ 39119	173%
EM	EM	EM	EM	EM	$103133	457%

Key: H = Hartford Hospital system. H + ox + ind = Hartford Hospital system + oxidase + indole for *Escherichia coli*. E = Enterotube. A10BJ = API 10S (method of best judgement (13)). A20 = API 20E. A10 = API 10S (includes 20% API 20E repeats for improved accuracy). EM = Criteria of Ewing and Martin (6), Hugh and Gilardi (9). LNF = Lactose non-fermenters.

Microbiologists must constantly avoid the temptation to provide more complete and academically satisfying evaluation of specimens at the expense of producing information useful for patient care as quickly as possible. An immediate report indicating that a Gram-stained direct smear contains leukocytes and bacteria may be more valuable to the clinician than a report of the precise species and its antibiograms 2 days later. Similarly, the preliminary report indicating the presence of significant numbers of bacteria in the urine may result in the initiation of effective antimicrobial therapy days before complete identification and an antibiogram is reported. This may prevent potentially fatal septicemia or at the least should shorten hospitalization by earlier initiation of antimicrobial therapy.

Control of delay in transmission and evaluation of the quality of specimens will speed reporting of useful information and help concentrate use of laboratory resources on specimens which are most likely to produce useful information. So also will preliminary reporting and abbreviated workup of mixed cultures from specimens showing evidence of contamination. The gram stain is valuable for assessing the quality of specimens and the rapid reporting of presumptive evidence of potential pathogens.

Potential savings and more rapid identification and reporting may be achieved through use of prepackaged commercially available kit systems for microbial identification but their indiscriminate use on increasing numbers of isolates will incur additional cost. Studies should be conducted to evaluate the impact on patient care of limited identification when isolates are observed in mixed culture or fail to correlate with morphologic types observed in Gram-stained direct smears when there is evidence of superficial contamination. Objective criteria should be established to define the clinically necessary levels of statistical confidence which should be sought in applying criteria to microbial identification. We must continue to develop more objective criteria to establish priorities for allocation of resources to those procedures which offer the most benefits to the patient at the lowest cost.

LITERATURE CITED

1. Bartlett, R. C. A plea for clinical relevance in medical microbiology. *Am J Clin Pathol 61:*867–872, 1974.
2. Bartlett, R. C. *Medical Microbiology Quality Cost and Clinical Relevance*. New York: John Wiley & Sons, 1974.
3. Bartlett, R. C. Control of cost and medical relevance in clinical microbiology. *Am J Clin Pathol 64:*518–524, 1975.
4. Boyle, V. J., Fancher, M. E. and Ross, R. W., Jr. Rapid, modified Kirby-Bauer susceptibility test with single, high concentration antimicrobial disks. *Antimicrob Agents Chemother 3:*418–424, 1973.
5. Cherry, W. B., and Moody, M. B. Fluorescent antibody techniques in diagnostic bacteriology. *Bacteriol Rev 29:*222–250, 1965.
6. Ewing, W. H., and Martin, W. J. Enterobacteriaceae, pp. 189–219. In *Manual of Clinical Microbiology,* Ed. 2. E. H. Lennette, E. H. Spaulding, and J. P. Truant (Eds.). Washington, D. C.: American Society for Microbiology, 1974.
7. Griner, P. F., and Liptzin, B. L. Use of the laboratory in a teaching hospital; implications for patient care, education and hospital costs. *Ann Intern Med 75:*157–163, 1971.
8. Harkness, J. L., Hall, M., Ilstrup, D. M., and Washington, J. A., II. Effects of atmosphere of incubation and of routine subcultures on detection of bacteremia in vacuum blood culture bottles. *J. Clin Microbiol 2:*296–299, 1975.
9. Hugh, R., and Gilardi, G. L. Pseudomonas, pp. 250–269. *Manual of Clinical Microbiology,* Ed. 2. E. H. Lennette, E. H. Spaulding, and J. P. Truant (Eds.), Washington, D.C.: American Society for Microbiology, 1974.
10. Mitruka, B. M. *Gas Chromatographic Applications in Microbiology and Medicine*. New York: John Wiley & Sons, 1975.
11. Nord, C. E., Lindberg, A. A., and Dahlback, A. Evaluation of five test kits—API, Auxotab, Enterotube, PathoTec, and R/B-for identification of Enterobacteriaceae. *Med Microbiol Immunol 159:*211–220, 1974.
12. Report, Ad Hoc Committee, State of Connecticut, Department of Health: Medical Microbiology Laboratory Utilization. *Conn Med 39:*499–505, 1975.
13. Robertson, E. A., and MacLowry, J. D. Construction of an interpretive pattern directory for the API 10S kit and analysis of its diagnostic accuracy. *J. Clin Microbiol 1:*515–520, 1975.
14. Rytel, M. Counterimmunoelectrophoresis in diagnosis of infectious disease. *Hosp Practice 10:*75–82, 1975.
15. Smith, P. B., Tomfohrde, K. M., Rhoden, D. L., and Balows, A. API system: a multitube micromethod for identification of *Enterobacteriaceae*. *Appl Microbiol 24:*449–552, 1972.
16. Sutter, V. L., Vargo, V. L., and Finegold, S. M. *Wadsworth Anaerobic Bacteriology Manual,* Ed. 2. Los Angeles: Anaerobic Bacteriology Laboratory, Wadsworth Hospital Center, Veterans' Administration, 1975.

17. Thornsberry, C., and Kirven, L. A. Ampicillin-resistance in *Hemophilus influenzae* as determined by a rapid test for β-lactamase production. *Antimicrob Agents Chemother* *6:*653–654, 1974.
18. Vestal, A. L. *Procedures for the Isolation and Identification of Mycobacteria, Publication 73-8230*. Atlanta: U.S. Department of Health, Education and Welfare, Center for Disease Control, 1973.
19. Weaver, R. E., Tatum, H. W., and Hollis, D. G. *The Identification of Unusual Pathogenic Gram Negative Bacteria*. Atlanta: U.S. Department of Health, Education and Welfare, Center for Disease Control, 1972.
20. *A Workload Recording Method for Clinical Laboratories,* Ed. 2. Chicago: College of American Pathologists, 1972.

3

The Speciation Polemic: An Analysis of the Debate Largely from an Epidemiologic Point of View

DONALD B. LOURIA

Archibald MacLeish noted "The loyalty of science is not to humanity but to truth—its own truth and that law of science is not the law of the good, what humanity thinks of as good, meaning moral, decent, humane, but rather the law of the possible. What is possible for science to know, science must know. What is possible for technology to do, technology will have done. The frustration, and it is a real and debasing frustration, in which we are mired today will not leave us until we believe in ourselves again, assume the mastery of our lives, the management of our means."

That elegant statement of man's growing battle with his burgeoning technological capabilities is perhaps too strong and comprehensive to be applied to speciation in microbiology, but it is a marvelous statement of one of the major dilemmas of our time—how to use our growing technology most effectively without having it obfuscate or confuse.

The polar views of the recent technological advances in microbiological speciation can be expressed as follows. On the one side is the view that we have gone too far, that speciation is usually unnecessary, and the most important data are simple identification and antibiotic sensitivity data. This view, in essence, rejects the new technology as cumbersome and largely unnecessary.

The contrary view holds that it is highly desirable to always have the maximum amount of information, and that speciation, as well as biotyping, should be carried out uniformly. This view says that the more information available to the physician, the better.

My own view is that neither polar view makes complete sense. There is a middle ground that is more desirable.

Clearly, there are instances in which speciation and biotyping is absolutely necessary. For example, several years ago, we were confronted with 3 cases of systemic moniliasis occurring on a single surgical ward over a short time span. We were justifiably concerned that we were dealing with a point source epidemic in which the yeast was being introduced from, for example, contaminated infusions. This concern was obviated by the speciation studies that showed that each of the 3 isolates was different; one was *Candida albicans,* one was *C. tropicalis* and one was *C. parapsilosis.* Here, speciation was absolutely necessary to define the epidemiological circumstances. Indeed, had speciation not been carried out, a series of judgmental errors could easily have been made.

There are other situations in which speciation (using the term in its broadest context to include biotyping and other efforts at the most precise definition of the isolate) is clearly beneficial. Maki and Martin (3) investigated a nationwide epidemic of septicemia caused by contaminated infusion products and found that the implicated organisms were *Enterobacter cloacae* and *Enterobacter agglomerans.* Speciation in that case was essential if the extent of the epidemic and its point source were to be defined precisely. Similarly, Adler and his colleagues (1) studied the epidemiological pattern of proteus infections by the use of proticine typing; they found that proticine types 6 and 9 were hospital acquired. Other studies have suggested that clinical proteus infections are caused by spread of organisms from the patient's own gastrointestinal tract. In point of fact, the latter observation can be made only by biotyping, proticine typing, or other similar techniques; to show merely that a *Proteus* species is in the stool and in the urine, wound, or blood provides limited information. If, however, a specific proticine type or biotype is found in both stool and blood, and it can be shown that stool colonization occurs earlier, then the association of stool carriage and deep-seated infection is reasonably documented. Without such specific information, the epidemiologic pattern of hospital-acquired infections cannot be defined adequately.

Another unequivocal example of the benefit of speciation relates to group D streptococcal infections. Group D includes enterococci as well as *Streptococcus bovis* and *Streptococcus equinus.* Until relatively recently, most laboratories confined themselves to determining for the clinician whether a given organism was a group D streptococcus, presumably an enterococcus. This had certain implications for the clinician. It is well known that enterococci usually resist penicillin alone. Some data indicated that perhaps 20–40% of enterococci were susceptible to penicillin and, in those cases, penicillin alone might well be perfectly adequate as therapy. But the majority of clinicians, mindful of the convincing data showing that the recovery rate in enterococcal endocarditis was greatest if a combination of penicillin and streptomycin was used, would administer the two in combination, the penicillin for 4–6 weeks in large dosage intravenously. Only a few clinicians would risk treating enterococcal endocarditis with penicillin alone; if they did, they would likely be criticized by their colleagues, regardless of in vitro sensitivities. In point of fact, many of the penicillin-sensitive "enterococci" were probably *S. bovis,* but the clinician could not take a chance on undertreating an enterococcus and so the enterococcal regimen was utilized. Penicillin alone is the treatment of choice for most *S. bovis* infections, and, since *S. bovis* is ordinarily very sensitive to penicillin, the dosage need not be nearly as great as for enterococci, and this in turn means the antibiotic can be administered intramuscularly or even by mouth (4).

If patients are treated unnecessarily with the combination of large amounts of penicillin and streptomycin for *S. bovis* infection, the patient is avoidably exposed to potentially toxic antimicrobials, and to the risk of potentially lethal superinfection such as that caused by species of *Candida* (4). If a laboratory fails to speciate a group D streptococcus, it is mandatory for the physician to assume the isolated organism is an enterococcus requiring treatment with the currently accepted regimens of either penicillin or ampicillin, plus an aminoglycoside. Here again, speciation is vital if the clinician is to treat the infection most intelligently with the smallest risk to the patient.

There are many other instances in which the maximum available information is helpful. For example, Noriega and colleagues (5) have reported recently that isolation of *C. guilliermondii* or *C. parapsilosis* from the blood stream suggests that the patient suffers from Candida endocarditis and this in turn compels the physician to look particularly for the epidemiologic circumstance most often associated with Candida endocarditis occurring outside the hospital, namely, intravenous use of illicit drugs. Speciation in such cases offers a great deal to the clinician. We have just observed a small epidemic of staphylococcal endocarditis in a population of drug users in Newark. Coincident with this outbreak, the heroin-using population was shifting from the white to the more available brown heroin, smuggled from Mexico and the Far East. One possibility to explain the outbreak was contamination of the herion with a specific staphylococcus; the other was coincidence. More definition was needed than *Staphylococcus aureus,* coagulase positive. One helpful technique, of course, is serotyping; this was carried out, but most of the isolated strains were nontypable. As an alternative, we attempted to determine if the strains were identical by antibiotic sensitivity testing using the disc method. All but one of the eight strains were penicillin-resistant, but sensitive to other antibiotics tested. Even a single deviation from the general pattern was in this case disconcerting. Was this a significant difference, this single deviation on antibiotic testing? I think probably it was not. There are obvious limitations to making epidemiologic inferences from antibiotic sensitivity tests.

If the equivalent of biotyping were available for staphylococci, or if we had available to us better techniques for serotyping, the information derived would be of immense importance in defining the epidemiologic pattern of the outbreak.

Another example of the advantages of speciation relates to Pseudomonas infections. If a laboratory reports a Pseudomonas in blood cultures, this ordinarily tells the clinician to treat with drugs such as gentamicin or carbenicillin, but if the organism is *Pseudomonas cepacia,* these drugs are worthless. If *P. cepacia* is isolated, the agents of choice are chloramphenicol or trimethoprim-sulfamethoxazole (2). Furthermore, because of the growing association of this organism with endocarditis among intravenous users of illicit drugs, the clinician would promptly investigate the possibility that the patient used such drugs parenterally. Here then, speciation is absolutely crucial.

The advantages of speciation seem clear, but there are disadvantages, too. Rennie and Duncan (7) studied Klebsiella strains by biotyping, and they were able to increase the number of demonstrably different strains from 62 to over 100. They found that in each serotype there was more than 1 biotype and in every biotype there was more than 1 serotype (4). The implications are mind-boggling. The more sophisticated the analysis, the more confusing it may be to the clinician. If a clinician is told that, from the blood stream of 3 patients in the same ward, 3 different klebsiellas have been isolated, one biotype 22-64-31, one biotype 22-63-32, one biotype 22-63-33, what is he to think? Are these significant differences? Is he to think there are multiple environmental sources for the observed infections, or should he assume these are small differences in biotype of no clinical importance?

To me, this is a major issue. It is well known that aberrant microbial forms (and these can be created by a variety of influences) may take many months to regain the biochemical characteristics of the parent organism (8). Additionally, mutants arising in any colony may result in change in the biochemical characteristics of a given subculture of the strain. All that is fine for the microbiologist, but it does not help the clinician. He needs to know how to interpret the clinical data in light of the various pieces of laboratory data provided to him. At times, our laboratories are getting so sophisticated they are not providing the simple bits of information that are of great clinical significance. For example, despite the marriage for antigenic reasons of low and high Klebsiella serotypes, there are vast virulence differences between the old Friedlander's bacilli (Klebsiella types A–F or 1–6) and the higher serotypes. If 10 low serotype organisms are injected into the peritoneum of mice, they will multiply rapidly, ordinarily killing the mouse within 48 hr. In striking contrast, injection of 1,000,000–1,000,000,000 organisms of higher serotypes is required to produce lethality. Thus, the low and higher serotypes may be antigenically similar, but they are dissimilar in animal virulence. And there are major differences in disease patterns in the human lung between low and high serotypes; the low serotypes often produce a far more virulent, progressive, and necrotizing pneumonia. To my way of thinking, the most important piece of information a laboratory can supply is whether the isolated Klebsiella is sticky, a characteristic of the highly encapsulated low serotypes. Although many infectious disease authorities would disagree, my own feeling is that in pneumonia caused by sticky, low serotypes, the proven regimens are tetracycline or chloramphenicol plus streptomycin with kanamycin as an alternative. Gentamicin is probably satisfactory but I personally would not treat clinical Friedlander's pneumonia with cephalosporins alone, regardless of in vitro sensitivity tests. Here then, at least in my view, is an instance in which the most important information is not sophisticated serotyping or biotyping but rather the simple characteristic of colony stickiness.

Listed in Tables 3.1 and 3.2 are the disc and tube sensitivities of *Salmo-*

TABLE 3.1
Disc sensitivities of *Salmonella heidelberg* isolates*

Source	Chloramphenicol	Ampicillin	Kanamycin	Streptomycin	Tetracycline	Cephalothin	Polymixin	Gentamicin	Trimethoprim-Sulfamethoxazole
Nosocomial									
Case 1	R	R	S	R	R	S	S	S	R
Case 2	S	R	S	S	R	S	S	S	–
Case 5	R	R	S	R	R	S	S	S	SS
Case 9	R	S	S	R	R	S	S	S	–
Kitchen worker	R	R	S	R	R	S	R	R	–
Practical nurse	S	S	S	R	S	S	S	S	–

* R, resistant; S, sensitive; –, SS, slightly sensitive.

TABLE 3.2
Test tube sensitivities of *Salmonella heidelberg* isolates

Isolate Identification: Nosocomial Case	Isolated From	Concentrations of Antibiotic Inhibiting Growth at 24 hr by Gross Observation			
		Chloramphenicol	Ampicillin	Gentamicin	Cephalothin
		μg/ml	μg/ml	μg/ml	μg/ml
1	Sputum	>50	>50	>50	50
2	Urine	>50	>50	25	50
3	Stool	>50	>50	1.5	–
4	Blood	>50	25	>50	12.5
5	Stool	>50	50	>50	25

nella heidelberg isolated from the stools or blood of patients during a recent epidemic among patients hospitalized at the Martland Medical Center in Newark. There are certain discrepancies suggesting that all of a sudden we had multiple strains of *S. heidelberg* causing severe clinical disease in our hospital.

Would one assume, on the basis of the sensitivities, that the kitchen worker or the nursing aid were not part of the epidemic? If the *S. heidelberg* strains were biotyped and more than one biotype was found, would this indicate more than one source and more than one organism? I think not. Chance variation in test outcome or genetic modification of the single causative agent would be more likely. If the microbiologist or the laboratory results of the antibiotic testing or biotyping or subtyping within a serotype led the clinician to believe he was dealing with two or more nosocomial pathogens, he might make some erroneous clinical and epidemiologic judgments with regard to disease patterns and disease control.

Similarly, by speciation techniques, Overdorf, Wilkins, and Ressler (6) found 90 strains of *Providencia stuartii* in a burn unit. Of these strains, 5% were sensitive to gentamicin, 35% were sensitive to tobramycin, but they were identical in resisting kanamycin, ampicillin, carbenicillin, tetracycline, and chloramphenicol. Are these really different strains, or would it have been better for the clinician to know merely that *P. stuartii* was causing problems in the burn unit and that amikacin was the antibiotic of choice. From the epidemiologic viewpoint, if there were really 90 different invading strains, it would be an epidemiologic nightmare.

What this suggests to me is that speciation is valuable and that data on speciation and biotyping should be accumulated and stored via a computerized system for recall months or years later. Unless this is done, we will not be able to make proper judgment about epidemiological patterns as well as the relative virulence of different species and biotypes. But speciation can also confuse and thus be a disservice to the clinician if carried too far.

The argument against speciation says, in essence, that the physician wishes to know the nature of the organism in the broadest sense and the sensitivity of the isolated organism but regards additional information as superfluous. Thus, one could argue that it would be adequate in many cases for the physician to know only that a possible pathogen was isolated and what agents are effective against it in vitro. For example, one might report: "blood culture positive; enterobacter group; sensitive gentamicin, chloramphenicol, kanamycin, and tobramycin."

In part, the polemic has arisen because the nomenclature and categories in which organisms are placed change with such regularity and unpredictability. It may well be that the unending changes are indeed indicated on the basis of biochemical tests and antigenic composition. However, that does not help the physician. He finds it difficult to make much clinical sense out of the separation of klebsiellas from nonpigmented serratias, enterobacters, citrobacters, or arizonas. The clinician finds he cannot (and does not want to) keep track of specific genuses, let alone individual species within the group. And it is not only the clinician who is confused. The infectious disease staff of most hospitals has, for the most part, an inadequate familiarity with the latest version of the speciation nomenclature and even of the changes in genus categorization.

Is speciation necessary in most cases? The answer is that from the point of view of the individual case and the individual practicing physician, it is not. What the physician most often needs is relatively simplistic identification accompanied by enough antibiotic sensitivity data to permit the initiation of a therapeutic program and to provide enough data to enable a sensible plan of action if the patient does not respond to the treatment program initially administered.

If speciation (including biotyping) is to be useful, the lines of communication between laboratory and clinician must be kept open. Perhaps the best way to achieve this is to make sure that the makeup of every hospital's infection committee includes the microbiologist, the infection control officer (or hospital epidemiologist), and the infectious disease clinician. Of course, often the latter two will, in point of fact, be the same person. The committee must work to make sure the clinician receives the most useful information and does not get inextricably mired in the complexities of biotyping.

To return to my initial point: our technology permits us to define a given organism ever more precisely. Now we have to make sure that this same technology contributes to the well-being of individual patients and helps the physician.

LITERATURE CITED

1. Adler, J. L., Burke, J. P., Martin, D. F., and Finland, M. L. Proteus infections in a general hospital. *Ann Intern Med 75:*517–531, 1971.
2. Giardi, G. L. Infrequently encountered Pseudomonas species causing infection in humans. *Ann Intern Med 77:*211, 1972.
3. Maki, D. B., and Martin, W. T. Nationwide epidemic of septicemia caused by contaminated infusion products. *J Infect Dis 131:*267, 1975.
4. Moellering, R. C., Watson, B. K., and Kunz, L. J. Endocarditis due to group D streptococcus. *Am J Med 57:*239, 1974.
5. Noriega, E. R., Rubinstein, E., Simberkoff, M., and Rahal, J. J. Subacute and acute endocarditis due to *Pseudomonas cepacia* in heroin addicts. *Am J Med 59:*29, 1975.
6. Overdorf, G. D., Wilkins, J., and Ressler, R. Emergence of resistance of *Providentia stuartii* to multiple antibiotics; speciation and characterization of providentia. *J Infect Dis 129:*353, 1974.
7. Rennie, R. P., and Duncan, I. B. R. Combined biochemical and serological typing of clinical isolates of Klebsiella. *Appl Microbiol 28:*534, 1974.
8. Zierdt, C. H., and Westlake, P. T. Transitional forms of *Corynebacterium acnes* in disease. *J Bacteriol 91*(47):327, 1944.

4

Blood Cultures—Merits of New Media and Techniques—Significance of Findings

JOHN A. WASHINGTON II

Whereas there is little question about the clinical significance of bacteremia, because of the degree of morbidity and mortality associated with it, there remain a number of questions about the optimal means of its detection in the clinical laboratory. Many of our currently recommended procedures reflect arbitrary but unproven practices, and it has been only in very recent years that some of these procedures have become objects of scientific scrutiny. Among the variables known to influence the detection of bacteremia are the manner of collection and transport of blood, the types and composition of media employed and their atmosphere of incubation, the timing and frequency with which media are examined, and the timing of routine or "blind" subcultures. Because of the known influence of such variables on the detection of bacteremia, it behooves those following the literature on this subject to study carefully those portions of published reports describing the materials and methods employed in any investigation of variables or comparisons of systems. In many instances, studies purporting to demonstrate the superiority of a new system of blood culture have used so-called conventional methods as a basis for comparison which, in my opinion, are scarcely adequate and compared with which almost any other system would have to be an improvement! A homemade blood-culture bottle, for example, is ordinarily not comparable to its commercially prepared counterpart containing the same medium, because few, if any, clinical laboratories have the equipment or expertise available in industry to bottle such media under vacuum and with CO_2 and, hence, to prepare a bottle containing media with an oxidation-reduction potential sufficiently low for the cultivation of anaerobic or "microaerophilic" bacteria.

In the discussion that follows, I will attempt to examine some of the variables known to influence the detection of bacteremia and to point out those areas about which we remain ignorant.

SKIN ANTISEPSIS

Preparation of the skin for venipuncture is an often overlooked variable. It has considerable significance, however, because of its direct bearing on the incidence of contamination of cultures with microorganisms that, although usually regarded as harmless commensals, have been increasingly recognized as causing infectious diseases in patients with implanted prosthetic material. The isolation, for example, of a *Staphylococcus epidermidis* or of a *Propionibacterium acnes* from such a patient cannot be disregarded casually. Contamination rates of blood cultures with such microorganisms relate closely to the care exercised in the preparation of the venipuncture site. Blood for culture at the Mayo Clinic and its affiliated hospitals is routinely collected by trained venipuncture teams; however, as a quality control measure, we maintain a chart showing the number of presumed contaminants—that is *Alcaligenes, S. epidermidis, P. acnes, Corynebacterium,* and *Bacillus*—isolated from single sets of blood cultures only on a daily basis (Fig. 4.1). Any marked increase in the isolation rates of these types of organisms is an indication for us to reemphasize aseptic techniques to our venipuncturists. As a rule, such organisms are isolated from approximately 2% of all cultures collected in a month's time.

In general, preparation of the skin with alcohol followed by tincture of

Figure 4.1

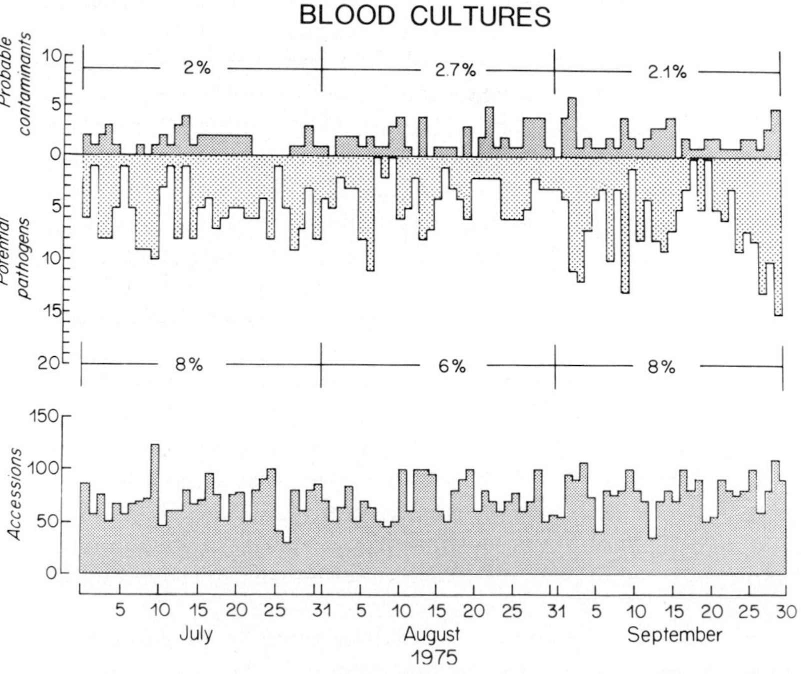

Quality control record during 3 months of 1975 illustrating number of blood-culture sets accessioned daily and number of probable contaminants and potential pathogens isolated daily. Monthly isolation rates in each category are also shown.

iodine or an iodophor should provide satisfactory antisepsis, provided that a suitable period of time (1–2 min) is allowed to elapse for the iodine or iodophor to be effective on the skin and that the venipuncturist's fingers are similarly prepared should palpation of the vein become necessary.

NUMBER OF CULTURES AND VOLUME OF BLOOD TO BE COLLECTED

Studies in our laboratory of bacteremic patients without endocarditis from whom at least four separate sets of cultures were collected within a 24-hr interval have shown that in approximately 80% the organisms were detected with the first set of cultures, in approximately 90% within the first two sets of cultures, and in 99% within the first three sets of cultures (Fig. 4.2) (36). In other words, it is rarely necessary to collect more than three separate sets of blood cultures per 24-hr interval, and we now require the clinician to consult with our laboratory staff before more than three sets of blood cultures are to be collected within 24 hr. Conversely, however, we believe that collection of only a single set of blood cultures is hazardous. In the first place, such a practice reflects ignorance about the general intermittency of bacteremia, other than that associated with intravascular infection; secondly, it makes it nearly impossible to evaluate the clinical significance of isolation of the types of organisms that are presumed contaminants but that may be causing disease in patients with, for example, implanted prosthetic material. The repeated isolation of such organisms from two or three separate sets of blood cultures would certainly be regarded with a high degree of suspicion. In contrast, their isolation from the only set of blood cultures collected from a patient raises more questions than it resolves. The isolation of viridans streptococci from the blood of

Figure 4.2

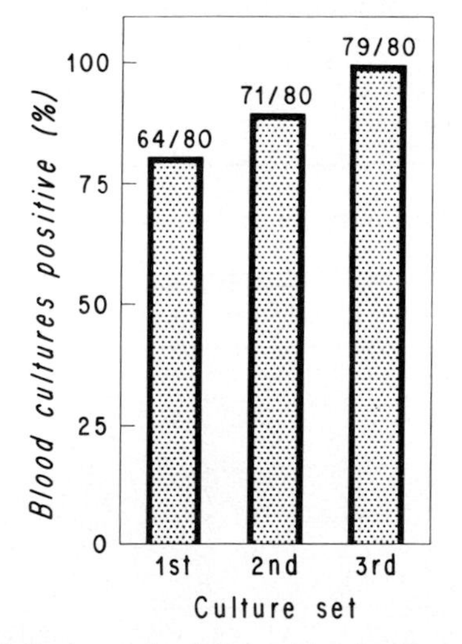

Cumulative rates of positivity in each of three blood-culture sets from 80 bacteremic patients. (From J.A. Washington II: *Mayo Clin Proceedings 50:*91–98, 1975. Reproduced with permission.)

patients undergoing a large variety of manipulations ranging from dental procedures to sigmoidoscopy has been well documented in the literature. The isolation of such an organism from the blood of a patient with underlying cardiac valvular disease certainly has important implications. Its isolation from a single set of cultures could represent transient bacteremia resulting from any one of a variety of procedures or activities, including mastication; however, it could also represent intravascular infection. For this reason, we have established a policy whereby an order for a single set of blood cultures is automatically doubled by our venipuncture team, and two separate cultures are routinely collected.

Werner et al. (39) have demonstrated that, in nearly all patients with endocarditis, the etiologic agent was isolated from the first two sets of blood cultures collected. We have confirmed these findings. The isolation of bacteria from patients with endocarditis is commonly and erroneously regarded as being difficult, and this leads many house officers to collect an excessive number of blood cultures. Before we established a policy limiting the number of cultures collected from our hospitalized patients to three per 24 hours, we had several patients in our records with 17–19 positive sets of blood cultures.

It can honestly be said that the ideal volume of blood to be collected for culture is unknown. Because of the fact that the order of magnitude of bacteremia is low (Tables 4.1 and 4.2) (33, 39), there must obviously be a minimum volume of blood which should be collected for culture. However, that minimum has never been established. Recommendations of 5–10 ml in adults and of 1–5 ml in infants and children are commonly made (1, 9, 25, 36) and should probably be followed until evidence to the contrary becomes available.

In my opinion, the number of separate blood cultures collected is as important as the volume of blood collected. The intervals between collections of blood can be arbitrarily set at 1 hr but must be sufficiently flexible to accommodate to clinical circumstances. In the critically ill patient in

TABLE 4.1
Order of magnitude of bacteremia in patients with endocarditis*

Colony-forming *Units/ml*†	Streptococci		Staphylococci	
	no.	%	*no.*	%
1–10	98	24	7	41
11–20	73	18	5	29
21–30	50	12	0	0
31–40	30	7	1	6
41–50	30	7	0	0
51–100	63	15	2	12
>100	71	17	2	12

* Data adapted from A.S. Werner, C.G. Cobbs, D. Kaye, et al.: *Journal of the American Medical Association, 202:*199–203, 1967 (39).

† Data do not include cultures positive in broth only.

TABLE 4.2
Order of magnitude of bacteremia after genitourinary tract manipulation*

Organism	Average No. of Colony-forming *Units/ml*†
Staphylococcus aureus	7.6
S. epidermidis or *Micrococcus*	0.2
Streptococcus	
Group A	0.8
Group D	1.4
Viridans group	0.8
Peptostreptococcus	0.2
Escherichia coli	8.2
Klebsiella pneumoniae	84.3
Enterobacter aerogenes	0.2
Proteus mirabilis	1.1
Pseudomonas aeruginosa	10.3

* Data adapted from N.M. Sullivan, V.L. Sutter, W.T. Carter, et al.: *Applied Microbiology, 23:*1101–1106, 1972 (33).
† Data do not include cultures positive in broth only.

whom parenteral antimicrobial agents are being started with some degree of urgency, it obviously becomes necessary to shorten the intervals between collections substantially. It remains important, however, to collect blood for cultures from separate venipuncture sites, rather than to try to collect a large volume of blood from one site and distribute it into several sets of blood-culture bottles. The reason for this should be obvious but is commonly overlooked under the guise of simplicity or convenience. Contamination of a large volume of blood will yield many sets of positive cultures, the interpretation of which may be erroneous and may lead to inappropriate therapy.

In our hospitals, blood is inoculated at the bedside directly into the blood-culture bottles. In some institutions, blood is transported to the laboratory in tubes containing an anticoagulant, sodium polyanetholsulfonate (SPS), and is then inoculated into blood-culture media (7). To the best of my knowledge, no direct comparison has been made of these two approaches with clinical specimens, and I am hesitant to recommend transport in SPS as a routine procedure until results of such studies become available.

MEDIA

The variety of media employed for blood culture is almost infinite. Although we have compared many media in terms of detection rates and time intervals to detection of bacteremia, there are many more media that we have not tested and probably will never test because of the sheer magnitude of such projects. Briefly summarizing our experiences, we have demonstrated that the performance in vacuum blood-culture bottles with SPS of soybean-casein digest broth (for example, Tryptic or Trypticase soy) is as good as, and with some species superior to, that of thioglycollate broth, Thiol broth, and Columbia broth (36). For example, our studies have demonstrated significantly fewer isolations of *Pseudomonas aeruginosa* from Thiol than from Tryptic soy broth (TSB); conversely, anaerobes were not isolated more frequently from thioglycolate or Thiol than from TSB (Table 4.3) (12). It would, therefore, be a serious mistake to rely exclusively on Thiol for detection of bacteremia. We also demonstrated in a comparative study that there were significantly fewer isolations of *Staphylococcus aureus* from Columbia broth than from TSB (Table 4.4) (13). We have, therefore, continued to employ TSB as our standard blood-culture medium and as the medium against which other media or blood culture systems are evaluated.

There has been much interest in recent years in the role of hypertonic media in enhancing bacterial detection. Whereas Sullivan et al. (33), Rosner (30, 31), and Henrichsen and Bruun (15) did demonstrate increased recovery of various groups of bacteria in media made hypertonic by the

TABLE 4.3
Isolates in positive cultures, by medium*†

Organism	No. Positive in			Total Positive‡	P§	Adjusted Percent positive¶
	TSB and Thiol	TSB only	Thiol only			
Bacillus	0	7	7	14	NS	–
Clostridium	2	1	1	4	NS	0.8
Corynebacterium	8	70	20	98	<0.001	–
Escherichia	87	22	10	119	<0.1	25.1
Salmonella	2	0	0	2	NS	0.4
Citrobacter	2	2	1	5	NS	1.1
Klebsiella	35	12	5	52	NS	10.9
Enterobacter	6	3	2	11	NS	2.3
Proteus	11	3	2	16	NS	3.4
Haemophilus	5	5	0	10	<0.1	2.1
Streptococci						
S. pneumoniae	14	3	3	20	NS	4.2
Group A	6	1	1	8	NS	1.7
Group D	27	5	4	36	NS	7.6
Other groups	0	0	1	1	NS	0.2
Viridans	30	4	5	39	NS	8.2
Acinetobacter	0	4	1	5	NS	1.1
Alcaligenes	1	1	0	2	NS	0.4
Bacteroidaceae	23	10	3	36	<0.1	7.6
Staphylococci						
S. aureus	45	20	11	76	NS	16.0
S. epidermidis	7	19	15	41	NS	–
Peptostreptococcus	0	1	0	1	NS	0.2
Peptococcus	0	2	0	2	NS	0.4
Pseudomonas	2	25	0	27	<0.001	5.7
Candida	0	2	0	2	NS	0.4
Torulopsis	0	2	0	2	NS	0.4
CDC group IIIA	1	8	4	13	NS	–

* From M. Hall, E. Warren, and J. A. Washington II: *Applied Microbiology, 27:*187–191, 1974 (12); by permission of the American Society for Microbiology, Washington, D.C.
† TSB = Tryptic soy broth.
‡ Total = 642.
§ By chi square analysis, for difference between media.
¶ Based on total positive minus 166 presumed contaminants equals 475.

TABLE 4.4
Numbers of isolates in positive cultures, by medium*†

Organism	TSB and Columbia	TSB Only	Columbia Only	Total Positive	P‡
Bacillus	0	7	24	31	<0.01
Clostridium	6	0	2	8	NS§
Corynebacterium	23	55	40	118	NS
Lactobacillus	2	0	1	3	NS
Escherichia	79	27	21	127	NS
Salmonella	0	0	1	1	NS
Citrobacter	2	0	0	2	NS
Klebsiella	37	5	5	47	NS
Enterobacter	2	2	1	5	NS
Serratia	10	1	0	11	NS
Proteus	11	8	7	26	NS
Haemophilus	10	1	0	11	NS
Listeria	2	0	2	4	NS
Streptococcus					
S. pneumoniae	3	4	2	9	NS
Viridans	36	5	4	45	NS
Group A	0	0	1	1	NS
Group B	0	2	2	4	NS
Group D	19	5	1	25	NS
Eubacterium	0	1	0	1	NS
Acinetobacter	0	2	1	3	NS
Alcaligenes	0	1	3	4	NS
Flavobacterium	1	0	0	1	NS
Bacteroidaceae	18	11	15	44	NS
Micrococcus	0	0	1	1	NS
Staphylococcus					
S. aureus	60	25	5	90	<0.01
S. epidermidis	25	33	19	77	NS
Peptostreptococcus	3	1	0	4	NS
Peptococcus	0	0	1	1	NS
Veillonella	1	0	0	1	NS
Pseudomonas	35	15	6	56	0.05
Aeromonas	3	1	1	9	NS
Candida	6	1	1	8	NS
Torulopsis	0	3	0	3	NS

* From M. Hall, E. Warren, and J. A. Washington II: *Applied Microbiol, 27*:699–702, 1974 (13). By permission of the American Society for Microbiology, Washington, D.C.
† TSB = Tryptic soy broth.
‡ For difference between media.
§ NS = not significant.

addition of sucrose, we were unable to confirm their findings in a comparison of TSB with and without 15% sucrose (Table 4.5) (37). Because the media and conditions employed in each of these studies varied considerably, the reasons for the differences in results are certainly unclear, and one's decision whether or not to use hypertonic media would have to be guided by the similarity of one's techniques to those employed in any of these studies. Our studies, at least, clearly demonstrated that the addition of 15% sucrose to TSB with SPS in vacuum blood-culture bottles did not enhance bacterial detection and, in some instances, actually impaired it

TABLE 4.5
Numbers of isolates in positive cultures, by medium*†

Organism	Both Media	TSB with SPS Only	TSB with SPS and Sucrose Only	*P*
Bacillus	1	7	23	<0.01
Clostridium	0	2	0	NS‡
Corynebacterium	17	32	37	NS
Escherichia	37	7	13	NS
Klebsiella	13	4	4	NS
Enterobacter	2	1	0	NS
Serratia	7	0	2	NS
Proteus	13	3	3	NS
Cardiobacterium	4	3	1	NS
Haemophilus	3	7	0	<0.05
Streptococcus				
S. pneumoniae	9	1	0	NS
Viridans group	22	5	5	NS
Group A	5	4	0	NS
Group D	6	1	3	NS
Other groups	0	0	1	NS
Alcaligenes	0	3	9	NS
Neisseria	0	1	0	NS
Bacteroidaceae	13	7	0	<0.05
Staphylococcus				
S. aureus	36	13	3	<0.05
S. epidermidis	13	22	11	NS
Peptostreptococcus	0	0	1	NS
Veillonella	0	1	0	NS
Pseudomonas	14	5	2	NS

* From J. A. Washington II, M. M. Hall, and E. Warren: *Journal of Clinical Microbiology, 1:*79–81, 1975 (37). By permission of the American Society for Microbiology, Washington, D.C.

† TSB = Tryptic soy broth; SPS = sodium polyanetholsulfonate.

§ NS = not significant.

(Table 4.5). Again, careful attention must be paid to the materials and methods employed in blood-culture studies.

ATMOSPHERE OF INCUBATION

This represents an area of considerable recent interest. In vitro studies have demonstrated delayed and decreased recovery of pseudomonads and yeasts from unvented vacuum blood-culture bottles and, conversely, delayed and decreased recovery of Bacteroidaceae from vented vacuum blood-culture bottles (10, 18). These data have been substantiated by Blazevic et al. (3) and by Harkness et al. (14) in clinical studies comparing vented and unvented vacuum blood-culture bottles containing Columbia broth and TSB, respectively. It has, therefore, been established that, for the time being at least, it is necessary to use two vacuum blood-culture bottles and to vent one of these but not the other. It is recommended that venting of the one bottle be transient only—that is, sufficiently long to release the vacuum but not the CO_2 usually present in such bottles. No further venting appears to be necessary.

Incidentally, although it has been clearly shown that venting vacuum blood-culture bottles increases the rate of recovery of yeasts (3, 10, 14) and that there are no significant differences between recovery rates of yeasts from vented vacuum blood-culture bottles and bottles containing biphasic brain-heart infusion (BHI) medium (28), it has also been shown that recovery of yeasts in the biphasic BHI occurs significantly more rapidly than in vented TSB (Table 4.6).

Anaerobic bacteremia is no longer a rare phenomenon, and most clinically significant anaerobic bacteremias are due to the Bacteroidaceae (40). The detection of anaerobic bacteremia can be made reliably in commercially available vacuum blood-culture bottles, the oxidation-reduction potential of which is usually very low. We did not find supplemented, prereduced, anaerobically sterilized media to be superior to unvented TSB in the recovery of anaerobes (Table 4.7) (38), but we did find that a supplemented peptone broth was distinctly inferior to unvented TSB in the

TABLE 4.6
Time intervals to detection of yeasts in matched pairs of positive blood cultures*

Medium	Mean Time Interval to Detection of Positivity, Days	*P*
Biphasic brain-heart infusion (BHI), vented	2.6	<0.01†
Tryptic soy broth (TSB), vented	5.2	
Biphasic BHI, vented	3.2	0.028‡
TSB, unvented	9.0	

* Data adapted from G. D. Roberts, C. Horstmeier, M. Hall, et al.: *Journal of Clinical Microbiology 2:*18–20, 1975 (28).
† By the sign test.
‡ By the delta-*t* test.

TABLE 4.7
Numbers of anaerobic isolates in positive cultures*†

Organism	No. Positive					
	By medium			By combination of media		
	TSB	Thio	BHI	TSB + Thio	Thio + BHI	TSB + Thio + BHI
Bacteroides fragilis	18	19	19	1	1	15
B. melaninogenicus	2	1	1			1
Fusobacterium fusiforme	1					
Eubacterium cylindroides	1					
Peptococcus	1	1	1			1
Peptostreptococcus	1	1		1		

* From J. A. Washington II, and W. J. Martin: *Applied Microbiology, 25:*70–71, 1973 (38). By permission of the American Society for Microbiology, Washington, D.C.
† TSB = Tryptic soy broth; Thio = thioglycollate; BHI = brain-heart infusion.

recovery of anaerobes and other clinically significant bacterial isolates from blood (Table 4.8) (12).

EXAMINATION OF CULTURES

It is our standard practice to inspect blood-culture bottles later on the day of collection and daily thereafter for 7 days. Those cultures that have remained negative are then reincubated for an additional 7 days, inspected once more, and then discarded. Broth cultures should be examined closely for signs of turbidity, gaseousness, hemolysis, or colony formation, and Gram-stained smears should be made whenever there is the slightest indication of the presence of these macroscopic signs.

Our decision to examine blood cultures daily for the first 7 days and then again after the 14th day of incubation is somewhat arbitrary and also reflects our role as a referral center. It would seem entirely reasonable for a community hospital to retain blood cultures routinely for 7 days only, because at least 95% of clinically significant bacterial isolates are detected within the first 5–7 days after collection of the culture (Table 4.9). Patients with subacute endocarditis due to unusual or fastidious bacteria requiring longer than a week for their detection are frequently seen at the Mayo Clinic; hence, our blood cultures are retained for a longer time. One must, therefore, attempt to assess the characteristics of a hospital's patient population in deciding how long to retain blood cultures routinely and

TABLE 4.8
Comparison of Tryptic soy broth, Thiol broth, and aerobic and anaerobic Vacutainer culture tubes wit supplemented peptone broth*

Organism	Tryptic Soy Broth		Thiol Broth		Supplemented Peptone Broth in Vacutainer Tubes				*P*†
					Aerobic		Anaerobic		
	No. positive	Days (mean)	No. positive	Days (mean)	No. positive	Days (mean)	No. positive	Days (mean)	
Corynebacterium‡	51	8.2	13	12.5	13	8.4	11	7.4	<0.01
Escherichia coli	87	2.0	81	1.8	34	2.1	36	1.8	<0.01
Enterobacter	12	1.0	12	1.9	2	1.0	3	1.3	<0.01
Haemophilus	4	9.8	2	12.0	0		0		–§
Streptococci									
Viridans	17	1.8	14	2.4	6	1.7	7	1.6	<0.01
Group A	15	1.7	13	2.7	6	1.3	10	1.5	<0.01
Group D	6	1.5	6	1.3	2	2.0	3	1.7	–§
Alcaligenes	8	2.3	1	2.0	2	5.0	1	3.0	–§
Bacteroidaceae	24	2.3	26	2.5	9	3.4	8	3.4	<0.01
Staphylococci									
S. aureus	52	2.9	38	4.1	23	3.1	20	3.6	<0.01
S. epidermidis	41	4.5	30	5.4	17	3.2	16	4.2	<0.01
Pseudomonas	28	4.0	3	6.7	12	3.9	7	5.4	<0.01
Candida	6	3.2	2	4.0	2	4.5	1	2.0	–§

* From M. Hall, E. Warren, and J. A. Washington II: *Applied Microbiology, 27:*187–191, 1974 (12). B permission of the American Society for Microbiology, Washington, D.C.

† For hypothesis that proportions of positives are the same in all four media.

‡ Includes *Propionibacterium*.

§ Although *P* <0.05 in these instances, the sample sizes were too small for determination of significance

BLE 4.9
umulative percentage positive of some commonly isolated species, by medium*

Organism	Tryptic Soy Broth						Thiol Broth					
	By day (%)					No. positive	By day (%)					No. positive
	1	2	3	4	7		1	2	3	4	7	
cillus			14	43		7					43	7
ostridium	67	100				3	67	100				3
rynebacterium				3	41	78	4				21	28
cherichia	70	88	91	92	96	109	55	93	96		99	97
bsiella	43	79	81		96	47	48	88		90	98	40
terobacter	67	78			100	9	75	100				8
oteus	58	86	93		100	14	38	85		92	100	13
emophilus		20	90	100		10		20		40	100	5
reptococci												
S. pneumoniae	59	100				17	29	94	100			17
Viridans	32	59	71	74	97	34	34	69	83		94	35
Group A	29	100				7	67	100				7
Group D	72	97			100	32	61	94	97			31
cteroidaceae	6	39	73		94	33	8	50	77		90	26
aphylococci												
S. aureus	26	65		71	86	65	14	45	70	73	93	56
S. epidermidis	8	31	62	69	96	26	9	18	23	32	68	22
eudomonas	4	44	81	93	96	27	50	100				2

* From M. Hall, E. Warren, and J. A. Washington II: *Applied Microbiology, 27:*187–191, 1974 (12). By rmission of the American Society for Microbiology, Washington, D.C.

should avoid, if possible, making such a decision on the basis of incubation space available in the laboratory.

Routine preparation of a Gram-stained smear of macroscopically negative cultures has been recommended by Blazevic et al. (2), who initially detected 23% of their positive cultures by this means. These findings could not be substantiated by Hall et al. (13), who concluded that a routine Gram-stained smear of macroscopically negative cultures was not justified. The reasons for the differences between the results of our studies are speculative and have never been satisfactorily resolved. Obviously, in some people's hands the routine Gram-stained smear has been found to be helpful.

No one disagrees about the necessity for a Gram-stained smear of a culture macroscopically suspected to be positive, and it is absolutely essential that the technologist maintain an extraordinarily high degree of suspicion while inspecting blood cultures.

SUBCULTURE

The selection of media to be used in the subculture of a positive culture is guided by the Gram morphology of the organism. One principle bears emphasis and is based on the fact that between 6 and 10% of all clinically significant bacteremias are polymicrobic—that is, are due to two or more organisms. It is, therefore, mandatory that suitable differential media be used in subcultures of positive cultures. Since at least one of the two or more organisms present may be anaerobic bacteria, it is also imperative to include an extra blood-agar plate to be incubated anaerobically along with the other media used for the isolation of aerobic and facultatively anaerobic bacteria. Examining this issue from a different perspective, it has been our experience that 32% of clinically significant anaerobic bacteremias are polymicrobic (40). Polymicrobial bacteremia is, therefore, an important problem that demands careful laboratory attention.

Despite the problems posed by mixed cultures, it is nonetheless worthwhile to include in the subcultures performed such biochemical or other tests that are appropriate for identification of the organism seen in the Gram-stained smear. Moreover, it is of value to proceed with direct antimicrobial susceptibility tests. When the turbidity of the inoculum can be adjusted to correspond with that of the barium sulfate standard, the results of direct and conventional tests correlate closely. Nevertheless, the antimicrobial susceptibilities should be confirmed by the conventional test when isolated colonies are available.

The importance of the role of a routine or "blind" subculture of macroscopically negative blood cultures is beyond dispute. Our studies (13) and those of Blazevic et al. (2) established that routine subculture 24 hr after blood collection and again several days later provided the initial means of detection of at least 12% of positive cultures, particularly those containing *Haemophilus*, pathogenic species of *Neisseria, Moraxella, P. aeruginosa,* and yeasts. More recently, Harkness et al. (14) have reported that a routine subculture onto chocolate blood agar carried out on the day of blood collection provided the means of detection of nearly half of the positive blood cultures (Table 4.10). Similar data have been reported by Todd and Roe (35), who found that 85% of all significant blood culture isolates were detected within 24 hr of incubation. Of particular interest in their study, conducted in a pediatric hospital, was the finding that most of the cultures exhibiting growth on early subculture represented true bacteremias, in contrast to presumed contaminants that infrequently grew on early subculture. Our procedure currently, therefore, is to subculture routinely on the day of blood collection (usually between 3 and 18 hr after collection) and then to resubculture routinely 48 hr later. We perform no further routine subcultures of macroscopically negative bottles. Obviously, pour plates performed at the time of blood collection or shortly thereafter serve somewhat the same function as the early subculture; however, since pour plates impose staffing requirements that may not exist around the clock in

TABLE 4.10
Means of detection of 128 positive blood cultures*

Cultures	Routine Subculture (Days)			Macroscopically Positive	Total
	<1	1	5		
Number	61	7	1	59	128
Percent	48	5	1	46	100

* From J. L. Harkness, M. Hall, D. M. Ilstrup, et al.: *Journal of Clinical Microbiology, 2:*296–299, 1975 (14). By permission of the American Society for Microbiology, Washington, D.C.

all hospitals, and since they do not utilize chocolatized blood, it is our opinion that the early subculture to chocolate blood agar is the more convenient and better procedure. In some hospitals, the house staff has been recruited to perform pour plates and other microbiologic tests at odd hours; however, it has been my personal observation that the conditions and performance in the vast majority of house staff laboratories are sufficiently atrocious to warrant their being dismantled in any modern hospital in the interest of improved patient care. At least, if they cannot be dismantled, they should be subject to the same standards of quality assurance and performance as are the hospital's regular clinical laboratories.

It has been our experience that routine anaerobic subcultures are not warranted. The majority of clinically significant anaerobic bacteria isolated from blood are Bacteroidaceae (40), which are promptly detected macroscopically.

MISCELLANY REGARDING COLLECTION AND PROCESSING

Anticoagulants

Most laboratories currently use 0.025–0.05% SPS for anticoagulation. This substance is also antiphagocytic and anticomplementary and has represented one of the major contributions to the area of blood cultures in recent years. Its addition to blood-culture media has significantly enhanced bacterial recovery. Its principal disadvantages appear to be limited to its inhibitory effects on the growth of gonococci (8) and of *Peptostreptococcus anaerobius* (11). The latter phenomenon has been limited to in vitro studies and has not been substantiated in clinical studies (29, 30, 32). This latter disadvantage has, nonetheless, spawned considerable interest in a new polyanion, sodium amylosulfate (SAS), which has been shown not to be inhibitory to *P. anaerobius* (19, 20). We have recently evaluated SAS by comparing it in unvented TSB with unvented TSB containing SPS and have found no significant differences between the two in nearly 6000 cultures. In other words, we did not suddenly encounter a large number of bacteremias due to *P. anaerobius*. Since SPS is likely to be cheaper than SAS in the foreseeable future, there is as yet little or no clinical evidence to suggest that a change to SAS is warranted.

Other Additives

The value of the addition of penicillinase to blood cultures from patients receiving penicillins remains controversial. Since the advent of media under vacuum and CO_2 and containing SPS, there have been no studies establishing the value of adding penicillinase. There are numerous studies in the literature of "pseudosepticemia" wherein lots of penicillinase, contaminated either by the manufacturer or during use, have led to spuriously positive blood cultures. Although there may be an indication for its addition to blood cultures of patients receiving penicillins parenterally, it is essential that the laboratory add it to only one of the two bottles of blood-culture media inoculated and that the penicillinase solution itself be concurrently cultured to establish its sterility. Because of the conventional 10–20% vol/vol dilution of blood in broth, there are few indications for the addition of penicillinase to blood cultures.

CULTURE FOR OTHER ORGANISMS

Brucellae

It has become reasonably established practice to inoculate blood from patients suspected of having brucellosis into a bottle with biphasic media (Castaneda principle). In this system, the blood broth mixture can be tipped so as to coat the agar slant. Such bottles should contain CO_2 or be incubated in an atmosphere of increased CO_2, and they should be incubated for a minimum of 4 weeks. Because of the difficulty in seeing colonies of brucellae, it is recommended that the agar slants in such bottles be routinely scraped and that Gram-stained smears be prepared to ensure that the organism is not present before the culture is discarded.

Fungi

It has already been pointed out that a biphasic BHI is recommended for the prompt recovery of yeasts from blood. Such bottles should be incubated at 30 C for a minimum of 4 weeks to ensure recovery of filamentous fungi as well. It has been our practice to vent such bottles chronically with a sterile, cotton-plugged needle. Obviously, this medium is also suitable for the growth of most bacteria encountered in blood.

Leptospires

Unfortunately, the possibility of a diagnosis of leptospirosis most frequently occurs at 2 a.m. when a beleaguered house officer is confronted with a patient who gives a 6-month history of jaundice and prior exposure to rodents. At that point, the house officer requests blood and urine cultures for leptospires. Equally unfortunate is the fact that leptospires are seldom, if ever, recovered from blood beyond the first week after onset of the acute illness or from the urine beyond the first 2–3 weeks after onset of this disease. Except for the acutely ill patient with leptospirosis, therefore, blood cultures for this organism are worthless and serology must be relied on to establish the diagnosis.

NEW PROCEDURES: RADIOMETRY

The evolution of $[^{14}C]CO_2$ from radioactively labeled substrates by bacteria has been utilized in principle for the detection of bacteremia. Reports in the scientific literature and otherwise have provided conflicting results (4, 27, 34) for reasons that should be obvious. These relate—to reemphasize what is perhaps an overworn point—to what is contained in the materials and methods segments of these studies, namely to what was the basis for comparison of the radiometric method. Under conditions similar to those described in this review, we were able to document earlier detection of bacteremia by radiometric means in only 17% of instances (27). Differences between our results and those reported by others in the scientific literature can readily be ascribed to methodologic differences associated with their conventional techniques. One conclusion is certain, and that is the lesser capability of radiometric techniques in detecting *Haemophilus* (21) and anaerobic bacteria (27, 34). It is evident, therefore, that radiometric techniques currently cannot be relied on exclusively for detection of bacteremia. Moreover, they do not obviate routine subculturing (5). Whether or not they are useful or practical as supplementary techniques obviously remains in question.

REPORTS

It behooves the laboratory to report positive findings in blood culture by phone and in written reports to the patient's physician so that appropriate modifications in antimicrobial therapy can be made. As has already been pointed out, there are inherent difficulties in trying to ascribe clinical significance to all isolates. MacGregor and Beaty (23) found that pneumococci, group A streptococci, Enterobacteriaceae, Bacteroidaceae, *Haemophilus*, and *Candida* were associated uniformly with clinically significant bacteremia. Encountered both in clinically significant bacteremias and as contaminants were staphylococci, viridans streptococci, group D streptococci, *Klebsiella*, and *Pseudomonas*. As contaminants only, they encountered *Corynebacterium, Bacillus*, and nonhemolytic streptococci. Although their findings provide rough guidelines for the significance of isolates from blood cultures, each case must be considered individually, and the laboratory is in a poor position to assess the clinical significance of isolates. Certainly the repeated isolation of nearly any organism from several sets of blood cultures helps in ascribing significance. It is therefore important for the laboratory to see to it that more than one culture per patient per 24 hr is collected to reduce the confusion associated with the isolation from a single culture set of an organism usually regarded as a normal commensal.

ORGANISMS IN BACTEREMIA

The types and distribution of bacteria isolated from blood cultures vary according to hospital size, location (urban, suburban, or rural), level of care (primary, secondary, or tertiary), and services rendered (acute, chronic, subspecialty). In their analysis of Gram-negative bacteremia, Dupont and Spink (6) demonstrated differences in the distribution of various Gram-negative bacilli according to hospital service. *Proteus* was commonly isolated from patients on the urology service and this reflected a urinary tract portal of entry, whereas *Pseudomonas* was commonly isolated from the pediatrics service because of the many children with burns and those with leukemia. In newborns with septicemia, *Escherichia coli* was the most frequent pathogen encountered, and the Klebsielleae were most often found in newborns after surgery. Bacteroidaceae were more often isolated from patients on the gynecologic and surgical services. Ninety percent of bacteremias due to Klebsielleae, *Pseudomonas, Proteus*, and of those due to multiple organisms were found by DuPont and Spink (6) to have been hospital-acquired. Bacteremias due to *E. coli* were less often (72%) hospital-acquired but represented the most frequent nonhospital-acquired type of Gram-negative bacillemia.

Ledger et al. (22) reviewed the bacteremias occurring on an obstetric-gynecologic service and found that the three most frequently encountered organisms were *E. coli*, enterococcal group D streptococci, and β-hemolytic streptococci other than groups A or D. *Peptostreptococcus, Peptococcus*, and *Bacteroides* were the principal anaerobes recovered, and these, as a group, accounted for 29% of all of the microorganisms encountered.

In a study of bacteremia following genitourinary tract manipulation, Sullivan et al. (33) found a 31% incidence of bacteremia in patients undergoing transurethral prostatic resection, 17% in patients undergoing cystoscopy, and 24% in patients undergoing urethral dilatation. Enterococcal group D streptococci and *Klebsiella pneumoniae* were the most frequent species isolated from this patient population.

Myerowitz et al. (26) found that the organisms most often associated with bacteremia in renal homotransplant recipients were *S. aureus, E. coli, Serratia*, and *Pseudomonas*. *P. aeruginosa, E. coli,* and *S. aureus* were reported by Hughes (16) to be the principal etiologic agents of infection in 199 patients dying with leukemia.

As is obvious from these studies and from those summarized by Klainer and Beisel (17), there is a strong association between certain predisposing factors and infection with specific microorganisms. All of these variables, however, affect the types and distribution of bacterial isolates from blood and limit a statistical comparison of the incidence of the types of bacteremias seen among various medical institutions. With these qualifications in mind, there are listed in Table 4.11 the numbers of patients by organism group for the Mayo Clinic and affiliated hospitals during 1974. The majority of the nearly 200,000 outpatients seen at the clinic and of the inpatients occupying the nearly 1,700 beds in affiliated hospitals during 1974 represented referrals from the Upper Midwest.

From all but six patients, *Bacillus* was isolated from single sets of cultures only. *Corynebacterium, S. epidermidis*, and *Alcaligenes* were isolated from single sets of cultures only from all but 23 patients, 27 patients, and 1 patient, respectively. These organisms presumably represented contaminants and are counted as such for control purposes (Fig. 4.1). The number of these organisms isolated during 1974 exceeded that ordinarily recorded in our laboratory, primarily because of an industry-wide problem with contaminated lots of blood-culture bottles. The 1,505 patients shown in Table 4.1 represented 2,559 cultures out of a total of 22,566 blood-culture sets received by our laboratory during 1974. There were 67 patients with polymicrobial bacteremia, exclusive of those with presumed contaminants.

For purposes of presenting recent blood-culture data from at least one other medical center, but being mindful of the difficulties inherent in trying to compare such data, there are listed in Table 4.12 the number of patients with bacteremia by organism group at the Boston City Hospital in 1972 (24). Isolates of *S. epidermidis*, diphtheroids, *Bacillus*, and others considered as contaminants were excluded from the study, as were anaerobic bacteria because adequate methods for their isolation were not used during this period of time. The smaller bed capacity of Boston City Hospital, its urban location, and the more acute nature of its practice are but a few of the variables that limit a comparison between these data and our own.

As may be seen in Figure 4.1, the percentage of positive cultures, exclusive of presumed contaminants, varied during the given 3-month interval from 6 to 8%. These fluctuations are common and reflect such variables as changes in the composition of a hospital's patient population, the ordering habits of physicians, patient selection, and other nuances. Be that as it may, such fluctuations do, for the most part, preclude any sequential statistical analysis of blood-culture media or system performance. It is not uncommon for laboratories to use this type of analysis to try to prove the superiority or inferiority of a blood-culture system's performance. Such analyses simply lack statistical validity, and comparisons of performance of systems must be based on parallel studies of two or more systems. In our own studies, a soybean-casein digest broth, Tryptic soy (Difco Laboratories), has served as the standard medium against which others are compared.

TABLE 4.11
Number of patients with bacteremia by organism group – Mayo Clinic and affiliated hospitals, 1974

Organism	No. of Patients	Organism	No. of Patients	Organism	No. of Patients
Bacillus	119	*Providencia*	1	*Bacteroides*	67
Clostridium	24	*Haemophilus*		*Fusobacterium*	3
*Corynebacterium**	316	*H. influenzae*	9	*Micrococcus*	2
Escherichia	159	*H. parainfluenzae*	5	*Staphylococcus*	
Salmonella	5	*H. aphrophilus*	2	*S. aureus*	113
Citrobacter		*Streptococcus*		*S. epidermidis*	231
C. freundi	1	*S. pneumoniae*	34	*Peptostreptococcus*	3
C. diversus	5	Viridans	76	*Peptococcus*	2
Klebsiella	68	Group A	18	*Veillonella*	1
Enterobacter		Group B	10	*Pseudomonas*	
E. aerogenes	4	Group D	25	*P. aeruginosa*	50
E. cloacae	13	Group F	2	*P. maltophilia*	10
E. agglomerans	2	Other	4	*P. fluorescens*	1
Serratia	29	*Eubacterium*	2	*P. cepacia*	3
Proteus		*Acinetobacter*	7		
P. vulgaris	4	*Alcaligenes*	36		
P. mirabilis	19	*Pasteurella multocida*	1		
P. morganii	9	*Neisseria gonorrhoeae*	1		
P. rettgeri	6	Neisseria spp.	3		

* Includes *Propionibacterium*.

TABLE 4.12
Number of patients with bacteremia by organism groups at the Boston City Hospital, 1972*†

Organism Group	No. of Patients	Organism Group	No. of Patients
Escherichia	97	*Streptococcus*	
Klebsiella	75	*S. pneumoniae*	108
Enterobacter	19	Viridans	63
Proteus	32	β-Hemolytic (not D)	45
Pseudomonas aeruginosa	22	Group D	38
		Staphylococcus aureus	87
	Other 91		

* Data adapted from J. E. McGowan, Jr., M. W. Barnes, and M. Finland: *Journal of Infectious Disease 132:*316–335, 1975 (24).

† Presumed contaminants (e.g., *Staphylococcus epidermidis, Bacillus,* diphtheroids) and anaerobes excluded from study.

CONCLUSIONS

We are undoubtedly entering a phase in which the various factors associated with the detection of bacteremia are finally undergoing scrutiny. Newer techniques employing physical or chemical principles, or both, are being studied. By the time this manuscript is published, there will have been changes in established techniques based on new data. Fortunately, the field is no longer static. It is, however, of utmost importance for the clinical microbiologist to be as discriminating as possible about new developments in the field. As has already been pointed out, the basis for comparison of a new system for detection of bacteremia must be carefully scrutinized. If the conventional system is poor or deficient in one or more ways, its comparison with a new system loses validity. The virtues of newer physicochemical techniques remain unproven in comparison with satisfactory conventional methods.

LITERATURE CITED

1. Bartlett, R. C., Ellner, P. D., and Washington, J. A., II. ASM Cumitech No. 1. In *Cumulative Techniques and Procedures in Clinical Microbiology: Blood Cultures,* J. C. Sherris. (Ed.). Washington D.C.: American Society for Microbiology, 1974.
2. Blazevic, D. J., Stemper, J. E., and Matsen, J. M. Comparison of macroscopic examination, routine Gram stains, and routine subcultures in the initial detection of positive blood cultures. *Appl Microbiol 27:*537–539, 1974.
3. Blazevic, D. J., Stemper, J. E., and Matsen, J. M. Effect of aerobic and anaerobic atmospheres on isolation of organisms from blood cultures. *J Clin Microbiol 1:*154–156, 1975.
4. Brooks, K., and Sodeman, T. Rapid detection of bacteremia by a radiometric system: a clinical evaluation. *Am J Clin Pathol 61:*859–866, 1974.
5. Caslow, M., Ellner, P. D., and Kiehn, T. E. Comparison of the BACTEC system with blind subculture for the detection of bacteremia. *Appl Microbiol 28:*435–438, 1974.
6. DuPont, H. L., and Spink, W. W. Infections due to gram-negative organisms; an analysis of 860 patients with bacteremia at the University of Minnesota Medical Center, 1958–1966. *Medicine 48:*307–332, 1969.
7. Ellner, P. D. System for inoculation of blood in the laboratory. *Appl Microbiol 16:*1892–1894, 1968.

8. Eng, J. Effect of sodium polyanethol sulfonate in blood cultures. *J Clin Microbiol 1:*119–123, 1975.
9. Franciosi, R. A., and Favara, B. E. A single blood culture for confirmation of the diagnosis of neonatal septicemia. *Am J Clin Pathol 57:*215–219, 1972.
10. Gantz, N. M., Swain, J. L., Medeiros, A. A., et al. Vacuum blood-culture bottles inhibiting growth of *Candida* and fostering growth of *Bacteroides*. *Lancet 2:*1174–1176, 1974.
11. Graves, M. H., Morello, J. A., and Kocka, F. E. Sodium polyanethol sulfonate sensitivity of anaerobic cocci. *Appl Microbiol 27:*1131–1133, 1974.
12. Hall, M., Warren, E., and Washington, J. A., II. Detection of bacteremia with liquid media containing sodium polyanetholsulfonate. *Appl Microbiol 27:*187–191, 1974.
13. Hall, M., Warren, E., and Washington, J. A., II. Comparison of two liquid blood culture media containing sodium polyanetholesulfonate: Tryptic soy and Columbia. *Appl Microbiol 27:*699–702, 1974.
14. Harkness, J. L., Hall, M., Ilstrup, D. M., et al. Effects of atmosphere of incubation and of routine subcultures on detection of bacteremia in vacuum blood culture bottles. *J Clin Microbiol 2:*296–299, 1975.
15. Henrichsen, J., and Bruun, B. An evaluation of the effects of a high concentration of sucrose in blood culture media. *Acta Pathol Microbiol Scand [B] 81:*707–710, 1973.
16. Hughes, W. T. Fatal infections in childhood leukemia. *Am J Dis Child 122:*283–287, 1971.
17. Klainer, A. S., and Beisel, W. R. Opportunistic infection; a review. *Am J Med Sci 258:*431–456, 1969.
18. Knepper, J. G., and Anthony, B. F. Diminished growth of *Pseudomonas aeruginosa* in unvented blood-culture bottles. *Lancet 2:*285–287, 1973.
19. Kocka, F. E., Arthur, E. J., and Searcy, R. L. Comparative effects of two sulfated polyanions used in blood culture on anaerobic cocci. *Am J Clin Pathol 61:*25–27, 1974.
20. Kocka, F. E., Magoc, T., and Searcy, R. L. New anticoagulant for combating antibacterial activity of human blood. *Proc Soc Exp Biol Med 140:* 1231–1234, 1972.
21. Larson, S. M., Charache, P., Chen, M., et al. Automated detection of *Haemophilus influenzae*. *Appl Microbiol 25:*1011–1012, 1973.
22. Ledger, W. J., Norman, M., Gee, C., et al. Bacteremia on an obstetric-gynecologic service. *Am J Obstet Gynecol 121:*205–212, 1975.
23. MacGregor, R. R., and Beaty, H. N. Evaluation of positive blood cultures; guidelines for early differentiation of contaminated from valid positive cultures. *Arch Intern Med 130:*84–87, 1972.
24. McGowan, J. E., Jr., Barnes, M. W., and Finland, M. Bacteremia at Boston City Hospital; occurrence and mortality during 12 selected years (1935–1972), with special reference to hospital-acquired cases. *J Infect Dis 132:*316–335, 1975.
25. Minkus, R., and Moffet, H. L. Detection of bacteremia in children with sodium polyanethol sulfonate; a prospective clinical study. *Appl Microbiol 22:*805–808, 1971.
26. Myerowitz, R. L., Medeiros, A. A., and O'Brien, T. F. Bacterial infection in renal homotransplant recipients; a study of fifty-three bacteremic episodes. *Am J Med 53:*308–314, 1972.
27. Renner, E. D., Gatheridge, L. A., and Washington, J. A., II. Evaluation of radiometric system for detecting bacteremia. *Appl Microbiol 26:*368–372, 1973.
28. Roberts, G. D., Horstmeier, C., Hall, M., et al. Recovery of yeast from vented blood culture bottles. *J Clin Microbiol 2:*18–20, 1975.
29. Rosner, R. Effect of various anticoagulants and no anticoagulant on ability to isolate bacteria directly from parallel clinical blood specimens. *Am J Clin Pathol 49:*216–219, 1968.
30. Rosner, R. A quantitative evaluation of three blood culture systems. *Am J Clin Pathol 57:*220–227, 1972.
31. Rosner, R. A quantitative evaluation of three different blood culture systems, pp 61–75. In *Bacteremia: Laboratory and Clinical Aspects*, A. C. Sonnenwirth (Ed.). Springfield, Ill.: Charles C Thomas, 1973.
32. Rosner, R. Comparison of recovery rates of various organisms from clinical hypertonic blood cultures by using various concentrations of sodium polyanethol sulfonate. *J Clin Microbiol 1:*129–131, 1975.
33. Sullivan, N. M., Sutter, V. L., Carter, W. T., et al. Bacteremia after genitourinary tract manipulation; bacteriological aspects and evaluation of various blood culture

systems. *Appl Microbiol 23:*1101–1106, 1972.

34. Thiemke, W. A., and Wicher, K. Laboratory experience with a radiometric method for detecting bacteremia. *J Clin Microbiol 1:*302–308, 1975.
35. Todd, J. K., and Roe, M. H. Rapid detection of bacteremia by an early subculture technic. *Am J Clin Pathol 64:*694–699, 1975.
36. Washington, J. A., II. Blood cultures; principles and techniques. *Mayo Clin Proc 50:*91–98, 1975.
37. Washington, J. A., II, Hall, M. M., and Warren, E. Evaluation of blood culture media supplemented with sucrose or with cysteine. *J Clin Microbiol 1:*79–81, 1975.
38. Washington, J. A., II, and Martin, W. J. Comparison of three blood culture media for recovery of anaerobic bacteria. *Appl Microbiol 25:*70–71, 1973.
39. Werner, A. S., Cobbs, C. G., Kaye, D., et al. Studies on the bacteremia of bacterial endocarditis. *JAMA 202:*199–203, 1967.
40. Wilson, W. R., Martin, W. J., Wilkowske, C. J., et al. Anaerobic bacteremia. *Mayo Clin Proc 47:*639–646, 1972.

5

Surgical Infections

PAUL D. ELLNER

Surgical infections comprise a variety of soft tissue infections that may occur subsequent to surgery, and which may require surgical treatment (2). The occurrence of a local infection must be preceded by the introduction of one or more species of microorganisms into the normally sterile tissues together with some alteration in the host's resistance.

Factors which diminish the resistance of the patient and predispose to infection include local ischemia, hematomas, dead spaces, necrotic tissue, the presence of foreign bodies such as sutures, prosthetic devices or implants; underlying disease such as diabetes, leukemia, hypogammaglobulinemia, and treatment with steroids.

Other major determinants in wound infections are the age of the patient, the presence of cardiac disease or hypertension, and the type of surgical procedure. Surgical factors include the anatomic area, the extent of the incision, and the duration of the operation (4).

Once the mechanical barrier of the intact skin or mucous membrane has been broached by the incision, resistance to infection depends upon the action of the phagocytic cells, the presence of specific immunoglobulins, and nonspecific blood and tissue substances such as lysozyme.

The introduction of bacteria into a previously sterile site such as a surgical wound is termed contamination. Many contaminating bacteria find the tissues of the wound a hostile environment, and succumb. Those organisms able to survive begin to actively multiply, and the wound may be said to have been colonized. Colonization does not always lead to infection, but infection is always preceded by colonization. The determination of which of the colonizing bacterial species eventually emerge as the actual etiologic agent(s) of infection depends upon their number (dose), their virulence, the wound environment (pH and oxidation-reduction potential), and the selective effect of antibiotics.

Surgical wounds may become infected during the operative procedure, or subsequently during the healing process. Exogenous sources of organisms may be aerosol droplets from the respiratory tract of doctors and nurses that are generated during talking, coughing, or sneezing. Improperly sterilized instruments, dressings, or other materials are rarely an actual cause of infections. Cross-infection from other patients in the ward is not uncommon and is most often transferred by the hands of personnel, frequently during the process of changing dressing. Airborne transmission of organisms from patient to patient is effective only over relatively short distances and is considerably less important than the direct transfer of the infective agents.

Clean surgical wounds may become infected hematogenously during bacteremic episodes secondary to indwelling intravenous catheters.

The source of the organism occurring at the time of operation is frequently the patient himself. Endogenous sources of the infecting organisms are the inhabited areas—the skin, upper respiratory, gastrointestinal, and genitourinary tracts.

Normal skin flora is predominantly Gram-positive and consists largely of *Propionibacterium acnes, Corynebacterium* spp., *Micrococcus* spp., *Staphylococcus epidermidis,* and *Staphylococcus aureus*. Approximately 50% of all wound infections are caused by *S. aureus*. In one study of 50 patients, *S. aureus* was cultured from the wound by the end of the operative procedure in 92% of the cases, and from the air of the operating room in 68% of the cases. Bacteriophage typing of the isolates revealed that approximately half of these organisms originated from the patients (3).

Group A streptococci are involved in 2–5% of surgical wound infections. The remaining 48% of infections are caused by enteric organisms such as *Escherichia coli, Proteus, Klebsiella, Enterobacter, Serratia, Pseudomonas,* and *Bacteroides* species, enterococci, *Peptostreptococcus,* and *Clostridium perfringens*. Many wound infections are polymicrobic and contain from 2 to 8 bacterial species, 1 to 5 of which may be strict anaerobes.

TYPES OF SURGICAL INFECTIONS

A common soft tissue infection is the "stitch abscess" occurring at the suture line. This infection usually appears before the fifth postoperative day and is almost always due to *S. aureus*, although occasionally coagulase-negative staphylococci are isolated in pure culture from the exudate.

Erysipelas, a rapidly spreading cellulitis due to hemolytic streptococci, is seldom seen since the advent of antibiotics. Necrotizing crepitant infections of subcutaneous tissue may occasionally occur within 1–4 days of relatively minor trauma. This infection may be caued by *Peptostreptococcus, Clostridia* (invariably *C. perfringens*), or a mixture of bacteria; often a *Bacteroides* sp. and *Staphylococcus* or *Streptococcus* (12).

Necrotizing fasciitis is a surgical infection that complicates surgical or traumatic wounds. It occurs most often following abdominal operations and is characterized by a rapidly spreading necrosis of the superficial fascia. This is a mixed synergistic infection from which Gram-positive cocci (hemolytic or anaerobic streptococci or staphylococci) and Gram-negative rods (*Pseudomonas, Bacteroides, Proteus,* or other Enterobacteria) may be recovered.

Postoperative gangrene, a progressive infection involving all layers of tissue including the musculature, may be caused by *Clostridia,* streptococci, or *Bacteroides*. A type known as Meleney's synergistic gangrene is caused by a microaerophilic streptococcus, together with a staphylococcus or Gram-negative rod (usually Proteus) (9).

Infections of prosthetic devices often appear many months after their implantation. When cardiac valve prostheses become infected the clinical presentation is similar to that seen in endocarditis, with fever, changing murmurs, and embolic phenomena. The organism most often involved is *S. epidermidis*, but infections due to *S. aureus, Serratia, Nocardia,* and *Candida* also occur. Other vascular prostheses, such as arterial grafts, may also become infected with the subsequent development of aneurysms, thrombosis, septic emboli, septicemia, and metastatic abscesses (7). A particularly disturbing problem, whose cause is not yet clear, is the delayed infection sometimes occurring 6–18 months after the installation of a prosthetic hip replacement. The organism invariably recovered is *S. epidermidis*.

OBTAINING SPECIMENS FOR CULTURE

In order to facilitate the recovery of any anaerobic bacteria that might be present, the clinical material should be protected from prolonged exposure to air. Aspirated pus or fluid may be left in the syringe and, after expressing any air, the entire syringe is sent to the laboratory. Alternatively, the fluid may be injected into a gassed-out tube for transport to the laboratory.

Smaller amounts of liquid material may be collected on a swab which should be promptly placed in a non-nutritive transport medium such as Stuart's or Amies. Formulations including charcoal should not be used since the particles are often confused with bacteria on direct microscopic examination.

The rubbing of a swab over the surface of a wound often merely reproduces the skin flora and may fail to reflect the actual agents of infection within the wound. Three-dimensional data obtained by wound biopsy are more meaningful than culture of wound surfaces or exudates. Swabs taken from the surface of a wound may provide qualitative information as to the various species of bacteria that may be present – but special techniques are required to obtain quantitative data.

All visible medications and superficial detritus should be carefully wiped away with a sponge soaked in sterile saline solution. Alternatively, the wound surface may be cleansed with alcohol. A biopsy is taken under sterile conditions with a dermal punch, or the center and edge of the wound may be scraped with a 3-mm curette. Approximately 0.1 ml of tissue is obtained. The tissue or the punch is placed in a sterile tube and sent to the laboratory where it is processed within the hour.

ROLE OF THE LABORATORY

A direct smear should be prepared for a Gram stain. The specimen should be plated on selective and nonselective media appropriate for the recovery of aerobic organisms – a suitable combination would be CNA agar with 5% blood, chocolate agar and EMB or MacConkey agar. In addition, direct primary anaerobic plating should be done in preference to placing the specimen in chopped-meat-glucose broth or thioglycolate medium. The use of reducible plates of kanamycin-vancomycin blood agar with hemin and vitamin K; neomycin-blood agar; CNA, and sheep blood agar plates can be recommended. These anaerobic plates should be promptly placed in an anaerobic jar such as the Gas-Pak for incubation.

The futility of attempting to perform direct antibiotic susceptibility studies on wound exudate has been clearly demonstrated (6), and the microbiologist should not permit himself to be coerced into this practice.

For quantitative studies, the tissue is aseptically weighed and ground with 0.4 ml of sterile broth. Serial dilutions are made with sterile broth (10^{-2}, 10^{-3}, 10^{-4}, 10^{-5}, 10^{-6}) and pour plates (aerobic and anaerobic) are prepared. Incubation is for 24 hr. A smear may be made with 0.02 ml of the 1:10 dilution, and Gram stained (11).

It is most important that the laboratory report preliminary results by 24 hr. Presumptive identification of one to three species on aerobic plates may be made based upon colonial morphology and a few rapid tests (13). Similarly, growth occurring only on anaerobic plates may be Gram stained and a presumptive identification made. (These plates should be promptly returned to an anaerobic environment and reincubated for an additional 24 hr.) If the 24-hr plates show four or more organisms such as Streptococcus viridans group, enterococci, non-hemolytic streptococci, *S. epidermidis, Micrococcus* sp., diphtheroids, or nonpathogenic Neisseria, the culture may conveniently be reported as "mixed saprophytic flora," rather than enumerating the various species. No further work or susceptibility tests need be done on such specimens.

The extent to which speciation is carried out should relate directly to its diagnostic, prognostic, therapeutic, and epidemiologic value. In most cases identification to the genus level is sufficient for a preliminary report, but inadequate for a final report. For example, the speciation of staphylococci and streptococci often provide important clues as to the origin and etiologic significance of the organisms. In other cases identification to species level e.g. *Proteus* or *Bacteroides* is helpful in selecting antimicrobial agents. For epidemiologic purposes it may be necessary to go beyond speciation and determine serologic, bacteriocin, or phage type.

SIGNIFICANCE OF QUANTITATIVE STUDIES

It has been shown that there is a direct relationship between the number of bacteria present in the wound and the rate of wound closure and healing (8), the survival of skin grafts (1), and the development of sepsis.

It would appear that irrespective of the species of bacteria present, the critical number is 10^5 organisms per gram (or milliliter) of tissue. A study of decubitus ulcers and infected open wounds showed that when the number of organisms in the wound exceeded 10^5 per ml, healing was markedly delayed. A study by Noyes et al. (10) showed that only when the wound exudate contained more than 10^6 organisms per ml did invasive process and infection occur. Sepsis is seldom a problem if the bacterial count is less than 10^5 per ml; when the count exceeds 10^5 per ml, invasive infection and sepsis often occur.

The laboratory diagnosis of surgical infections depends on close cooperation between the surgeon and the microbiologist. Adequate specimens, promptly delivered to the laboratory, permit the most rapid and accurate results (5).

LITERATURE CITED

1. Bacchetta, C. A., et al. Biology of infections of split thickness skin grafts. *Am J Surg 130:*63–67, 1975.
2. Baxter, C. R. Surgical management of soft tissue infections. *Surg Clin North Am 52:*1483–1499, 1972.
3. Burke, J. F. Identification of the sources of staphylococci contaminating the surgical wound during operation. *Ann Surg 158:*898, 1963.
4. Cohen, L. S., Fekety, F. R., Jr., and Cluff, L. E. Studies on the epidemiology of staphylococcal infection. VI. Infections in the surgical patient. *Ann Surg 159:*321, 1964.
5. Dineen, P. Laboratory diagnosis of surgical infections. *Surg Clin North Am 52:*1379–1383, 1972.
6. Ellner, P. D., and Johnson, E. Unreliability of direct antibiotic susceptibility testing on wound exudates. *Antimicrob Agents Chemother 9:*355–356, 1976.
7. Fry, W. J. Vascular prosthesis infection. *Surg Clin North Am 52:*1419–1424, 1972.
8. Lyman, I. R., Tenery, J. H., and Basson, R. P. Correlation between decrease in bacterial load and rate of wound healing. *Surgery 130:*616–621, 1970.
9. Meleney, F. L. *Treatise on Surgical Infections*. New York: Oxford University Press, 1945.
10. Noyes, H. E., et al. Delayed topical antimicrobials as adjuncts to systemic antibiotic therapy of war wounds. Bacteriologic Studies. *Milit Med 132:*461–468, 1967.
11. Robson, M. C., and Heggers, J. P. Bacterial quantification of open wounds. *Milit Med 134:*19–24, 1969.
12. VanBeek, A., et al. Nonclostridial gas-forming infections. *Arch Surg 108:*552–557, 1974.
13. Wasilauskas, B. L., and Ellner, P. D. Four-hour presumptive identification of certain bacteria from blood cultures. *J Infect Dis 124:*499–504, 1971.

6

The Diagnostic Microbiology Laboratory in the Care of the Immunosuppressed Patient

DONALD ARMSTRONG

Exact speciation and rapid identification of clinical microbial isolates would seem unquestionably just and necessary to most microbiologists. Clinicians are frequently more interested in the results of rapid antibiotic susceptibility tests. Hospital epidemiologists should want both. In immunosuppressed patients, these needs are even more important than in a general hospital population. In many instances, the infection follows a particularly fulminant course in the immunosuppressed patient, and, therefore, at the least, rapid sensitivity tests are desirable. A single source epidemic within a hospital caring for many patients who are immunosuppressed may spread especially rapidly and, therefore, exact speciation and rapid identification become important. In this chapter, I will discuss the problem by region of infection, present some exemplary clinical problems that we have seen at Memorial Sloan-Kettering Cancer Center, and draw some conclusions as a result.

THE IMMUNOSUPPRESSED PATIENT

A patient may be immunosuppressed because of a basic underlying disease or because of the treatment received (4, 8, 9, 23) (Table 6.1). It is immunosuppressive to have an indwelling intravenous catheter allowing otherwise innocuous organisms access to the blood stream. It is immunosuppressive to have Hodgkin's disease or a congenital immunodeficiency, and it is immunosuppressive to receive therapy with corticosteroids as well as cytotoxic agents. There are multiple examples of ways of affecting the immune responses, including more than one of the immune mechanisms, but usually one is affected more than the others, resulting in predictable types of infections according to the specific immune defense(s) altered (Table 6.1).

OROPHARYNGEAL, SINUS, EAR INFECTIONS

A bacterial pharyngitis in immunologically intact persons in the United States is most always due to *Streptococcus pyogenes,* with few exceptions.

Although diptheria has almost disappeared, gonorrheal pharyngitis is increasing. In contrast, in the neutropenic patient, a great number of organisms may be responsible. *Pseudomonas aeruginosa* or other Gram-negative enteric bacteria as well as staphylococci may be responsible for a primary pharyngitis in any neutropenic patient, and *Candida albicans* or other *Candida* species may also be the cause. Identification and tests for sensitivities are important on throat cultures under these circumstances because (a) the throat isolate should be identified because the infecting organism may well appear in the blood cultures, and it is important to document that these are the same and the throat is the source, as the duration of therapy may be shorter in such superficial infections; and (b) the appropriate antibiotic can be selected after sensitivity testing. Multiple antibiotics should be used initially to cover a broad spectrum of Gram-positive cocci and aerobic Gram-negative rods (3), and one or more may be stopped when sensitivity test results are available.

In the case of *Candida* sp. infection, a smear and Gram stain, in addition to culture, can be helpful for semiquantitation and the evaluation of pseudohyphae, which suggest, but do not definitely indicate, significant tissue invasion. Speciation of Candida isolates is important, for if a blood culture is positive for one species and the throat for another, then we know

ABLE 6.1
Alterations in immune responses in immunosuppressed patients

Immune System Altered	Organisms Most Frequently Causing Opportunistic Infection			
	Bacteria	Fungi	Parasites	Viruses
ntact integument	*Streptococcus pyogenes* *Staphylococcus aureus* Enteric bacilli	*Candida* sp.		Vaccinia
nfusion infections	Enteric bacilli *Salmonella* sp.	*Candida* sp.	Malarial species *Toxoplasma gondii*	Hepatitis B Cytomegalovirus
Surgical procedures	*S. pyogenes* *S. aureus* Enteric bacilli *Bacteroides fragilis* *Salmonella* sp.	*Candida* sp.		Herpes simplex
Polymorphonuclear phagocyte	*S. pyogenes* *S. aureus* Enteric bacilli	*Candida* sp. *Aspergillus* sp. *Mucor* sp.		
Mononuclear phagocyte	*Listeria monocytogenes* *Salmonella* sp. *Nocardia asteroides* *Mycobacterium tuberculosis*	*Cryptococcus neoformans* *Candida* sp. *Histoplasma capsulatum* *Coccidioides immitis*	*Pneumocystis carinii* *T. gondii* *Strongyloides stercoralis*	Cytomegalovirus Varicella-zoster Herpes simplex Vaccinia virus
Humoral factors	*Streptococcus pneumoniae* Enteric bacilli		*P. carinii*	Vaccinia virus
Reticuloendothelial system	*S. pneumoniae* *Haemophilus influenzae*			

to look for other sources for the blood culture isolate, such as an indwelling intravenous catheter or the kidneys.

In the patient who is not neutropenic, or suspected of having Candida, a culture can be obtained only to rule out a *S. pyogenes*. We have seen a patient with Hodgkin's disease with a *S. pyogenes* septicemia and without a pharyngitis, but who was a pharyngeal carrier of the organisms. Three of her family members were also carriers and gave histories of repeated sore throats (22). All family members were treated and the sore throats stopped.

It must be stressed that, while in the general population in the United States throat cultures of *S. pyogenes* and *Neisseria gonorrhea* are all that may be necessary, in the neutropenic patient, predominant organisms of any kind should be identified and sensitivity tests performed.

Sinus and ear infections should be handled just as with any patient, since they are essentially closed space infections in many instances and they may be caused by a variety of single organisms or by a mixture of aerobic and anaerobic bacteria. The latter are usually not important to identify since they respond to the drainage that made it possible to obtain the pus to isolate them.

It is important to consider *Mucor* species in any pharyngeal or sinus infection in the heavily immunosuppressed patient and to do a wet mount as well as room and 37°C temperature cultures on Sabaroud's medium (31). *Aspergillus* species may also cause a rhinocerebral syndrome (49). Black necrotic tissue in the nasopharynx or sinuses should suggest either of these fungi and, in addition, *P. aeruginosa* as the invading organism.

Stomatitis may be due to herpes simplex, which can be rapidly isolated and presumptively identified by cytopathic effect within 24 hr. Another 48–72 hr is usually necessary for absolute identification by demonstrating the presence of herpes simplex antigen by the complement fixation (CF) test. Where the reagents are available, it should require only an hour to document the presence of antigen by indirect immunofluorescence (IF). Since herpes simplex may be in oral secretions without causing illness, and even if the stomatitis is due to this virus, it may become rapidly superinfected with bacteria or Candida. These latter possibilities, which are treatable, must be pursued. Inclusions suggesting invasive herpes simplex infection may be seen on smears with Giemsa or Papanicolaou stains, but the identification of the organism or the demonstration of mucosal cell reactions is probably worthwhile only in a research setting.

CENTRAL NERVOUS SYSTEM INFECTION

As with any central nervous system (CNS) infection, it is imperative to isolate and identify the offending organism. In the immunosuppressed host, the variety of possible organisms to be expected on a regular basis (Table 6.2) is much greater than in the general population (14, 15), and the sensitivities may vary greatly so that accurate and rapid results are essential. *Listeria monocytogenes* may appear Gram-negative or resemble Gram-positive cocci on smear of the cerebrospinal fluid (CSF), and positive

ABLE 6.2
rganisms most frequently causing central nervous system infection according to immune defect

Immune System Altered	Bacteria	Fungi	Parasites	Viruses
ononuclear phagocyte	*Listeria monocytogenes* *Nocardia asteroides*	*Cryptococcus neoformans*	*Toxoplasma gondii*	Varicella-zoster Herpes simplex
olymorphonuclear phagocytes	*Pseudomonas aeruginosa* *Escherichia coli*	*Aspergillus* sp. *Mucor* sp.		
tegument	*Staphylococcus aureus* *E. coli* *P. aeruginosa* *Klebsiella* sp.			
ncertain	*Streptococcus pneumoniae*	*Candida* sp.		Papovavirus

identification, including tumbling motility and passage on blood agar to detect hemolysis, are necessary as immediate steps for presumptive identification. Sensitivity determinations are particularly important in the penicillin-allergic patient. In addition, every CSF sample from an immunosuppressed patient should be tested for cryptococcal antigen, since this infection is one of the most frequent causes of meningitis in this population (14, 15, 28). In every case where we have found cryptococcal antigen by the latex agglutination test, we have been able to isolate and identify the organism by obtaining large volumes of CSF on subsequent lumbar punctures (10). We would prefer to do this since sensitivity tests to amphotericin B, 5-fluorocytosine, and miconazole should be done and will hopefully be more readily available in the future.

In the case of a brain abscess, when the organisms are available, all should be identified and sensitivities determined including those of anaerobes. Some of the latter, such as *Bacteroides fragilis* or *Bacteroides melaninogenicus,* may be resistant to penicillin, and antibiotics are more important in anaerobic CNS infections where sustained adequate drainage may be difficult to maintain.

Nocardia asteroides infections involving the CNS usually occur as abscesses and do not shed organisms in the CSF (48). Serological tests have not been consistently reliable (24), but deserve further study.

Aspergillus CNS infections are obviously becoming more frequent and are usually diagnosed at autopsy (14, 32, 49). In one of our cases, a brain biopsy revealed the organism by culture and histopathology, but treatment was unsuccessful, probably because the diagnosis was made too late. A serological test (vida infra) which we have developed (39, 50) should allow earlier treatment, hopefully before spread to the CNS occurs.

Mucor species invading the CNS are also usually found at autopsy, and early diagnosis will depend on either serological methods or brain biopsy.

Reports of *Candida* species CNS infections have been increasing, but again, diagnosis is made at autopsy. In only 1 of 10 cases did we isolate the organism from the CSF antemortem (15). Serological tests in our hands have been of limited value (18, 50). In one patient who was a narcotics addict with a persistent aseptic meningitis, strong precipitin bands and a high agglutinin titer did stimulate us to continue attempts at isolation until hypertonic media cultures yielded *C. albicans* (30).

Viruses need identification for epidemiologic purposes. Varicella-zoster (V-Z), especially as herpes zoster of the cranial nerves, can involve the CNS and early in the clinical illness resemble herpes simplex. Cultures on human embryonic lung (HEL) cells, (V-Z) and rabbit kidney (RK) cells (herpes simplex) will rapidly differentiate the viruses by the rapidity and type of cytopathic effect (CPE) and lack of growth of V-Z on RK cells. It is important to establish whether the infection is due to V-Z virus, for the patient should be isolated to protect other immunosuppressed patients (40). There is no evident need to isolate patients with herpes simplex or papovavirus infections; documentation of the latter is usually made at autopsy and is presently only academically valuable.

The diagnosis of CNS toxoplasmosis is discussed subsequently under "Serology"; in some cases we have resorted to brain biopsy and mouse inoculation in addition to histopathology and serology. The organisms have been grown in cell culture, but most workers prefer intraperitoneal inoculation of the mouse.

PULMONARY INFECTIONS

Among the most difficult infections to diagnose and treat in the immunosuppressed patient, pneumonias remain the leading cause of infectious deaths. Representative sputum specimens are infrequently obtained because of lack of neutrophils and/or debilitation of the patient. When the specimen is a good one, identification is important to rule out a nosocomial source or hematogenous spread to the lungs from another source, although this may be difficult to document. In the neutropenic patient, the isolate may be any of the usual causes of pneumonia, such as *Streptococcus pneumoniae, Klebsiella pneumoniae,* or even *Haemophilus influenzae;* but, just as often, the cause may be a Gram-negative enteric bacillus which is highly antibiotic-resistant, such as *Pseudomonas aeruginosa* (13, 42) (Table 6.3). It is therefore mandatory to isolate, do antibiotic sensitivity tests, and identify the organism. If adequate specimens are not available, I urge maneuvers which will produce them, ranging from transtracheal aspiration to open lung biopsy. These resistant infections may appear de novo, or superinfect a patient on broad spectrum antibiotic therapy, or occur as a result of respiratory assistance equipment. Since the latter may be a threat to others, it is important to identify the organism as precisely as possible in case the same organism subsequently appears in other patients or in the respiratory assistance equipment. There are no studies of which I am aware that demonstrate spread of a bacteria from patient to patient resulting in pneumonias in the immunosuppressed hosts, other than by respiratory assistance equipment. Nevertheless, there are excellent examples of patients, after entering the hospital, becoming colonized with Gram-negative enteric organisms such as *P. aeruginosa* or *K. pneu-*

TABLE 6.3
Organisms most frequently causing pneumonias according to immune defect

Immune System Altered	Bacteria	Fungi	Parasites	Viruses
Polymorphonuclear phagocytes	*Streptococcus pneumoniae* *Streptococcus pyogenes* *Staphylococcus aureus* *Klebsiella pneumoniae* Enteric bacilli	*Aspergillus* sp. *Mucor* sp.		
Mononuclear phagocytes	*Nocardia asteroides* *Mycobacterium tuberculosis* *Mycobacterium* sp.	*Histoplasma capsulatum* *Coccidioides immitis* *Cryptococcus neoformans*	*Pneumocystic carinii* *Toxoplasma gondii* *Stronglyloides stercoralis*	Cytomegalovirus Herpes simplex Varicella-zoster Measles
Humoral	*S. pneumoniae* *Haemophilus influenzae* Enteric bacilli		*P. carinii*	
Integument	Enteric bacilli			

moniae, and subsequently developing various types of infections with these organisms (17, 41). Since there are so many strains of these Gram-negative aerobic enteric organisms, it may be difficult to pinpoint person-to-person spread precisely, but with new developments, such as the immunotyping of *P. aeruginosa,* more specific markers for these strains should become available (19).

Communicable cases of *Nocardia asteroides* pneumonia have not been described (48), but further observations should be made. Therefore, such a potentially treatable organism needs identification for epidemiologic means and sensitivity testing, because of its notable poor response to therapy in some instances (11).

Clusters of infections with *Aspergillus* species have been reported in relation to building construction in the environment and require identification of the organisms to document this type of hazard to the immunosuppressed patient (1, 37). Whether surveillance methods could detect this type of hazard remains to be proven. We have recently observed an increase (32) and then a decrease in Aspergillus infection. In 1971, the number of patients with an Aspergillus infection seen at autopsy peaked at 28 and then fell to 9 in 1974. There are multiple factors which may have affected this. One possible influence was that, in November of 1973, we moved to a new hospital, and major construction projects were stopped. Another factor was that a serological test was under study for the early diagnosis of Aspergillosis, and patients were treated early and successfully (39). Once the serological test was accepted, patients were treated on the basis of clinical syndromes and a conversion from negative to positive immunodiffusion (ID) bands, and some even on the basis of the clinical syndrome alone. It may be that this was the reason the number of patients diagnosed by histopathology decreased. We are therefore not certain that the decrease in cases of Aspergillosis seen at autopsy represents a true decrease in incidence due to an epidemiologic change. We are presently trying to evaluate this and to observe for a possible new increase in prevalence associated with renewed construction.

Clusters of *Pneumocystis carinii* pneumonia have also been described using histopathological identification of this microorganism which is not readily grown in vitro (43). In the case of microorganisms such as *P. carinii* that have a long and variable incubation period, communicability may be difficult to detect, so that exact identification of every case is extremely important to avoid missing any links in the chain.

Acute diffuse pneumonias are a common problem in the immunosuppressed host (6). The causes are so many (Table 6.4) that a specific diagnosis is of utmost importance to avoid empiric therapy with multiple toxic therapeutic agents. In addition, the clinical course may progress very rapidly, and an early as well as specific diagnosis is mandatory. If less invasive means of obtaining an adequate specimen (including bronchoscopy with bronchial brushings or transbronchial biopsy) are unavailable, I recommend early open lung biopsy. The path from the biopsy site to the laboratory seems to be strewn with obstacles, and a rigid protocol should be established and adhered to in every instance, if the specimen is to receive proper handling. We recommend the procedures in Table 6.5 be done on each specimen, so that it is evident that an adequate sized specimen will be necessary, as well as a coordinated approach involving the surgeons, pathologists, and microbiologists involved. We have found the safest procedure is to have an infectious disease fellow as the representative of the diagnostic microbiology laboratory in the operating room at the time of surgery, so that the appropriate steps are ensured.

CARDIAC AND INTRAVASCULAR INFECTIONS

Endocarditis in the immunosuppressed patient is not appreciably different from that infection in any other individual, and the importance of

TABLE 6.4
Acute diffuse pneumonia in immunosuppressed patient

Bacteria	Fungi	Parasites	Viruses	Others
A. Infectious				
Streptococcus pyogenes	*Cryptococcus neoformans*	*Pneumocystis carinii*	Cytomegalovirus	*Mycoplasma pneumoniae*
Staphylococcus aureus	*Candida* spp.	*Toxoplasma gondii*	Herpes simplex	*Chlamydia psittaci*
Enterobacteraceae	*Aspergillus* spp.	*Strongyloides stercoralis*	Measles	
Pseudomonas aeruginosa	*Coccidioides immitis*			
Nocardia asteroides	*Histoplsma capsulatum*			
B. Noninfectious				
Undetermined etiology	Neoplastic infiltrate	Congestive heart failure		Drug toxicity
	Hemorrhage	Uremia or shock		Oxygen Toxicity

isolation, identification, and sensitivity teting is obvious in any case (38). Identification suggests the source—e.g., *Streptococcus viridans* group, the oropharynx; Enterococcus or *Streptococcus bovis,* the gastrointestinal or genitourinary tract; and *Candida* species, indwelling intravenous catheters. Some feel that the association between *S. bovis* endocarditis and carcinoma of the colon is sufficient to indicate workup for the latter in the face of endocarditis due to the former (36).

Myocardial abscesses may be found in disseminated Candida or Aspergillus infections (32), but are usually a small part of the syndrome and do not offer the microbiology laboratory any help in identifying the responsible organism. Toxoplasmosis may cause a myocarditis, especially in the immunosuppressed patient, and the diagnosis of this infections is discussed below, under "Serology."

GASTROINTESTINAL TRACT INFECTIONS

A common complication of immunosuppressive therapy is esophagitis due to herpes simplex, *Candida albicans,* or, less often, enteric bacilli (Table 6.6). Esophograms prove only inflammation of the esophagus, not an etiologic agent. A specific microbial diagnosis is usually difficult for esophagoscopy with washings, or biopsy is necessary. Many of the patients who are susceptible to such infections are thrombocytopenic, so that this procedure carries the risk of bleeding, and cases have been described where Pseudomonas bacteremia and death followed esophagoscopy of a patient with a Pseudomonas esophagitis (20). Mouth or throat cultures do not necessarily reflect what is causing an esophagitis, and only a specimen from the involved area is valid. Speciation of Candida isolates is important

TABLE 6.5
Management of specimens for diagnosis of acute diffuse pneumonia

Smears*	Cultures	Pathology	Serology†
Gram stain	Bacteria	Gram-Weigert	Cryptococcal antigen
Gram-Weigert stain	Fungi	Methenamine silver	Aspergillus antibody
Cytology	Mycoplasma	Gram stain	Cold agglutinins
Acid fast stain	Virus	PAS	
		Acid fast	
		Mucicarmine	

* Includes impression smears from biopies and frozen sections.
† If base-line (preacute) serums are available, should do battery of tests available for agents listed in Table 6.4. A significant rise in titer might be apparent.
Reprinted with permission from *Transplantation Proceedings*, Vol. VIII, No. 4, 1976 by Grune & Stratton, Inc.

TABLE 6.6
Organisms showing predilection for severe gastrointestinal or intraabdominal infections in immunosuppressed patient

Bacteria	Fungi	Parasites	Viruses
Salmonella sp.	*Candida* sp.	*Strongyloides stercoralis*	Cytomegalovirus
Clostridia sp.	*Aspergillus* sp.	*Giardia lamblia*	Herpes simplex
Bacteroides sp.	*Mucor* sp.		
Enteric bacilli			

in this instance, just as it is from any other heavily infected site, such as the mouth (vide supra), for if a different *Candida* species is isolated from the blood, another source for its invasion should be sought. Usually, in patients with esophagitis, we do not get an adequate specimen or a positive blood culture, and the clinician must treat empirically.

Instrumentation of the gastrointestinal tract should also be suspected as a source of infection as noted above for esophagoscopy. We recently had an outbreak of ampicillin, chloramphenicol-resistant *Salmonella oslo* infections, which, in some instances, we attributed to gastroscopy with common bile duct cannulation (16). Not only the exact speciation, but the sensitivity patterns served as markers for this unusual isolate. Following the isolation of two *Salmonella* species identified as *S. cubana,* we found that both patients had received carmine red dye markers, an association previously described (46). Thereafter, we have autoclaved our carmine red dye.

Clostridial sepsis has been associated with instrumentation, and the speciation of an isolate as *Clostridia septicum* has been suggested as an indication for a search for an underlying neoplasm (2, 47).

Upper gastrointestinal signs along with fever, with or without pulmonary lesions, should raise the question of "hyperinfection" with *Strongyloides stercoralis,* especially in the patient whose disease or treatment will affect his mononuclear phagocyte function (8). Although stools should be examined, a duodenal aspirate is more likely to reveal the larvae and/or eggs of the organism.

A Gram stain of the stool may be very important in suggesting a presumptive diagnosis until cultures can be evaluated semiquantitatively, predominant organisms identified, and sensitivities determined. We have seen persistent diarrhea for over a week in a patient who had lymphosarcoma and whose stools showed predominantly Gram-positive rods and yielded *Clostridium perfringens* on culture. The diarrhea subsided on penicillin therapy. Other causes of diarrhea in these patients suggested by a Gram stain are *Candida albicans* and *Staphylococcus aureus*. Ordinarily, nonpathogenic enteric organisms that have caused ulcerative gastrointestinal lesions and diarrhea include *P. aeruginosa*. A Gram stain would not be helpful, nor would identification, unless the organism was clearly in predominance or pure culture. Giardiasis may be a cause of persistent or recurrent diarrhea, especially in patients with IgA deficiencies. The cysts and trophozoites will usually be found in the stool in the presence of diarrhea, but, if not, duodenal aspirates would be necessary to rule out *Giardia lamblia* as the offending agent.

INTRAABDOMINAL INFECTIONS

Intraabdominal infections (Table 6.6) in the immunosuppressed host are not much different from those in the general population, with one notable exception. *Clostridia* sp. infections, which start in the bowel wall of patients with infiltrating or obstructing tumor, may be manifest early by cellulitis in the flanks, which rapidly goes on to ecchymosis and crepitation (47). Aspiration of the cellulitis should be cultured for anaerobes as well as aerobes, and, in fact, the Gram stain will frequently show large Gram-positive rods suggesting *Clostridia* sp. Since we have seen cellulitis due to Clostridia from an intraabdominal source appear first in the axilla, any aspirate of a cellulitis should be cultured anaerobically. This would also be helpful in identifying some of the organisms which we see causing a mixed synergistic gangrene in patients with tumors of the lower bowel where fistulas may be the route for bowel flora to reach the subcutaneous tissues or the fascia. In most other instances of infections of an intraabdominal source, the offending organisms are aerobic Gram-negative bacilli, usually *Escherichia coli,* Klebsiella, Pseudomonas, or Proteus. Neutropenia, or immunosuppression of any kind, does not appear to predispose the patient to infection with other bowel flora, particularly anaerobes. Following the extensive bowel or gynecological surgery done frequently on patients with neoplastic disease, postoperative infections may well include anaerobes such as *Bacteroides fragilis* (26), *Peptostreptococcus* sp., and *Peptococcus* sp. Once good drainage has been achieved, the identification and sensitivity patterns of these organisms become far less important, because drainage alone tends to cure the infection, and after drainage, many may not even require antibiotic therapy. An astute clinician can determine with his nose, well before the laboratory can tell by isolation and identification, that anaerobes are present in a wound, and that the wound needs better drainage, and if this cannot be achieved, antibiotics such as clindamyicn, chloramphenicol, or tetracycline may be needed. Liver or splenic abscesses in the immunosuppressed host appear to be due to the same organisms that predominate in their bowel flora or circulate in the blood. The majority are the same as those in the normal population with a few exceptions. *Aspergillus* sp., *Mucor* sp., and *Candida* sp. are more commonly found in liver and spleen abscesses, most apparently arriving by the hematogenous route, others questionably by the biliary tract. We have all too often seen hepatic abscesses due to *Candida* sp. at postmortem, and have all too infrequently made the diagnosis antemortem. It is not only important to isolate and identify the organism and to speciate it for epidemiological purposes, but, as more antifungal agents become available, sensitivity tests can be done. If 5-fluorocytosine or miconazole were as effective as amphotericin B against a *Candida* species tested in vitro, either compound appears far less toxic and therefore preferable to amphotericin B.

URINARY TRACT INFECTIONS

The importance of speciating isolates from urine samples is obvious from various studies in general hospitals, where a nosocomial source was first suspected by the identification of the organisms, then documented by epidemiological investigations, and prevented by corrective measures. The isolation from the urine of *Serratia marcescens, Citrobacter* sp., *Pseudomonas* sp., or any other environmental or water borne organism should suggest to the microbiologist and clinician a nosocomial source, and it should be sought.

The only remarkable difference between urinary tract infections in the immunosuppressed patient and those in other patients is the prevalence of infections due to *Candida* sp. in those that are immunosuppressed. Although a Gram stain is helpful in evaluating urinary tract infections in any sort of patient, it may be even more helpful in the immunosuppressed patient where a rapid answer is most important because of the threat of impending sepsis from any type of infection. The Gram stain may suggest an enterococcus, which would alert the clinician that he should increase his dose of ampicillin over that ordinarily used for the usual *E. coli* urinary tract infection, or that he should indeed include ampicillin in the regimen, which frequently is made up of a cephalosporin or a semisynthetic penicillin and an aminoglycoside, without including ampicillin. Another Gram-positive coccus, the Staphylococcus, may be suggested by the Gram stain, and prompt, appropriate therapy for that organism. Finally, *Candida* sp., which so often infect the immunosuppressed patient, can be suspected immediately on Gram stain. In the absence of other organisms and the presence of pyuria, the clinician may want to be alerted to this laboratory finding in order to treat early for presumptive bladder or renal candidiasis. Although we do not routinely do Gram stains on all urine samples, we do them on request, and I have them done on all my immunosuppressed patients who appear the least toxic with presumptive urinary tract infections.

Speciation of Candida isolated from the urine is important because, if a different species is isolated from the blood, a different source than the urinary tract should be sought and corrected when possible. We have isolated *Candida albicans* in significant amounts from the urine, and *Candida tropicalis* from the blood. We then removed all intravascular catheters and found the *C. tropicalis* on one of them, and this was the only documented source. In addition, since *C. tropicalis* may be relatively resistant to amphotericin B, and sensitivity tests to this agent are not routinely done; the recognition of this species from a significant source should prompt efforts to determine sensitivities. The isolation of a *Candida* species, even from the blood, is so often questionably significant, and so many different species are potentially pathogenic, that it is important to speciate isolates from different body sites to try to determine the source of fungemia in case a simple curative procedure, such as the removal of a catheter, can be done.

PERINEAL INFECTIONS

Infected rectal fissures are frequent in patients with leukemia or lymphomas who are neutropenic; they may be the source for lethal infections. Although they frequently start out as predominantly *E. coli* infections, with antibiotic therapy they become quickly superinfected with *Klebsiella pneumoniae* or *Pseudomonas aeruginosa*. The latter organism may cause cellulitis of the perineum, of the scrotum or the vulva, and go on to form necrotic areas which eventually slough. In the perirectal area, the scrotum, or the vulva, cultures are usually contaminated with normal bowel or vaginal flora. Aspiration of an edge of an area of cellulitis surrounding the fissure or vulvitis has been helpful on some occasions. If the microbiologist is aware of the importance of such a procedure and of identifying the organisms from such an aspirate, the clinician may benefit by this information. Both the microbiologist and the clinician should keep in mind that the cellulitis is usually due to *P. aeruginosa,* and the earlier they are treated as such, the better. Significant Candida infections do not appear to occur in the rectal area, but do so in the vagina. Again, since the organism is normal flora, it is difficult to make an estimate as to whether it is invasive or not, even in the presence of pseudohyphae on smear. The role of serology in this evaluation will be discussed below.

SKIN INFECTIONS

The importance of aspirating skin lesions in patients who are immunosuppressed cannot be overstressed. These lesions may be the only source for isolating the offending organism and examples have already been cited above. Since many of the skin lesions are secondary to bacteremia, fungemia, or viremia from another source, identification of the organism may suggest a source, and the isolate may be the only one on which to do sensitivity tests. Table 6.7 lists organisms which the microbiologist can anticipate isolating from skin lesions. Certain isolates may have epidemiological implications, such as *S. pyogenes, S. aureus, Salmonella* sp., or varicella-zoster virus. We have seen cellulitis due to groups B, C, and G streptococci occurring from an endogenous source rather than a communicable one, and find it worthwhile to group streptococci as a result (7). Since groups B, C, and G streptococci do not occur in epidemics in adults, we have been able to avoid considerable unnecessary epidemiologic work by identifying them. The epidemiology of *S. aureus* infections requires phage typing facilities, absent in many microbiology laboratories. These are usually not necessary, but are helpful when a common source outbreak seems possible or likely.

When an aspirate of a skin lesion is unrevealing, then a biopsy should be the next step. The specimen should be large enough to smear, culture, and fix for sectioning. Precious time may be wasted by dividing up a specimen into multiple inadequate pieces. We feel that biopsy specimens are best handled by one individual – preferably from the infectious disease service,

TABLE 6.7
Organisms to be anticipated in skin lesions in immunosuppressed patients

Bacteria	Fungi	Parasites	Viruses
Pseudomonas aeruginosa	*Candida* sp.	*Toxoplasma gondii*	Varicella-zoster
Aeromonas hydrophila	*Cryptococcus neoformans*		Herpes simplex
Klebsiella pneumoniae	*Aspergillus* sp.		Vaccinia
Streptococcus pyogenes	*Mucor* sp.		Cytomegalovirus
Staphylococcus aureus	*Histoplasma capsulatum*		
Clostridium perfringens			

who first takes the specimen to the microbiology laboratory, removes a sterile portion, and then delivers the rest to the pathology department.

BLOOD CULTURE ISOLATES

All blood culture isolates should be speciated and have sensitivity determinations—in any group of patients, let alone the immunosuppressed patient. In an instance where the organism identified is usually not a pathogen, it should not be reported as a contaminant, even if the laboratory suspects that it is. The clinician must decide whether a specific organism is a contaminant or not. A *Bacillus* sp. in an immunosuppressed patient may be life threatening (25), and the physician who is caring for the patient must decide whether it is significant or not. A "diphtheroid" may be *Listeria monocytogenes* or even a true *Corynebacterium* species, such as *C. equi,* which causes life-threatening infections in the immunosuppressed patient (12). Mixed bowel flora or even skin flora found in a blood culture may reflect the source of the infection, such as a bowel perforation or abscess, or an infected intravenous catheter. The neutropenic patient is particularly likely to circulate multiple organisms which require treatment from sources such as superficial bowel or skin ulcers (44, 45).

SEROLOGY

Serological Tests to Identify Infections

There are a number of infections in immunosuppressed patients which defy ready clinical diagnoses. An organism may be so plentiful, as with *C. albicans,* that its isolation, even in large numbers, is of uncertain significance. In contrast, an agent such as Aspergillus may cause flagrant pulmonary disease and not appear in the sputum. Only 12% of our patients with proven pulmonary aspergillosis had positive sputum tests by smear or culture (32). Compounding the problem were two leukemic patients in the same series who had Aspergillus in their sputum and no pulmonary aspergillosis. Other examples of opportunistic infections in which the organism grows slowly or is rarely or never isolated in the microbiology laboratory are nocardiosis, tuberculosis, mucormycosis, toxoplasmosis, and *P. carinii* pneumonia. Specific, sensitive, and rapid serological tests to document infections with causative agents are necessary. In invasive candidiasis, we have found up to 50% false negative immunodiffusion (ID) reactions, or nondiagnostic agglutination titers in patients with neoplastic disease (18). On the other hand, up to 100% of various groups of patients followed carefully and with no evidence of invasive candidiasis had either positive ID tests or diagnostic rises or falls in agglutination titers. This type of test cannot be used as a major factor in evaluating the possibility of invasive Candida infection in an individual who is immunosuppressed.

In contrast, we found an ID test for Aspergillus antibody far more reliable (39). Using a microtemplate on agarose (50), and concentrating the serum 3 times, we were able to demonstrate conversions from negative to positive ID reactions in 7 of 10 patients with proven aspergillosis. Subsequently, using a fluorescent antibody (FA) test (50), serum of the 3 patients with negative ID reactions were tested and showed a 4-fold rise or fall in titer (51). We now include the FA test along with the ID test in evaluating patients for invasive aspergillosis. The FA test has the advantage of yielding results in a matter of hours, while the ID test takes 3 days. There has been no reliable serological test developed for mucormycosis as yet. Serological tests for the common fungi infecting both the intact and immunosuppressed patient are outlined in Table 6.8.

Nocardia asteroides grows slowly and is often difficult to find on Gram stains of sputum samples. *Mycobacterium tuberculosis* – or other mycobacterial infections – may progress rapidly in the immunosuppressed patient (27), or disseminate to liver, spleen, and bone marrow, without clinically involving the lungs. Serological tests for both of these infections would be helpful to the clinician but have been found rather insensitive in studies thus far (24, 29, 35). It would be desirable to develop both more sensitive and specific tests for nocardiosis and tuberculosis because they appear to share common antigens, and, in addition, patients with atypical mycobacterial infections tend also to react with *M. tuberculosis* antigens (29, 35).

Serologic tests for toxoplasmosis are multiple, and so-called significant titers vary from laboratory to laboratory. The only diagnostic test is one including acute and convalescent serum samples, demonstrating a 4-fold rise in titer. Table 6.9 contains a list of tests and titers which should raise the clinicians' suspicion. Since therapeutic action may have to be taken before a convalescent serum sample is available, or the patient is tested well into his illness, and an acute specimen is not available, antibody levels suggesting recent infection are also listed in Table 6.9.

TABLE 6.8
Serology for common fungal infections

A Four-Fold Rise in Titer is Diagnostic*
Titers Listed Suggest Recent Infection

	CF	ID	Agglutination	Others†
Histoplasmosis	≥1:8 ‡	+ M‡ old infection + H recent infection + C cross reacting	Latex – recent infection	IFA Hemagglutination
Coccidioidomycosis	≥1:8	+ recent infection	Latex – recent infection	Precipitin – recent
Blastomycosis	≥1:8			IFA
Aspergillosis	–	+ recent or old infection		Latex agglutination, IFA
Candidiasis	–	+ recent or old infection	Recent or old infection	Latex agglutination, IFA, CF

* Since infections are common in endemic areas, a diagnostic rise in titer must be carefully correlated with the clinical picture. Dual infections do occur.

† Various tests are under study in different laboratories where interpretations must be obtained.

‡ Influenced by skin test – draw blood before skin test.

CF = Complement fixation. ID = immunodiffusion. IFA = indirect fluorescent antibody.

TABLE 6.9
Toxoplasma serology

Test	Titers Suggesting* Recent Infection
Sabin-Feldman dye test (SFDT)	≥1:1024
Complement-fixation test (CFT)	≥1:8
Hemagglutination test (HA)	≥1:1024
Immunodiffusion test (ID)	+ band(s)
Indirect immunofluorescent test (IFA)	≥1:1024
Total serum immunoglobulins (IgS)	
Immunoglobulin G (IgG)	≥1:1024
Immunoglobulin M (IgM)	≥1:128

* A 4-fold rise in titer is diagnostic of ongoing infection between the serums tested. Conversion from a negative to positive ID test is strongly suggestive.

It would be desirable to have a sensitive and specific serologic test for every microorganism infecting man, especially in the immunosuppressed individual who so often becomes prey to one or more of many organisms that are normal flora and that can be frequently cultured from body sites in the absence of invasive disease. A serological test correlating with invasive disease would be most helpful in choosing specific antimicrobial therapy. The least we can do at present is to store serums on all patients before we immunosuppress them and at regular intervals during periods of immunosuppression, so that we have acute or "preacute" serums available when needed for tests that are available and valid.

The detection of circulating specific antigen as an indicator of invasive infections has been a great step forward in diagnostic microbiology. This was first demonstrated by *Cryptococcus neoformans*, subsequently with serum hepatitis B, and then with *Haemophilus influenzae, Neisseria meningitidis*, and *Streptococcus pneumoniae*. Table 6.10 lists the organisms where free or circulating antigen has been detected in the presence of invasive infection.

Detection of specific components of the organism in specific quantities is another method of attempting to demonstrate invasive infection with a commonly isolated agent such as *Candida albicans*. Demonstration of certain patterns of polysaccharides and lipids by gas-liquid chromotogra-

TABLE 6.10
Organisms where free or circulating antigen has been detected in presence of invasive infection

Bacteria	Fungi	Parasites	Viruses
Streptococcus pneumoniae	*Cryptococcus neoformans*	—	Serum hepatitis
Neisseria meningitidis	*Candida* sp.		
Haemophilus influenzae			
Pseudomonas aeruginosa			
Streptococcus pyogenes			

phy (GLC) has correlated with invasive Candida infections in one study (33), and this will hopefully be confirmed and extended because GLC is a laboratory tool which can yield rapid results. There is some experience to suggest that this method can also be applied to the detection of selected bacteria (34).

The detection of specific antigen or cell components or metabolites of microorganisms in body fluids is one of the most promising areas in diagnostic microbiology for the early diagnosis of invasive infection.

Antimicrobial Sensitivity Testing

Microbiology laboratories in hospitals which care for large numbers of immunosuppressed patients have a special problem with antimicrobial sensitivity testing. Invasive infection with various fungi are common, and antifungal sensitivity testing to agents such as amphotericin B or 5-fluorocytosine is done in very few laboratories. As new antifungal agents, such as clotrimazole and miconazole are used, valid and rapid sensitivity tests which can be done in any community hospital microbiology laboratory should become available. The pharmaceutical companies which produce the antifungal agent, a responsible government agency (Center for Disease Control or Federal Drug Administration), or the American Society of Microbiology should assume responsibility. The production of any antimicrobial should be accompanied by valid methods to test sensitivity, and, if the antimicrobial is toxic, then methods for serum assay which can be done in community hospital microbiology laboratories should be marketed at the same time. It is my opinion that the pharmaceutical company marketing the drug should be responsible.

SUMMARY

The need for rapid sensitivity testing of significant isolates from an immunosuppressed patient is even more pressing than in the general population. Infections move more rapidly and are more frequently lethal in the immunosuppressed patient. Every advantage the clinician can have in choosing the appropriate antibiotic and the appropriate dose may be necessary to treat a patient successfully when the immune defenses are altered. Precise identification of the organism is also necessary in the immunosuppressed patient where the source may not be obvious, but the identification of the organism may suggest a potential source; for example, *Escherichia coli, Pseudomonas aeruginosa, Klebsiella, Enterobacter,* or *Proteus* species from the gastrointestinal tract. At times, if the organism suggests a source which can be drained, this can be extremely helpful, frequently even more so than choosing the appropriate antibiotic. The precise identification of the organism also helps in hospital epidemiology, and nosocomial infections are more frequent in the immunosuppressed patient who may be hospitalized for prolonged periods of time and receive multiple supportive measures, including catheters of various types and respiratory assistance equipment—all excellent foci for nosocomial infections. Thus, the isolation of a *Serratia marcescens* from more than one patient's urine and blood might suggest not only a nosocomial infection but a single source of nosocomial infection, such as has been seen with contamination of lubricant jellies. Finally, precise identification can help in evaluating whether an organism that has been isolated is truly a potential pathogen or a contaminant.

The best diagnostic microbiology laboratory can help to render extraordinary patient care if they not only follow the rules of scrupulous definitive microbiology, but they also must communicate with the clinician who is directly responsible for the patient's care. The directors, supervisors, and technologists must all seek to speak a common language with the clinicians, including educating them, if necessary, but ever so gently, for they may be oversensitive, particularly in areas where they lack specific knowledge. Likewise, the clinician must communicate with the laboratory and remember that the best microbiologist is rigid in certain areas both by nature and by training, and usually rather cautious and conservative. He is particularly sensitive when he cannot supply the information that the clinician obviously wants. Close cooperation and fluid communication is necessary between the medical microbiologist and the clinician to render the finest patient care in infectious diseases, especially in the immunosuppressed patient.

ACKNOWLEDGMENT

The author acknowledges the technical and supervisory staff of the Diagnostic and Special Studies Microbiology Laboratory of Memorial Sloan Kettering Cancer Center. Their devotion to precise and definitive medical microbiology has made possible first, excellent patient care and second, the collection of data which made possible studies such as those indicated in the bibliography.

LITERATURE CITED

1. Aisner, J., Schimpff, S. C., Bennett, J. E., Young, V. M., and Wiernik, P. A. Aspergillus infections in cancer patients. Association with fireproofing materials in a new hospital. *JAMA 235:*411–412, 1976.
2. Alpern, R. J., and Dowell, V. R. Clostridium septicum infections and malignancy. *JAMA 209:*385–388, 1969.
3. Armstrong, D. Life threatening infections in cancer patients. *CA 23:*138–150, 1973.
4. Armstrong, D. Infectious complications in cancer patients treated with chemical immunosuppressive agents. *Transplant Proc 5:*1245–1248, 1973.
5. Armstrong, D. Infections in patient with lupus erythematosus disseminatus. *Arthritis Rheum 17:*285–286, 1974.
6. Armstrong, D. Interstitial pneumonia in the immunosuppressed patient. *Transplant Proc* (in press).
7. Armstrong, D., Blevins, A., Louria, D. B., Henkel, J. S., Moody, M. D., and Sukany, M. Groups B, C, and G Streptococcus infections in a cancer hospital. *Ann NY Acad Sci 174:*511–522, 1970.
8. Armstrong, D., Chmel, H., Singer, C., Tapper, M. L., and Rosen, P. P. Non-bacterial infections associated with neoplastic disease. *Eur J Cancer 11*(Suppl.): 79–94, 1975.
9. Armstrong, D., Young, L. S., Meyer, R. D., and Blevins, A. H. Infectious complications of neoplastic disease. *Med. Clin North Am 55:*729–745, 1971.
10. Armstrong, D., Yu, B. Unpublished observations.
11. Bach, M. C., Monaco, A. P., and Finland, M. Pulmonary nocardiosis. Therapy with minocycline and with erythromycin plus ampicillin. *JAMA 224:*1378–1381, 1973.

12. Berg, R., Chmel, H., Mayo, J. B., Armstrong, D. *Corynebacterium equi* infection complicating neoplastic disease. *Am J Clin Pathol* (in press).
13. Bode, F. R., Pare, J. A. P., and Fraser, R. G. Pulmonary diseases in the compromised host. *Medicine 53:*255–293, 1974.
14. Chernik, N. L., Armstrong, D., and Posner, J. B. Central nervous system infections in patients with cancer. *Medicine 52:*563–581, 1973.
15. Chernik, N. L., Armstrong, D., and Posner, J. B. Central nervous system infections in patients with cancer: Changing patterns. Cancer (in press).
16. Chmel, H., and Armstrong, D. *Salmonella oslo:* A focal outbreak in a hospital. *Am J Med 60:*203–208, 1976.
17. Eichoff, T. C. Nosocomial infections due to *Klebsiella pneumoniae*: Mechanisms of intrahospital spread, pp. 117–122. In *Proceedings of the International Conference on Nosocomial Infections,* Center for Disease Control, August 1970.
18. Filice, G., Yu, B., and Armstrong, D. Candida immunodiffusion and Agglutination tests in patients with neoplastic disease: Inconsistent correlation with invasive infections. *J Infect Dis* (in press).
19. Fisher, M. W., Devlin, H. B., and Gnabasik, F. J. New immunotype schema for *Pseudomonas aeruginosa* based on protective antigens. *J Bacteriol 98:*835–836, 1969.
20. Greene, W. H., Moody, M., Hartley, R., et al. Esophagoscopy as a source of *Pseudomonas aeruginosa* sepsis in patients with acute leukemia. The need for sterilization of endoscopes. *Gastroenterology 67:*912–919, 1974.
21. Hart, P. D., Russell, E., and Remington, J. S. The compromised host and infection. I. Deep fungal infection. *J Infect Dis 120:*169, 1969.
22. Henkel, J. S., Armstrong, D., Blevins, A., and Moody, M. D. Group A β-hemolytic streptococcal bacteremia in a cancer hospital. *JAMA 211:*983–986, 1970.
23. Hersh, E. M., and Freireich, E. J. Host defense mechanisms and their modification by cancer chemotherapy. *Methods Cancer Res 2:*355, 1968.
24. Humphreys, D. W., Crowder, J. G., and White, A. Serological reactions to Nocardia antigens. *Am J Med Sci 269:*323–326, 1975.
25. Idhe, D. C., and Armstrong, D. Clinical spectrum of infection due to *Bacillus species*. *Am J Med 55:*839–845, 1973.
26. Kagnoff, M. F., Armstrong, D., and Blevins, A. Bacteroides bacteremia. Experience in a hospital for neoplastic disease. *Cancer 29:*245–251, 1972.
27. Kaplan, M. H., Armstrong, D., and Rosen, P. P. Tuberculosis complicating neoplastic disease. A review of 201 cases. *Cancer 33:*850–858, 1974.
28. Kaplan, M. H., Rosen, P. P., and Armstrong, D. Cryptococcosis in a cancer hospital. Clinical and pathological correlates in forty-six patients. *Cancer* (in press).
29. Kaplan, M. H., Yu, B., and Armstrong, D. Detection of antibody to *Mycobacterium tuberculosis* by a micro-immunodiffusion method (submitted for publication).
30. Louria, D. B., and Armstrong, D. Unpublished data.
31. Meyer, R. D., Roen, P., and Armstrong, D. Phycomycosis complicating leukemia and lymphoma. *Ann Intern Med 77:*871–879, 1972.
32. Meyer, R. D., Young, L. S., and Armstrong, D. Aspergillosis complicating neoplastic disease. *Am J Med 54:*6–15, 1973.
33. Miller, G. G., Witwer, M. W., Braude, A. L., and Davis, C. E. Rapid identification of *Candida albicans* septicemia in man by gas-liquid chromatography. *J Clin Invest 54:*1235–1240, 1974.
34. Mitruka, B. M., Jonas, A. M., and Alexander, M. Rapid detection of bacteremia in mice by gas chromotography. *Infect Immun 2:*474–478, 1970.
35. Parlett, R. C., and Youmans, G. P. An evaluation of the specificity and sensitivity of a gel double-diffusion test for tuberculosis. *Am Rev Respir Dis 80:*153–166, 1959.
36. Roberts, R. B. Personal communication.
37. Rose, H. D. Mechanical control of hospital ventilation and Aspergillus infections. *Am Rev Respir Dis 105:*306–307, 1972.
38. Rosen, P. P., and Armstrong, D. Infective endocarditis in patients treated for malignant neoplastic disease. *Am J Clin Pathol* 60:241–250, 1973.
39. Schaefer, J. C., Yu, B., and Armstrong, D. An Aspergillus immunodiffusion test in the early diagnosis of aspergillosis in adult leukemia patients. *Am Rev Respir Dis 113:*325–329, 1976.
40. Schimpff, S., Serpick, A., Stoler, B., Rumack, B., Millin, H., Joseph, J. M., and Block, J. Varicella-zoster infection in patients with cancer. *Ann Intern Med 76:*241, 1972.

41. Schimpff, S. C., Young, V. M., Greene, W. H., et al. Origin of infection in acute nonlymphocytic leukemia. Significance of hospital acquisition of potential pathogens. *Ann Intern Med 77:*707–714, 1972.
42. Sickles, E. A., Young, V. M., Greene, W. H., and Wiernik, P. H. Pneumonia in acute leukemia. *Ann Intern Med 79:*528–534, 1973.
43. Singer, C., Armstrong, D., Rosen, P. P., and Schottenfeld, D. *Pneumocystis carinii* pneumonia: A cluster of eleven cases. *Ann Intern Med 82:*772–777, 1975.
44. Singer, C., Kaplan, M. H., and Armstrong, D. Sepsis in leukemia and lymphoma. *Antibiot Chemother 21:*187–188, 1976.
45. Singer, C., Kaplan, M. H., and Armstrong, D. Bacteremia and fungemia complicating neoplastic disease: A study of 364 cases. *Am J Med* (in press).
46. Wolfe, M. S., Armstrong, D., Louria, D. B., and Blevins, A. Salmonellosis in patients with neoplastic disease. A review of 100 episodes at Memorial Cancer Center over a 13 year period. *Arch Intern Med 128:*546–554, 1971.
47. Wynne, J. W., and Armstrong, D. Clostridial septicemia. *Cancer 29:*215–221, 1972.
48. Young, L. S., Armstrong, D., Blevins, A., and Lieberman, P. *Nocardia asteroides* infection complicating neoplastic disease. *Am J Med 50:*356–367, 1971.
49. Young, R. C., Bennett, J. E., Vogel, C. L., Carbone, P. P., and DeVita, V. T. Aspergillosis. The spectrum of the disease in 98 patients. *Medicine 49:*147, 1970.
50. Yu, B., and Armstrong, D. Serological tests for invasive aspergillus and candida infections in patients with neoplastic disease. *Proceedings, VIth Congress of the International Society for Human and Animal Mycology*. University of Tokyo Press, Japan (in press).
51. Yu, B., Armstrong, D. Unpublished data.

7

Enterobacteriaceae: Clinical Significance of Speciation

WILLIAM R. MC CABE

The increasing complexity of delivery of health care has led to marked changes in hospital practices. The expansion of specialized and intensive care facilities and other changes have resulted in steadily progressive, almost astronomical, rises in hospital costs. These increasing costs, coupled with a diminution of local, state, and federal funding, have placed significant constraints upon hospital budgets. Recognition that attempts must be made to control constantly spiralling hospital costs has caused reexamination of the relative merits and cost-benefit ratios of many current hospital practices. These considerations are almost certainly responsible for the emphasis on the importance of detailed bacterial speciation to be considered in these sessions. Critical appraisal of the actual importance of various practices in the provision of quality patient care is often met with considerable opposition. Such examinations can serve an important role in defining the essential aspects of optimal medical care provided that they are carried out by authorities who are knowledgeable concerning requirments for patient care but who are not biased concerning the merits of various techniques or methods.

Before examining individual aspects of hospital practices, it is important to reexamine the purpose and function of a hospital. While this may appear to be obvious, there is ample evidence that various biases and the pressures of "special interests" may modify the primary purposes of hospitals. The only justification that can be found for the existence of hospitals is that they provide a site where intensive medical and nursing care and laboratory and roentgenologic diagnostic facilities are available. There can be little question that patient care has to be the major function of the hospital. One gains the impression, however, that other considerations have too often become of major importance to those who determine priorities in hospitals. Plush carpeting and private washrooms for hospital directors are not essential for optimal patient care – but neither are "ego gratification" for directors of microbiology laboratories and full employment for all microbiologists.

It is also important that assessment of hospital practices includes all activities and should not be limited just to appraisal of the practices of microbiology laboratories. The role of other aspects of hospital practice are beyond the scope of this presentation, however, and it will be necessary to limit these considerations to the importance of precise speciation of Enterobacteriaceae to patient care.

There is probably no area of clinical infectious diseases or related laboratory studies in which the work load has increased so markedly in the past 30 years as that of infections caused by the Enterobacteriaceae. Twenty-five years ago, most microbiologic teaching ignored Enterobacteriaceae, other than Salmonella and Shigella, as being unimportant causes of human disease other than urinary tract infection. Indeed, organisms such as *Escherichia coli*, Klebsiella, etc., were of concern to microbiologists primarily because they tended to make it more difficult to identify Salmonella or Shigella in stool cultures. The frequency of hospital-acquired infections caused by the Enterobacteriaceae and Pseudomonadeaceae has shown a progressive increase since the early 1950s. These organisms have replaced the Staphylococcus as the most frequent cause of wound infections. Other investigators have documented a steady increase in the frequency of hospital-acquired necrotizing pneumonia caused by Gram-negative bacilli (5, 8). The most convincing index of the increasing frequency and importance of infections caused by Gram-negative bacilli, however, is that of bacteremia caused by these organisms. Whereas, prior to 1950, bacteremia caused by Gram-negative bacilli was considered a clinical oddity, reports from several major medical centers have demonstrated that bacteremia with Gram-negative bacilli may occur as often as once per 100 hospitalized admissions (9, 10). Numerous other investigations have also added further documentation of the progressively increasing frequency of infections caused by this group of microorganisms. This increasing prevalence of Gram-negative bacillary infections has added substantially to the work loads of microbiology laboratories.

Another source of additional work for the laboratory has been the increasing frequency of colonization with Gram-negative bacilli, which is often a reflection of antimicrobial therapy. While it is often difficult to distinguish colonization from infection, the former is a much more frequent occurrence (11). Isolation of Gram-negative bacilli from pharyngeal and sputum cultures was relatively infrequent prior to the advent of antimicrobials but is a frequent occurrence today. An additional source of increased laboratory work load is the submission of poorly collected specimens which have been contaminated with Gram-negative bacilli. In one survey from our laboratory, as many as 25% of urine cultures were considered to have been contaminated, as manifested by the presence of large numbers of nonpathogens, diphtheroids, etc., or the presence of large numbers of more than one species of Gram-negative bacilli (6). The increasingly detailed speciation of Enterobacteriaceae has also added to the laboratories' "work loads." The past 15–20 years have been marked by frequent reclassification of Enterobacteriaceae and other Gram-negative bacilli (2–4). These taxonomic manipulations have greatly increased the number of tests performed per isolate—but have they contributed materially to the quality of medical care received by the patient? Taken in toto, the increasing number of infections and colonization by Gram-negative bacilli and the increasingly complex process of speciation of Gram-negative bacilli have markedly increased the work load and expense of the microbiology laboratory. A recent survey from our laboratory indicated that approximately 20% of the technician work time was devoted to isolation and speciation of Gram-negative bacilli (6). This is almost twice as much time as is spent in performing sensitivity determinations. Thus, since a significant proportion of the laboratory's time is devoted to detailed speciation of Gram-negative bacilli, it is important to be certain that this information contribute materially to the patient's care.

REASONS FOR SPECIATION OF ENTEROBACTERIACEAE

Some of the reasons which have been marshalled for the detailed and extensive speciation of Enterobacteriaceae are listed in Table 7.1. Before these can be discussed in detail, it is necessary to understand just how the work of the microbiology laboratory is utilized for clinical care. It is essential for laboratory personnel to have some understanding of the manner in which bacteriologic studies are employed since results that have no relevance to the immediate clinical situation are of little use in patient care irrespective of how carefully the work has been performed. The usual clinical problem with which the physician is faced is that of a patient in whom an infection is suspected on the basis of history, physical, and x-ray examination and examination of Gram-stained specimens. At this time, appropriate specimens are collected and sent for bacteriologic culture and antibiotic sensitivity determinations. Approximatley 36–48 h are required before the exact species and sensitivity results are known. Thus, in most instances, the results of antibiotic susceptibility testing are available at the same time that bacterial speciation is completed. More importantly, the physician is forced to treat the most critical and seriously ill patients primarily on the basis of clinical acumen for the first 24–36 hr with little assistance from the microbiology laboratory. It is those patients with the most severe and fulminant infections, in whom laboratory assistance would be of greatest assistance, that the laboratory serves only to confirm

TABLE 7.1
Reasons for speciation of Enterobacteriaceae

1. Assistance in diagnosis
   A. Identification of etiologic agent in infections
   B. Identification of "specific disease syndromes"
2. Assistance in determining treatment
3. Assistance in determining prognosis
4. Epidemiologic and research uses
5. Communication

or refute the clinical diagnosis and to provide information which allows the physician to select a more effective or less toxic antibiotic after 36–48 hr of treatment. Recognition of the clinical problems of treating patients with serious life-threatening infections makes it apparent that the development of more rapid diagnostic techniques in microbiology is urgently needed and is considerably more important than detailed speciation of microorganisms.

Assistance in Diagnosis

There are two ways, shown in Table 7.1, in which bacteriologic cultures and speciation of the etiologic agent may be of diagnostic assistance: (a) isolation and identification of the etiologic agent from sites of infection in patients with clinical illnesses and (b) identification of "disease syndromes" caused by specific bacterial species. Both of these features are interrelated and constitute a major basis for clinical diagnosis. As was described earlier, however, initial diagnosis and treatment of patients seriously ill with infection is dependent primarily on clinical judgment with little assistance from the microbiology laboratory until 24–48 hr after initiation of therapy. For this reason, speciation of the etiologic agent is of little value in the initial management of the acutely ill patient and serves primarily to confirm the clinical diagnosis.

Another way in which speciation may contribute to diagnosis in the patient with infections is by the association of "specific disease syndromes" with the isolation of individual species of Enterobacteriaceae. Once a sufficiently distinct constellation of clinical findings has been established to be produced by a specific agent, the clinician can then predict the etiologic agent in patients who present with these symptoms. Table 7.2 presents a list of those Enterobacteriaceae which produce relatively characteristic clinical syndromes. As may be seen, only a small number of distinctive clinical syndromes are produced by individual species of Enterobacteriaceae. Typhoid or enteric fever, which are rare in the United States today, are caused predominantly by *Salmonella typhi*, *S. paratyphi-A*, and *S. paratyphi-B* although other Salmonella occasionally may be responsi-

TABLE 7.2
Diagnostic assistance and syndrome identification

A. Specific clinical syndromes caused by enterobacteriaceae:
   1. Typhoid and enteric fever—*Salmonella typhi*, *Salmonella paratyphi-A*, and *Salmonella paratyphi-B*
   2. Gastroenteritis and diarrheal disease—other salmonella
   3. Diarrheal disease of infants—Enteropathogenic *Escherichia coli*
   4. Bacillary dysentery—*Shigella* sp.
   5. Friedlander's pneumonia—*Klebsiella pneumoniae*
B. Other Infections produced by multiple species of Enterobacteriaceae:
   1. Bacteremia
   2. Urinary tract infections
   3. Wound infections
   4. Pulmonary infections

ble. Friedlander's pneumonia also has relatively distinctive features and is produced by a limited number of capsular types of *Klebsiella pneumoniae*. Salmonella gastroenteritis and diarrhea, enteropathogenic *E. coli* diarrhea of infants, and bacillary dysentery caused by Shigella are often considered to have certain distinctive features. There is sufficient overlap in manifestations, however, that, in the individual patient, it is usually not possible to accurately predict the etiologic agent solely on clinical findings. Thus, while it is true that it has been possible to associate certain syndromes with individual species of Enterobacteriaceae in past studies, such diseases constitute only a small proportion of those infections caused by Enterobacteriaceae. In addition, the rather considerable experience with infections caused by Enterobacteriaceae indicate that it is highly unlikely that any new characteristic syndromes caused by specific bacteria will be identified.

Part B of Table 7.2 presents a list of the numerically most important types of infections caused by Enterobacteriaceae. The manifestations of infections in these sites are identical irrespective of the species of etiologic agent. Almost every species of Enterobacteriaceae has been isolated from patients with bacteremia and no differences in clinical findings can be detected as is described subsequently under "Assistance in Determining Prognosis." Similarly, all species of Enterobacteriaceae may produce urinary tract infections, hospital-acquired pulmonary infections, and wound infections, and the clinical features of each of these does not differ with different etiologic agents. A few clinical clues sometimes may result from knowledge of the species of etiologic agent. The finding of species other than *E. coli* in urine cultures suggests prior urinary infections, genitourinary manipulations, or structural genitourinary abnormalities and certain types of renal calculi may suggest infection with *Proteus mirabilis*. An accurate history will identify these features even more effectively, however.

Assistance in Determining Treatment

Another reason that has been given for detailed speciation of Enterobacteriaceae is to assist in the selection of an optimal therapeutic regimen. This proposed use assumed that knowledge of antibiotic sensitivity patterns of species of Enterobacteriaceae will allow the physician to select an appropriate antibiotic once the infecting agent has been identified. As pointed out earlier, however, in actual practice, the results of sensitivity testing are usually available by the time speciation is completed. Since the physician already has sensitivity results by the time the species of the etiologic agent is known, he routinely utilizes the sensitivity report for antibiotic selection.

Another potential use of speciation is in determining treatment in infections in which there is documented evidence that in vitro sensitivity results do not correspond with clinical efficacy. The sole instance of enterobacterial infection in which a distinct discrepancy between the results of susceptibility testing and clinical efficacy has been demonstrated is typhoid or "enteric" fever. There is clear documentation that, although the infecting Salmonella may be susceptible to a large number of antimicrobials, only chloramphenicol, ampicillin, and sulfonamide-trimethoprim are effective for the treatment of this syndrome.

Assistance in Determining Prognosis

Ability to predict the severity or outcome of infections once the infecting organism is known is another reason that has been advanced for careful and detailed speciation of Enterobacteriaceae. There is evidence that certain Salmonella species, e.g., *S. cholerae-suis*, are associated with more severe infection but, with the exception of diarrheal disease, there is no evidence that the species of the infecting agent influences the severity of other infections caused by Enterobacteriaceae. Extensive studies in our laboratory have demonstrated that the severity of the patient's underlying disease, at the onset of Gram-negative bacteremia, was the major determinant of fatality rates in this infection. As shown in Table 7.3, however, no significant differences in fatality rates of bacteremia caused by different species of Enterobacteriaceae could be detected when corrections were made for the severity of the host's underlying disease (7). The influence of the species of etiologic agent in bacteremia on the severity of the disease was also assessed using the frequency of occurrence of shock as another index of the severity of bacteremia as shown in Table 7.4. The frequency of shock in bacteremia caused by various species of Enterobacteriaceae did not differ significantly in patients with underlying disease of similar severity (7). Studies by Tillotson and Finland (11) of respiratory tract colonization and hospital-acquired pneumonia caused by Enterobacteriaceae demonstrated similar findings. No significant differences in infection rates (number of infections/number of colonizations) or fatality rates could be detected between various species of Enterobacteriaceae (11).

In addition, because of the 36–48 hr required to complete bacterial speciation, the physician is usually able to make a much better estimate of

TABLE 7.3
Fatality rates in relation to etiologic agent and underlying host disease in bacteremia

Etiologic Agent	Underlying Host Disease		
	Rapidly fata	Ultimately fatal	Nonfatal
Escherichia coli	7/10	20/53	5/77
Klebsiella-Enterobacter-Serratia	5/6	15/42	6/54
Pseudomonas aeruginosa	12/13	18/27	4/18
Proteus sp.	–	5/10	1/21
Bacteroides	–	3/12	2/7
Other	1/2	3/12	7/34
Mixed	1/3	27/44	16/28
Totals	26/34 (76%)	91/200 (46%)	41/239 (17%)

TABLE 7.4
Frequency of shock in relation to etiologic agent and underlying host disease in Gram-negative bacteremia

Bacterial species	Underlying host Disease		
	Rapidly fatal	Ultimately fatal	Nonfatal
Escherichia coli	1/2 (50%)*	18/41 (44%)	11/50 (22%)
Klebsiella-Enterobacter-Serratia	2/2 (100%)	8/21 (38%)	11/32 (34%)
Pseudomonas	1/3 (33%)	8/13 (62%)	5/10 (50%)
Proteus sp.	–	3/8 (38%)	3/14 (22%)
Other	0/2 –	4/8 (50%)	5/18 (28%)
Mixed	0/2 –	17/32 (53%)	14/23 (61%)
Bacteroides	–	7/12 (58%)	4/7 (57%)
Totals	4/11 (36%)	65/135 (48%)	53/154 (35%)

* Number of patients with shock/number of patients (% with shock).

the prognosis of the infection on the basis of the clinical course during this period than on knowledge of the species of the etiologic agent. Thus, knowledge of the exact species of Enterobacteriaceae producing infection is not of much assistance to the clinician in predicting the severity or prognosis of an infection.

Epidemiologic and Research Uses

A valid indication for detailed speciation of Enterobacteriaceae is for recognition and studies of hospital-acquired infections. The progressively increasing frequency of hospital-acquired infections caused by Enterobacteriaceae emphasized the importance of understanding the epidemiology of such infections and the development of control measures. Increased numbers or "clusters" of infection caused by individual species could not be recognized unless some method of bacterial identification were available. Even in these instances, however, bacterial speciation alone is not the sole criterion responsible for recognition of increased numbers of infections. An additional epidemiologic marker, such as resistance to a specific antibiotic, is usually required before individual species are recognized as a cause of an outbreak of infections. Thus, strains of *Pseudomonas aeruginosa*, *Proteus rettgeri*, and other bacteria are more apt to be considered a matter of concern if they are resistant to gentamicin and other aminoglycosides than if they are antibiotic susceptible. Such resistance markers are often more important to alerting us to outbreaks of infection and tracing their spread through the hospital than bacterial speciation. In addition, more sensitive methods, such as serotyping or bacteriophage typing, are usually required for precise identification of strains of species for detailed epidemiologic studies.

Marked changes in the classification of Gram-negative rods have occurred during the past two to three decades. Microorganisms such as Paracolobactrum, Herellea, and Mimae have disappeared and organisms such as Providencia and Acinetobacter have appeared (2–4). Some such changes have obvious merit but the value of other changes is less apparent. Irrespective of their merit, however, these numerous taxonomic changes have made it difficult to compare the prevalence of individual bacterial species as causes of various types of infections at various time periods. In addition, each delineation of a new species of Enterobacteriaceae is followed by a spate of publications describing these organisms as a cause of hospital-acquired infections and as a "new emerging pathogen" in hospitals. While this does expand bibliographies and support the already excessive number of journals, it really contributes little to our knowledge of Gram-negative bacillary infections to recognize that all Enterobacteriaceae are capable of producing hospital-acquired infections under appropriate circumstances. It is hardly surprising to anyone with an understanding of the pathogenesis of such infections to find that constituents of the fecal, skin, or environmental flora can traverse urinary catheters, tracheostomy tubes, etc., to produce infections.

Communication

Peculiarly enough, the most compelling reason for speciation is the necessity for communication between health care professionals. If we did not have specific names for microorganisms, communication of results between microbiologists and between microbiologists and physicians would not be possible. The complexity of attempting to communicate by the use of descriptive terms rather than by species names would seriously handicap medical care and research activities.

DISCUSSION

It is *not* the intent of this presentation to suggest that speciation of Gram-negative bacilli be discontinued. Its goal is to critically examine how detailed speciation of Enterobacteriaceae *does* and *does not* contribute to the care of the patient and other related hospital activities. It is hoped, however, that this review does indicate that the current zeal for detailed and comprehensive speciation of all clinical isolates of Enterobacteriaceae entails more expense and technician time than is justified by the clinical usefulness of the information supplied. While the cost-effectiveness of various hospital procedures was previously of limited concern, increasing hospital costs make it necessary to carefully examine both costs and utility of many hospital and laboratory practices. Others have also suggested that the expenses of microbiology laboratories could be reduced without materially reducing the quality or quantity of clinically useful diagnostic information (1). These suggestions emphasized that insistence on carefully collected clinical specimens rather than specimens contaminated with normal body flora, e.g., sputum uncontaminated with saliva, afforded a considerable reduction in clinically irrelevant bacterial identification (1). In addition, firm insistence that repetitious culturing or cultures of duplicate specimens not be performed reduces the laboratory's work load without affecting patient care. Other ways can also be found for simplification of laboratory procedures. The reading of all routine cultures, except blood cultures, from a single ward or hospital area daily by the same technician, is a helpful method for the early recognition of "clusters" of hospital-acquired infections. Also, identification schemes for speciation of Enterobacteriaceae can be simplified. There is a tendency to utilize excessively large numbers of tests initially for speciation to ensure that unusual species are identified rapidly. Since the vast majority of clinical isolates are *E. coli*, Klebsiella-Enterobacter-Serratia, Pseudomonads, or species of Proteus, these can usually be identified with a limited number of determinations plus morphology on selective media. More complex organisms can then be studied in more detail and the slight delay in identification of this small proportion of organisms would have little clinical impact for the reasons delineated above. Finally, the development of automatic methods may ultimately provide even more cost-effective techniques.

LITERATURE CITED

1. Bartlett, R. C. *Medical Microbiology Quality Cost and Clinical Relevance*. New York: John Wiley & Sons, 1974.
2. Buchanan, R. E., and Gibbons, N. E. *Bergey's Manual of Determinative Bacteriology*, Ed. 8, Baltimore: Williams & Wilkins Co., 1974.
3. Edwards, P. R. *Identification of Enterobacteriaceae*, Ed. 3. Minneapolis, Minn.: Burgess Publishing Co., 1972.
4. Edwards, P. R., and Ewing, W. H. *Identification of Enterobacteriaceae*, Ed. 2. Minneapolis, Minn.: Burgess Publishing Co., 1964.
5. Finland, M. Changing ecology of bacterial infections as related to antibacterial therapy. *J Infect Dis 122:*419, 1970.
6. Jacobs, A., and McCabe, W. R. Unpublished observations.
7. Kreger, B., Craven, D., and McCabe, W. R.: Gram-negative bacteremia (in preparation).
8. Mays, B. B., Thomas, G. B., Leonard, J. S., Jr., Southern, P. M., Jr., Pierce, A. K., and Sanford, J. P. Gram-negative bacillary necrotizing pneumonia; a bacteriologic and histopathologic correlation. *J Infect Dis 120:*687, 1969.
9. McCabe, W. R. Gram-negative bacteremia, in *Disease-a-Month*. Chicago: Yearbook Medical Publishers, 1973.
10. Myerowitz, R. L., Medeiros, A. A., and O'Brien, T. F. Recent experience with bacillemia due to gram-negative organisms. *J Infect Dis 124:*239, 1971.
11. Tillotson, J. R., and Finland, M. Bacterial colonization and clinical superinfection of the respiratory tract complicating antibiotic treatment of pneumonia. *J Infect Dis 119:* 597, 1969.

8

Pseudomonas–Identification Methods, Significance of Speciation, and Pathogenicity for Man

G. L. GILARDI

Among the members of the genus *Pseudomonas, P. aeruginosa, P. pseudomallei,* and *P. mallei* until recently were considered the only human pathogens. Implication of additional *Pseudomonas* species in infection has generated greater interest in their identification in the clinical laboratory and potential as human pathogens. The following is a review of recent developments in this area.

METHODS FOR IDENTIFICATION

Materials and methods for identification have been covered in detail elsewhere (16, 24) and are briefly summarized here. A two part identification system is sufficient for recognition of pseudomonads and other nonfermentative Gram-negative bacteria encountered in clinical bacteriology. The first part is a conventional Enterobacteriaceae setup which identifies most isolates and includes tests for urease (Christensen's urea agar); hydrogen sulfide, indol, motility (SIM medium); ONPG (ONPG test medium); phenylalanine deaminase (phenylalanine agar); arginine, lysine, ornithine decarboxylase (Moellers decarboxylase base medium); deoxyribonuclease (DNase test medium); oxidase (cytochrome oxidase strip); and susceptibility to penicillin and polymyxin.

Nonfermenters not identified at this point require a second setup which includes the flagella stain and tests for acid production from glucose, fructose, xylose, maltose, lactose, mannitol (OF basal medium); pyoverdin (Sellers differential agar); nitrogen gas (nitrate broth); esculin hydrolysis (Trypticase soy agar plus 0.1% esculin and 0.05% ferric citrate); gelatinase (nutrient gelatin); growth at 42°C; and acetate assimilation. A heavy inoculum is used for all enzymatic tests and a light broth suspension is used as inoculum for other tests. Age of the inoculum should be less than 48 hr. All cultures are incubated at 30 C for 24–48 hr.

PSEUDOMONADS IN CLINICAL SPECIMENS

The morphological and biochemical characteristics of pseudomonads and pseudomonas-like organisms encountered in clinical microbiology (Table 8.1) are extensively detailed elsewhere (24, 46, 49). Only the salient characteristics are covered here. Where appropriate, the significance of speciation is discussed. The clinical significance of each pseudomonad (excepting *P. aeruginosa, P. mallei, P. pseudomallei*) is reviewed.

Fluorescent Group (*P. aeruginosa, P. fluorescens, P. putida*)

Salient Features. Pyocyanin-producing strains of *P. aeruginosa* are easily identified in the laboratory, but apyocyanogenic strains must be differentiated from other pseudomonads. Apyocyanogenic strains can be recognized by several uniform characters, namely polar monotrichous flagella, arginine dihydrolase activity, growth at 42°C, and failure to oxidize dissacharides. The simple fluorescent pseudomonads (*P. fluorescens, P. putida*) are differentiated from apyocyanogenic strains of *P. aeruginosa* on the basis of polar multitrichous flagella, failure to grow at 42°C, and oxidation of disaccharides. *P. fluorescens* is distinguished from *P. putida* (*P. ovalis*) by gelatinase activity.

Two biotypes of *P. putida* and seven biotypes of *P. fluorescens* are recognized (46) on the basis of pigmentation, biochemical, and nutritional differences. Identification to biotype would yield epidemiological information, but in most circumstances the biotype need not be determined. There is some question whether nonproteolytic *P. putida* should be distinguished from *P. fluorescens*. Differentiation of the simple fluorescent pseudomo-

TABLE 8.1
Unusual Pseudomonas species encountered in clinical specimens and those species associated with infections

Pseudomonas Species Encountered	References Implicating the Species with Infections
P. putida	17, 27, 39
P. fluorescens	11, 17, 37, 40, 47
P. cepacia	2, 3, 5, 8–10, 17, 19, 20, 22, 29–32, 35, 36, 38, 41–45, 48, 51, 52
P. acidovorans	7
P. testosteroni	1, 4
P. alcaligenes	4, 34
P. pseudoalcaligenes	17, 25
P. stutzeri	4, 17, 18, 26
P. maltophilia	4, 12–14, 17, 21, 33, 34, 47
P. putrefaciens	17, 23, 50
P. mendocina	(No references for remaining bacteria)
P. diminuta	
P. vesiculare	
P. pickettii	
VA-2 group	
VE-1 group	
VE-2 group	
Xanthomonas	

nads from *P. aeruginosa* is necessary on the basis of the recognized pathogenicity of the latter species.

Clinical Significance. *P. fluorescens* and *P. putida* are usually environmental contaminants and rarely opportunistic pathogens for man. *P. putida* has been associated with a case of bacteriuria, wound infections (17), septic arthritis (27), and one case of septicemia in which the organism was recovered from blood and bone marrow cultures and from postmortem cultures of the spleen, an abdominal gland, and the lung (39).

P. fluorescens has been associated with four cases of empyema (37), urinary tract infections (40), septicemia (40, 47), a postoperative back abscess (17), and a febrile reaction following transfussion of contaminated blood (11). *P. fluorescens* and *P. putida* are generally susceptible to tetracycline, neomycin, kanamycin, polymyxin, and gentamicin, but resistant to carbenicillin (15).

Pseudomallei Group (*P. pseudomallei, P. mallei, P. cepacia*)

Salient Features. The phytopathogen *P. cepacia* (EO-1, *P. multivorans, P. kingii*) is genetically related to the animal pathogens *P. pseudomallei* and *P. (Actinobacillus) mallei*. Members of the pseudomallei group are polar multitrichous and utilize a wide range of organic compounds as carbon and energy sources. Many strains fail to produce or produce a slow and very weak indophenol oxidase reaction. Motility, growth at 42°C, and denitrification separate *P. pseudomallei* from *P. mallei*. Tests for ONPG, denitrification, arginine dihydrolase, and lysine decarboxylase separate *P. cepacia* from *P. pseudomallei*. Some isolates of *P. cepacia* produce a yellow-green, water-soluble, nonfluorescent, phenazine pigment; most strains are nonpigmented.

Clinical Significance. *P. cepacia* has been associated with endocarditis (22, 31, 32, 36, 42–44), septicemia (2, 10, 29, 35, 44, 52), pneumonia (8–10, 41, 51), urinary tract infections (10, 20, 22, 30, 38, 45), various wound infections and abscesses (3, 10, 17, 48), and chronic granulomatous disease (5, 9). *P. cepacia* is resistant to gentamicin and antibiotics of the polymyxin group. Only chloramphenicol has been shown to be effective against this species. Successful treatment of *P. cepacia* endocarditis with trimethoprim-sulfamethoxazole has been reported (36). The antibiogram of *P. cepacia* is quite similar to those of *P. pseudomallei* and *P. mallei* (28, 52).

Acidovorans Group (*P. acidovorans, P. testosteroni*)

Salient Features. Various names have been used for the organisms in this group including *P. acidovorans, P. testosteroni,* and *Comamonas terrigena*. Members of this group are polar multitrichous. The flagella are distinctive, resembling those of spirilla with a wavelength of 3.1 μm. Generally negative physiological features are associated with the members of this group. The acidovorans group contains in part species which are weakly saccharolytic and species which are alkali producers. *P. acidovorans,* but not *P. testosteroni*, produces weak acid from glucose, fructose, and mannitol. *P. testosteroni* may be regarded as a biotype of *P. acidovorans*.

Clinical Significance. Members of the acidovorans group are rarely etiologically significant. *P. testosteroni* was associated with conjunctivitis (4) and a case of septicemia in which the organism was isolated from peripheral blood and bone marrow (1). *P. acidovorans* was responsible for five cases of bacteremia due to contaminated transducers in patients undergoing cardiovascular monitoring (7). No single antibiotic is uniformly effective against members of the acidovorans group (15).

Alcaligenes Group (*P. alcaligenes, P. pseudoalcaligenes*)

Salient Features. Members of this relatively inert group are separated from the physiologically similar strains of the acidovorans group by polar monotrichous flagella. *P. pseudoalcaligenes* is nutritionally more active than *P. alcaligenes* and oxidizes glucose and fructose. The two physiological members of this group may be regarded as biotypes A and B, with biotype B corresponding to *P. pseudoalcaligenes*.

Clinical Significance. As in the case of the members of the acidovorans group, strains of the alcaligenes group are rarely opportunistic infectious agents. *P. alcaligenes* has been associated with empyema (34) and conjunctivitis (4). *P. pseudoalcaligenes* was the etiologic agent of pneumonitis, a postoperative knee infection (17), and was isolated from an infection of a pregnant uterus (25). No single antibiotic is effective against members of the alcaligenes group (15).

P. stutzeri

Salient Features. Strains of *P. stutzeri* (*Bacillus denitrificans* II) denitrify, tolerate 6.5% NaCl, produce extracellular amylase, utilize maltose, and are polar monotrichous. Freshly isolated strains produce colonies that are both wrinkled, tough, and coherent as well as smooth, with a buff to light brown intracellular pigment.

Clinical Significance. *P. stutzeri* is ubiquitous in soil and water and has been recovered from numerous clinical specimens but has been implicated only in otitis media, several wound infections sustained after trauma (17, 18), conjunctivitis (4), and septic arthritis (26). *P. stutzeri* belongs to the relatively antibiotic-susceptible group of pseudomonads demonstrating susceptibility to a wide spectrum of antibiotics (15).

P. maltophilia

Salient Features. *P. maltophilia* (*P. melanogena*) is polar multitrichous and produces lysine decarboxylase and extracellular deoxyribonuclease. It is unusual in not producing cytochrome oxidase.

Clinical Significance. *P. maltophilia* is the second most frequently encountered pseudomonad in clinical specimens and is occasionally an opportunistic pathogen for man. It has been associated with urinary tract infections including epididymitis (47), various wound infections and abscesses (14, 17, 34), bacteremia, pneumonia (13), acute mastoiditis (21), endocarditis (12), conjunctivitis (4), and meningitis (33). *P. maltophilia* is resistant to many antibiotics, with most strains demonstrating susceptibility only to chloramphenicol, nalidixic acid, and polymyxin (15). Trimethoprim-sulfamethoxazole was effective in treating a case of *P. maltophilia* endocarditis (12).

Diminuta Group (*P. diminuta, P. vesiculare*)

Salient Features. The flagella of the strains of this group are unusual, having a wavelength of only 0.6–0.98 μm. Members of this group are generally inert. *P. diminuta* fails to oxidize carbohydrates. *P. vesiculare* utilizes glucose, galactose, maltose, and esculin. *P. vesiculare* may be regarded as a biotype of *P. diminuta*. Accurate speciation of the biotypes in the diminuta group, as well as in the alcaligenes and acidovorans groups, would yield important epidemiological information. However, the members of these three groups are rarely, if ever, associated with infections, and under most circumstances this information would not be required.

Clinical Significance. Members of the diminuta group are infrequently recovered from human sources and appear to have no etiologic significance.

P. putrefaciens

Salient Features. *P. putrefaciens* (*P. rubescens*) is a polar monotrichous bacillus that produces hydrogen sulfide, ornithine decarboxylase, and extracellular deoxyribonuclease. A reddish-tan or pink water-soluble pigment is formed on most media. There are two physiological groups of strains which are distinguished by tolerance to 6.5% NaCl, as well as other different biochemical features (49). The salt-tolerant strains may be regarded as biotype 2. These strains, more so that biotype 1, are frequently recovered from human sources.

Clinical Significance. *P. putrefaciens* is known primarily as causing spoilage of food including butter, fish, and poultry. This organism has been isolated from a number of human sources but has been etiologically associated only with otitis media (17, 23, 50) and a tibia wound infection (17). This species demonstrates susceptibility to a wide range of antibiotics (15).

Other Pseudomonads

Several other *Pseudomonas* species and pseudomonas-like organisms have been recovered from clinical specimens but they appear not to have been implicated in infections (Table 8.1).

SUMMARY OF CLINICAL SIGNIFICANCE

The pseudomonads examined in this report are free-living and ubiquitous in the human body and in water, soil, or on plants. Although the majority of these isolates are present as saprophytes or contaminants and represent part of the transient flora or normal colonization of patients, these oganisms may serve as primary or opportunistic pathogens. An examination of the cases documented in the literature indicates that some of the manifestations of infection with pseudomonads are associated with certain predisposing factors. There is an increased incidence of infection in elderly patients (13) and in patients who have had serious underlying medical problems such as burns (52), trauma (10, 17), wounds following extensive surgery (10, 12, 17, 40, 44, 51), narcotic addiction (19, 27, 31, 32, 36), alcoholism (19), diabetes (10), malignancy (10, 34), aspiration pneumonia (13), or other major debiliting conditions (10). Infections may be of nosocomial origin in patients who have had prolonged hospitilization (13) with a history of prior instrumentation or a manipulative procedure, such as urethral cathcterizations (10, 13, 20, 34, 44) or cystoscopies (10, 13, 30); tracheostomy (13); instrumentation with contaminated disinfectant solutions (2, 3, 6, 20, 22, 44, 45), anaesthetics (41) or saline solutions (35); and intravenous infusions of contaminated medications and fluids (11, 29, 35). Contaminated pressure transducers have been incriminated as a source of nosocomial bacteremia in patients undergoing cardiovascular monitoring (35). Hospital-acquired infections with the pseudomonads examined in this report also occur in patients who have been placed on long term therapy with antibiotics (10, 13, 51). Prompt diagnosis of pseudomonas infections, identification of the species involved, and determination of specific therapy are necessary for a favorable prognosis.

LITERATURE CITED

1. Atkinson, B. E., Smith, D. L., and Lockwood, W. R., *Pseudomonas testosteroni* septicemia. *Ann Intern Med 83:*369–370, 1975.
2. Bassett, D. C. J., Dickson, J. A. S., and Hunt, G. H. Infection of holter valve by *Pseudomonas*-contaminated chlorhexidine. *Lancet 1:*1263–1264, 1973.
3. Bassett, D. C. J., Stokes, K. J., and Thomas, W. R. G. Wound infection with *Pseudomonas multivorans*. A water-borne contaminant of disinfectant solutions. *Lancet 1:*1188–1191, 1970.
4. Ben-Tovim, T., Eylan, E., Romano, A., and Stein R. Gram-negative bacteria isolated from external eye infections. *Infection 2:*162–165, 1974.
5. Bottone, E. J., Douglas, S. D., Rausen, A. R., and Keusch, G. T. Association of *Pseudomonas cepacia* with chronic granulomatous disease. *J Clin Microbiol 1:*425–428, 1975.
6. Burdon, D. W., and Whitby, J. L. Contamination of hospital disinfectants with *Pseudomonas* species. *Br Med J 2:*153–155, 1967.
7. Center for Disease Control. *Morbid Mortal Week Rep 24:*295, 1975.
8. Dailey, R. H., and Benner, E. J. Necrotizing pneumonitis due to the pseudomonad "eugonic oxidizer – group 1." *N Engl J Med 279:*361–362, 1968.
9. Denney, D., Bigley, R. H., Rashad, A. L., MacDonald, W. J., and Miller, M. J. Recurrent pneumonitis due to *Pseudomonas cepacia* – an unexpected phagocyte dysfunction. *West J Med 122:*160–164, 1975.
10. Ederer, G. M., and Matsen, J. M. Colonization and infection with *Pseudomonas cepacia*. *J Infect Dis 125:*613–618, 1972.
11. Felsby, M., Munk-Andersen, G., and Siboni, K. Simultaneous contamination of transfusion blood with *Enterobacter agglomerans* and *Pseudomonas fluorescens*, supposedly from the pilot tubes. *J Med Microbiol 6:*413–416, 1973.
12. Fischer, J. J. *Pseudomonas maltophilia* endocarditis after replacement of the mitral valve: A case study. *J Infect Dis 128:*S771–773, 1973.
13. Gardner, P., Griffin, W. B., Swartz, M. N., and Kunz, L. J. Nonfermentative gram-negative bacilli of nosocomial interest. *Am J Med 48:*735–749, 1970.
14. Gilardi, G. L. *Pseudomonas maltophilia* infections in man. *Am J Clin Pathol 51:*58–61, 1969.
15. Gilardi, G. L. Antimicrobial susceptibility as a diagnostic aid in the identification of nonfermenting gram-negative bacteria. *Appl Microbiol 22:*821–823, 1971.
16. Gilardi, G. L. Characterization of nonfermentative nonfastidious gram negative bacteria encountered in medical bacteriology. *J Appl Bacteriol 34:*623–644, 1971.

17. Gilardi, G. L. Infrequently encountered *Pseudomonas* species causing infection in humans. *Ann Intern Med 77:*211–215, 1972.
18. Gilardi, G. L., and Mankin, H. J. Infection due to *Pseudomonas stutzeri*. *NY State J Med 73:*2789–2791, 1973.
19. Hamilton, J., Burch, W., Grimmett, G., Orme, K., Brewer, D., Frost, R., and Fulkerson, C. Successful treatment of *Pseudomonas cepacia* endocarditis with trimethoprim-sulfamethoxazole. *Antimicrob Agents Chemother 4:*551–554, 1973.
20. Hardy, P. C., Ederer, G. M., and Matsen, J. M. Contamination of commercially packaged urinary catheter kits with the pseudomonad EO-1. *N Engl J Med 282:*33–35, 1970.
21. Harlowe, H. D. Acute mastoiditis following *Pseudomonas maltophilia* infection: Case report. *Laryngoscope 82:*882–883, 1972.
22. Henderson, A., and Byatt, M. E. *Pseudomonas cepacia* in the west of Scotland. *Health Bull (Edinb) 32:*100–102, 1974.
23. Holmes, B., Lapage, S. P., and Malnick, H. Strains of *Pseudomonas putrefaciens* from clinical material. *J Clin Pathol 28:*149–155, 1975.
24. Hugh, R., and Gilardi, G. L. *Pseudomonas*, p. 250–269. In E. H. Lennette, E. H. Spaulding, and J. P. Truant (Eds.), *Manual of Clinical Microbiology*, Ed. 2. Washington, D. C.: American Society for Microbiology, 1974.
25. Ledger, W. J., and Headington, J. T. Isolation of *Pseudomonas pseudoalcaligenes* from an infection of a pregnant uterus. *Int J Gynecol Obstet 10:*87–89, 1972.
26. Madhavan, T. Septic arthritis with *Pseudomonas stutzeri*. *Ann Intern Med 80:*670–671, 1974.
27. Madhavan, T., Fisher, E. J., Cox, F., and Quinn, E. L. *Pseudomonas putida* and septic arthritis. *Ann Intern Med 78:*971–972, 1973.
28. Mannheim, W., and Bürger, H. Über physiologische Merkmale und die Frage der systematischen Stellung des Rotz-Erregers. *Z Med Mikrobiol Immunol 152:*249–261, 1966.
29. Meyer, G. W. *Pseudomonas cepacia* septicemia associated with intravenous therapy. *Calif Med 119:*15–18, 1973.
30. Mitchell, R. G., and Hayward, A. C. 1966. Postoperative urinary-tract infections caused by contaminated irrigating fluid. *Lancet 1:*793–795, 1966.
31. Neu, H. C., Garvey, G. J., and Beach, M. P. Successful treatment of *Pseudomonas cepacia* endocarditis in a heroin addict with trimethoprim-sulfamethoxazole. *J Infect Dis 128:*S768–770, 1973.
32. Noriega, E. R., Robinstein, E., Simberhoff, M. S., and Rahal, J. J., Jr. Subacute and acute endocarditis due to *Pseudomonas cepacia* in heroin addicts. *Am J Med 59:*29–36, 1975.
33. Patrick, S., Hindmarch, J. M., Hague, R. V., and Harris, D. M. Meningitis caused by *Pseudomonas maltophilia*. *J Clin Pathol 28:*741–743, 1975.
34. Pedersen, M. M., Marso, E., and Pickett, M. J. Nonfermentative bacilli associated with man. III. Pathogenicity and antibiotic susceptibility. *Am J Clin Pathol 54:*178–192, 1970.
35. Phillips, I., Eykyn, S., Curtis, M. A., and Snell, J. J. *Pseudomonas cepacia* (*multivorans*) septicaemia in an intensive-care unit. *Lancet 1:*375–377, 1971.
36. Rahal, J. J., Jr., Simberkoff, M. S., and Hyams, P. J. *Pseudomonas cepacia* tricuspid endocarditis: Treatment with trimethoprim, sulfonamide, and polymyxin B. *J Infect Dis 128:*S762–767, 1973.
37. Riskó, T., and Nikodémusz, I. *Pseudomonas fluorescens*, mint kórokozó. *Népegészségügy 31:*106–108, 1950.
38. Roberts, J. B. M., and Speller, D. C. E. *Pseudomonas cepacia* in renal calculi. *Lancet 2:*1099, 1973.
39. Rogers, K. B. *Pseudomonas* infections in a children's hospital. *J Appl Bacteriol 23:*533–537, 1960.
40. Rutenberg, A. M., Koota, G. M., and Schweinburg, F. B. The efficacy of kanamycin in the treatment of surgical infections. *Ann NY Acad Sci 76:*348–362, 1958.
41. Schaffner, W., Reisig, G., and Verrall, R. A. Outbreak of *Pseudomonas cepacia* infection due to contaminated anaesthetics. *Lancet 1:*1050–1051, 1973.
42. Schiff, J., Suter, L. S., Gourley, R. D., and Sutliff, W. D. *Flavobacterium* infection as a cause of bacterial endocarditis. Report of a case, bacteriologic studies, and review of the literature. *Ann Intern Med 55:*499–506, 1961.
43. Sorrell, W. B., and White, L. V. Acute bacterial endocarditis caused by a variant of the genus *Herellea*. *Am J Clin Pathol 23:*134–138, 1958.

44. Speller, D. C. E. *Pseudomonas cepacia* endocarditis treated with co-trimoxazole and kanamycin. *Br Heart J 35:*47–48, 1972.
45. Speller, D. C. E., Stephens, M. E., and Viant, A. C. Hospital infection by *Pseudomonas cepacia*. *Lancet 1:*798–799, 1971.
46. Stanier, R. Y., Palleroni, N. J., and Doudoroff, M. The aerobic pseudomonads: A taxonomic study. *J Gen Microbiol 43:*159–271, 1966.
47. Sutter, V. L. Identification of *Pseudomonas* species isolated from hospital environment and human sources. *Appl Microbiol 16:*1532–1538, 1968.
48. Taplin, D., Bassett, D. C. J., and Mertz, P. M. Foot lesions associated with *Pseudomonas cepacia*. *Lancet 1:*568–571, 1971.
49. Tatum, H. W., Ewing, W. H., and Weaver, R. E. Miscellaneous gram-negative bacteria, p. 270–294. In E. H. Lennette, E. H. Spaulding, and J. P. Truant (Eds.), *Manual of Clinical Microbiology,* Ed. 2., Washington, D. C.: American Society for Microbiology, 1974.
50. von Graevenitz, A., and Simon, G. Potentially pathogenic nonfermentative, H_2S-producing gram-negative rod (1b). *Appl Microbiol 19:*176, 1970.
51. Weinstein, A. J., Moellering, R. C., Jr., Hopkins, C. C., and Goldblatt, A. Case report: *Pseudomonas cepacia* pneumonia. *Am J Med Sci 265:*491–494, 1970.
52. Yabuuchi, E., Miyajima, N., Hotta, H., and Ohyama, A. *Pseudomonas cepacia* from blood of a burn patient. *Med J Osaka Univ 21:*1–6.

9

Speciation of Anaerobic Bacteria: A Clinically Based Rationale

ROBERT M. SWENSON
BENNETT LORBER

In recent years there has been increasing interest in anaerobic bacteria as causes of human infections. Numerous studies have indicated that they are involved in a wide variety of human infections; and, unlike other infections where a single organism is isolated, in the majority of these situations multiple species of bacteria are recovered (6). Thus, the clinician is faced with the question of the significance of each isolate in the pathogenesis of the infection in question. In turn the microbiologist is faced with the question of "how far to go" in the isolation and identification of anaerobic bacteria.

The establishment of a Special Anaerobe Laboratory at Temple University Hospital in 1970 provided the opportunity for a detailed prospective study of the role of anaerobes in a wide variety of human infections. The results of our studies in 241 cases will be presented first to provide the groundwork for developing a clinically based rationale for the speciation of anaerobic bacteria.

MATERIALS AND METHODS

All patients included were hospitalized at Temple University Hospital during the period from January 1970 to December 1974, and had been seen by at least one member of the Section of Infectious Diseases. In all instances specimens were obtained with care to prevent contamination with anaerobic bacteria normally present in certain areas of the body. Thus, specimens of coughed sputum, bronchoscopy specimens, stool, vaginal, and cervical specimens were not processed since they either normally contained or were readily contaminated with anaerobes. In most instances specimens were collected from a closed space by aspiration using a syringe and an 18 gauge needle. Exceptions to this were: (a) sputum specimens which were obtained by transtracheal aspiration, (b) endometrial cultures which were obtained transcervically, (c) biopsy specimens, and (d) blood cultures. All materials collected by needle and syringe were immediately injected into an oxygen-free, medium-free tube for transportation to the laboratory. Biospy specimens were placed in sterile tubes without media and taken immediately to the laboratory. Blood cultures were obtained using a sterile vacutainer system containing supplemented peptone broth (Becton-Dickinson). In the Anaerobe Laboratory, specimens were passed into an anaerobic chamber where all subsequent manipulations were carried out (12). Anaerobic bacteria were identified by colonial morphology, reaction to Gram stain, growth on selective media, biochemical reactions, and gas-liquid chromatography of fermentation products. The criteria used were those outlined by the Anaerobe Laboratory at Virginia Polytechnic Institute (7). Facultative bacteria were isolated and identified using standard techniques (8). Patients were included in this study only if: (a) anaerobic bacteria were the only or predominant organisms isolated, and (b) the patient has received no antibiotics or received antibiotics for 24 hr or less.

RESULTS

Infection of Female Genital Tract

During the period of study a total of 241 patients fulfilled the criteria outlined for inclusion in the study. The largest single group was infections of the female genital tract. The sites of these infections are listed in Table 9.1. Ten cases of endomyometritis were a result of incomplete septic abortion while 11 cases followed prolonged rupture of fetal membranes. Cases of peritonitis represent 14 instances of recurrent pelvic inflammatory disease with marked signs of peritoneal irritation. The other 4 cases followed rupture of a pelvic abscess. In these cases specimens were obtained by culdocentesis. All of the specimens from pelvic abscesses were obtained at laparotomy to drain the lesion. All cases of vaginal cuff abscess followed abdominal hysterectomy and all wound infections followed operations.

TABLE 9.1
Infections of female genital tract

Infection	No.	Anaerobic Bacteria		
		Present[1]	Multiple[2]	Only[3]
Endomyometritis	21	18	11	8
Peritonitis, salpingitis[4]	18	12	8	6
Pelvic abscess	15	12	10	8
Vaginal cuff abscess[5]	21	16	13	11
Bartholin's abscess	17	12	8	5
Wound infection[6]	17	11	7	5
	109	81	57	43

[1] Number of cases in which anaerobic bacteria were isolated.
[2] Number of cases in which more than one species of anaerobic bacteria were isolated.
[3] Number of cases in which only anaerobic bacteria were isolated.
[4] Includes cases of peritonitis secondary to salpingitis.
[5] Following hysterectomy.
[6] Following surgical procedures on the female genital tract.

Anaerobic bacteria were isolated from 81 to 109 infections (74%). In 57 of 81 cases (70%) multiple species of anaerobic bacteria were found. Finally, in 43 of 109 instances (39%) only anaerobic bacteria were isolated. In general, the results were similar regardless of the site of infection.

A summary of the bacteria isolated from these infections is presented in Table 9.2. There was a total of 192 anaerobic isolates from these 109 infections. *Bacteroides* species were most commonly found, making up 43% of the total. *Bacteroides melaninogenicus* and *Bacteroides fragilis* were the most frequently isolated of the *Bacteroides* species.

There were 38 isolates of *Peptostreptococcus* species. *Peptostreptococcus anaerobius* and *Peptostreptococcus intermedius* were most frequently found. *Peptococcus* species were also frequently found. *Fusobacterium, Eubacterium,* and *Clostridium* species were infrequent isolates.

Facultative anaerobes or aerobes were isolated in 62 of the 109 cases (56%). There was a total of 88 such isolates. The Gram-negative bacilli, *Escherichia coli,* Proteus and Klebsiella, and streptococci made up the bulk of the isolates.

Intraabdominal Infections

A second major group was that of intraabdominal infections. A total of 64 such patients were studied. The types of infections and incidence of anaerobic bacteria are listed in Table 9.3. There were 26 cases of generalized peritonitis. The underlying diseases in these cases were penetrating wound (10 cases), ruptured appendix (6 cases), blunt trauma to the abdomen (4 cases), ruptured diverticulum (3 cases), and perforation of the gastrointestinal tract due to tumor (2 cases). In one case rupture of the colon appeared to be related to prolonged ileus secondary to narcotic ingestion.

TABLE 9.2
Bacteria isolated from infections of female genital tract

Anaerobes			Facultative Anaerobes or Aerobes		
Bacteroides melaninogenicus	25		*Escherichia coli*	28	
Bacteroides fragilis	22		*Streptococcus* spp.	25	
Bacteroides spp.	35		Group A 7		
		82	Group D 5		
Peptostreptococcus anaerobius	18		*Proteus* spp.	15	
Peptostreptococcus intermedius	11		*Klebsiella* spp.	6	
Peptostreptococcus spp.	9		*Neisseria gonorrheae*	6	
		38			
Peptococcus prevotii	15		*Staphylococcus aureus*	4	
Peptococcus magnus	7		*Hemophilus vaginalis*	3	
Peptococcus spp.	7		*Pseudomonas* spp.	1	
		29			88
Anaerobic streptococcus	11				
Fusobacterium spp.	9				
Eubacterium spp.	9				
Clostridium spp.	7				
Others	7				
		192			

TABLE 9.3
Intraabdominal infections

Infection	No.	Anaerobic Bacteria		
		Present	Multiple	Only
Peritonitis	26	24	22	10
Visceral abscess	15	11	7	6
Intraperitoneal abscess	11	9	6	4
Retroperitoneal abscess	4	2	2	1
Wound	8	6	4	3
	64	52	41	24

Visceral abscesses included 13 hepatic, 1 splenic, and 1 pancreatic abscess. Seven cases of liver abscess followed cholangitis or appendicitis. In the remainder there was no apparent antecedent infection. The splenic abscess occurred in a patient with pneumonia and empyema, and the pancreatic abscess followed an episode of acute pancreatitis.

Intraperitoneal abscesses were related to penetrating wounds (5 cases), ruptured diverticulum (4 cases), and ruptured appendix (2 cases). Two retroperitoneal abscesses followed penetrating wounds and two others were perinephric abscesses related to underlying pyelonephritis. All wound infections followed penetrating wounds or surgery to the gastrointestinal tract.

Anaerobic bacteria were isolated in 52 of 64 cases (81%). Multiple anaerobes were present in 41 of 64 cases (64%) and anaerobes were the only organisms isolated in 24 of 64 cases (38%).

A summary of the bacteria isolated is presented in Table 9.4. There was a total of 241 isolates. Of these, 153 were anaerobic bacteria. The remaining 88 were facultative anaerobes or strict aerobes.

Bacteroides fragilis was the most frequent isolate (37% of all anaerobes), and all *Bacteroides* species accounted for 61% of the anaerobes. *Peptococcus* species (12% of anaerobes) and *Peptostreptococcus* species (10%) were also frequently encountered. *Clostridium* species were isolated in only 5 cases.

Gram-negative bacilli made up 57% of the facultative anaerobes and strict aerobes isolated. *Escherichia coli* was most commonly isolated. *Streptococcus* species were isolated on 34 occasions. Eight of these were group D.

Pleuropulmonary Infections

Pleuropulmonary infections were also found to be frequently due to anaerobic bacteria. The majority of cases occurred in patients with poor dentition (43 of 45) who had a history of antecedent loss of consciousness (40 of 45). Characteristic putrid sputum or empyema fluid was present in 33 cases.

The types of infections in these cases are presented in Table 9.5. Lung abscess was most common. Cases of uncomplicated pneumonia and empyema occurred with equal frequency. Anaerobic bacteria were isolated in

TABLE 9.4
Bacteria isolated from intraabdominal infections

Anaerobes	No.		Facultative Anaerobes or Aerobes	No.
Bacteroides fragilis	57		*Escherichia coli*	36
Bacteroides melaninogenicus	24		*Streptococcus* spp.	34
Bacteroides spp.	13		Group D 8	
		94	*Proteus* spp.	7
Peptococcus prevotii	12		*Klebsiella* spp.	5
Peptococcus spp.	6		*Pseudomonas* spp.	2
		18		
Peptostreptococcus anaerobius	10		Others	4
Peptostreptococcus intermedius	3			88
Peptostreptococcus spp.	3			
		16		
Fusobacterium spp.	8			
Eubacterium spp.	6			
Clostridium perfringens	5			
Others	6			
		153		

TABLE 9.5
Pleuropulmonary infections

Infection	No.	Anaerobic Bacteria		
		Present	Multiple	Only
Pneumonia	11	10	9	8
Lung abscess	23	21	20	11
Empyema	11	9	6	5
	45	40	35	24

40 of 45 cases (89%), multiple anaerobic species in 35 (78%) and only anaerobes in 24 cases (54%).

The bacteriology of these infections is shown in Table 9.6. There were a total of 147 isolates. Of these, 106 were strict anaerobes. Anaerobic Gram-positive cocci were the most frequent isolates (63) followed by *Fusobacterium* species. *Bacteroides* species were isolated on 17 occasions but only 4 of these were *Bacteroides fragilis*. Facultative anaerobes were predominantly streptococci and *Neisseria* species.

Miscellaneous Infections

A variety of other infections was encountered infrequently (Table 9.7). There were 5 cases of hematogenous osteomyelitis. Three cases involved the vertebrae and one each the humerus and femur. In none of these patients was a primary focus of infection detected, nor was bacteremia detected. Four were due to *B. fragilis* and the other to *Fusobacterium gonidiaformans*. The remaining case of osteomyelitis of the humerus resulted from spread from a contiguous soft tissue abscess and involved multiple organisms.

Four of the 5 cases of cellulitis were due to multiple anaerobes, usually anaerobic cocci, *B. fragilis,* or clostridia. Many additional cases of cellulitis were encountered during the period of the study. However, organisms were not isolated so they were not included in this report.

Two of the three cases of septic arthritis were of hematogenous origin and were due to *B. fragilis*. Clinically these cases were monoarticular involving the elbow joint in one case and the knee in the other, and were indistinguishable from other forms of pyogenic arthritis. The remaining case resulted from spread from a contiguous soft tissue focus to involve the elbow joint.

A variety of soft tissue abscesses were encountered. Most of these occurred in narcotic addicts and were due to multiple anaerobes and facultative anaerobes or aerobes. Two cases of brain abscess involved multiple anaerobes. The single case of endocarditis followed gynecological surgery, involved the aortic valve, and was due to *B. fragilis*. The patient survived following prolonged antibiotic therapy and replacement of the valve. The case of meningitis occurred in a patient with an intraabdominal abscess secondary to a ruptured diverticulum with the abscess eventually extending into the subarachnoid space. *B. fragilis* and *E. coli* were isolated from the cerebrospinal fluid. The patient rapidly expired following the development of meningitis.

Bacteremia

The incidence of bacteremia in these 241 cases and the sites of origin are

TABLE 9.6
Bacteria isolated from pleuropulmonary infections

Anaerobic Bacteria			Facultative Bacteria	
Peptococcus prevotii	18		*Streptococcus* spp.	21
Peptococcus spp.	14		*Neisseria* spp.	8
		32	*Staphylococcus aureus*	4
Peptostreptococcus anaerobius	16		*Streptococcus pneumoniae*	4
Peptostreptococcus intermedius	8		*Klebsiella pneumoniae*	2
Peptostreptococcus spp.	7		*Escherichia coli*	2
		31		41
Fusobacterium fusiforme	12			
Fusobacterium nucleatum	6			
Fusobacterium spp.	4			
		22		
Bacteroides spp.	13			
Bacteroides fragilis	4			
Clostridium spp.	4			
		106		

TABLE 9.7
Miscellaneous infections

Infection	No.	Anaerobic Bacteria: Multiple	Anaerobic Bacteria: Only
Osteomyelitis	6	1	5
Cellulitis	5	4	1
Arthritis	3	1	2
Abscesses	5	4	1
Brain abscess	2	2	2
Endocarditis	1	0	1
Meningitis	1	0	0

shown in Table 9.8. Blood cultures were positive in 28 cases. In 7, more than one organism was isolated. Bacteremia was most commonly associated with intraabdominal and female genital tract infections. *B. fragilis* was the most frequent cause of bacteremia and accounted for 18 of 35 isolates from blood cultures.

DISCUSSION

Anaerobic Bacteria as Pathogens

The results presented here indicate that anaerobic bacteria are frequently involved in a wide variety of human infections. With the exception of rather uncommon infections such as hematogenous osteomyelitis, septic arthritis and endocarditis, these are usually mixed infections involving multiple anaerobic species and often other facultative anaerobes as well. We consider the anaerobes "clinically significant" if they are the only or predominant organisms isolated from appropriately collected and processed specimens. We realize this does not definitely answer the question of the role of each of the bacteria in these polymicrobic infections but feel that it is the most prudent approach to take. Using these criteria it seems clear that the anaerobic bacteria are frequent causes of intraabdominal, female genital tract, and pulmonary infections and, to a lesser extent, a wide variety of other infections such as cellulitis, postoperative wound infections, and brain abscesses.

Collection of Specimens

With the exception of clostridial infections most anaerobic infections are of endogenous origin occurring in proximity to the mucosal surfaces where anaerobic bacteria are normally present in large numbers, i.e., the colon, vagina, upper respiratory tract, and, to a lesser extent, the skin. Therefore, it is extremely important that specimens be collected in such a way as to avoid contamination from the nearby normal flora. Several collection methods are listed in Table 9.9. As can be seen most of these employ a needle and syringe to aspirate material from a closed space.

Properly collected specimens should be injected (without introducing

TABLE 9.8
Anaerobic bacteremia

Organism	Site of Origin					
	Abdomen	Genital tract	Respiratory tract	Other	Unknown	Total
Bacteroides fragilis	11	4		1	2	18
Bacteroides melaninogenicus	1	4				5
Bacteroides spp.	2	1	1			4
Fusobacterium fusiforme				1		1
Fusobacterium radiculosum			1			1
Peptostreptococcus intermedius			1			1
Clostridium spp.			1	4		5
	14	9	4	6	2	35

any air) into an oxygen-free transport tube. Such tubes are commercially available or can be prepared in the laboratory. Specimens transported in this manner have yielded large numbers of anaerobes after as long as a 4-hr delay in processing. A recent report has suggested that anaerobes may survive for significant periods of time in oxygen-containing transport vials (4). However, this study employed stock strains of anaerobes and they may not be applicable to recent clinical isolates. Purulent material itself is an adequate transport medium provided the volume is 2 ml or greater (3). Tissue specimens should be placed in a sterile tube and transported to the laboratory immediately for processing as soon as possible. An alternative method for transporting tissue is the anaerobic jar technique of Attebery and Finegold (2).

We have also found that many physicians are quite uninformed as to what specimens are appropriate for processing for anaerobic bacteria. For this reason the laboratory should establish its own criteria for what specimens should be handled this way. A list of acceptable clinical specimens is shown in Table 9.10.

Speciation of Anaerobic Isolates

There are several potential ways in which the speciation of bacteria may aid the clinician. We would now like to examine these in terms of their applicability to anaerobic bacteriology.

Diagnosis. As our data demonstrate, anaerobic bacteria are involved in a wide variety of human infections. The most important task for the microbiology laboratory is to identify those infections which involve anaerobes; i.e., have the capability to isolate anaerobic bacteria as quickly and

TABLE 9.9
Specimen collection methods

Source	Method
Closed space infection (abscess, empyema)	Aspiration by needle and syringe
Lower respiratory tract secretions	Percutaneous transtracheal aspiration
Uterine cavity, sinus tract	Aspiration using plastic intravenous catheter

TABLE 9.10
Acceptable clinical specimens for laboratory diagnosis of anaerobic infections

CENTRAL NERVOUS SYSTEM:	Cerebrospinal fluid, abscess material, tissue biopsy
PULMONARY:	Transtracheal aspirate, tissue biopsy, direct lung aspirate, pleural fluid
INTRAABDOMINAL:	Aspirate from loculated abscess, ascitic fluid, tissue biopsy
GENITOURINARY:	Aspirate from loculated abscess, aspirate collected by culdocentesis, tissue biopsy
OTHER:	Blood, tissue or fluid from any normally sterile site

accurately as possible. The *most important and fundamental* information needed by the clinician is whether anaerobic bacteria are PRESENT OR NOT. As indicated below THERE IS LITTLE NEED FOR DETAILED SPECIATION OF ANAEROBIC BACTERIA. A corollary of this is the importance of the Gram stain. Infections involving anaerobic bacteria are frequently mixed infections. Conversely, infections involving only facultative anaerobes or strict aerobes are usually monomicrobic infections; e.g., pneumococcal pneumonia, tuberculosis, gonorrheae, etc. Thus, a Gram stain of material showing multiple morphologically distinct microorganisms is strong presumptive evidence that this is an anaerobic infection.

Epidemiology. The precise identification of a Gram-negative facultative anaerobe as *Salmonella typhi* or *Brucella suis* has important epidemiological implications related both to identifying the patient's source of infection and instituting appropriate infection control measures. However, when we apply such epidemiological considerations to infections involving anaerobic bacteria the conclusions are somewhat different. As we have already pointed out, these are predominantly endogenous infections; i.e., patients are infected with organisms originally part of their own normal flora. Such infections are not contracted from any environmental source nor are they transmissible to other human contacts. Therefore, with the exception of *Clostridium botulinum,* there are no epidemiological reasons for detailed speciation of anaerobic bacteria.

Prognosis. Speciation of some facultatively anaerobic bacteria may be helpful in determining the prognosis of a given infection. For example, identification of Gram-positive cocci from blood cultures of a patient with infective endocarditis as *Staphylococcus aureus* implies a significantly worse prognosis than if that organism is identified as a Streptococcus viridans type. In the case of anaerobic bacteria there is little prognostic value accruing from detailed speciation. It has been suggested that *Clostridium perfringens* may cause more severe infections than other clostridial species (1). However, detailed studies of such infections do not bear this out (9). It would appear that *Bacteroides fragilis* may cause more severe infections (5), is more frequently associated with bacteremia (Table 9.8) and may produce metastatic foci of infection (Table 9.7). Thus, there is some prognostic significance to precisely identifying this organism, but there is little prognostic significance in precisely identifying other anaerobic bacteria.

Therapy. Probably the major reason for speciating bacteria is to provide a data base for rational selection of antimicrobial agents. For example, the information that the infecting organism is *Streptococcus pneumoniae* together with the physician's knowledge of the antimicrobial susceptibilities of this organism provide the basis for selecting penicillin for the treatment of this infection. However, in many instances involving bacteria results of in vitro susceptibility tests are often available before speciation is complete. In this situation, speciation is really not helpful in determining therapy.

Here again the situation is somewhat different with anaerobic bacteria.

TABLE 9.11
Susceptibilities of anaerobic bacteria*

Antibiotic	Percent of Strains Susceptible	
	All NSAB†	*Bacteroides fragilis*
Penicillin	78	39
Cephalothin	72	31
Tetracycline	64	67
Chloramphenicol	96	100
Lincomycin	72	66
Clindamycin	93	96

* Total of 432 strains.
† Nonsporeforming anaerobic bacteria.

Although methods for in vitro disc susceptibility tests have been described for anaerobes (13) we do not feel they are as yet sufficiently standardized for general use. In this situation the clinician must base his choice of an antimicrobial on the anaerobe(s) isolated and his knowledge of the in vitro susceptibilities of anaerobic bacteria. Results of such susceptibility tests are shown in Table 9.11. Essentially all clinical isolates are susceptible to penicillin except *Bacteroides fragilis*. Thus, in its simplest form, ALL THE CLINICIAN NEEDS TO KNOW IS: ARE ANAEROBIC BACTERIA PRESENT; AND, ARE ANY OF THEM *B. fragilis*? For this reason we feel that it is not necessary for clinical laboratories to employ gas-liquid chromatography and other complex techniques for extensive speciation of anaerobic bacteria. Rather they should focus on developing their capabilities to isolate anaerobic bacteria and employ simplified procedures for the rapid identification of the commonly encountered anaerobes, particularly *B. fragilis* (10, 11, 14).

This is not meant to minimize the need for larger research oriented laboratories to continue to study the role of anaerobic bacteria in human infections, develop standardized methods of susceptibility testing, and monitor antibiotic susceptibilities in order to rapidly detect any changes in kinds of anaerobic isolates or their antibiotic susceptibility patterns. If such changes do occur the clinical microbiology laboratory will have to adjust their procedures accordingly.

SUMMARY

The bacteriology and clinical aspects of anaerobic infections were studied prospectively in 241 patients. Special techniques for isolating anaerobic bacteria were employed along with standard microbiological methods. Infections of the female genital tract, other than those due to *Neisseria gonorrheae,* and intraabdominal infections almost always involved anaerobes. Bacteroides species, particularly *Bacteroides fragilis,* were most commonly found. Pleuropulmonary infections following aspiration of oropharyngeal contents also frequently involved anaerobes, usually fusobacteria and gram-positive cocci. A variety of other infections involved anaerobes.

These results are used as a basis for developing a rationale for more simplified procedures for the speciation of anaerobic isolates.

ACKNOWLEDGMENTS

The authors are indebted to Valerie Vargo, Myroslawa Korzeniowski, Sharon Brown, and Matthew Widomski for their excellent technical assistance, and Sharon Austin for her excellent secretarial assistance.

Supported in part by grants from the American Heart Association and the Upjohn Company.

LITERATURE CITED

1. Altemeier, W. A., and Fullen, W. D. Prevention and treatment of gas gangrene. *JAMA 217:*806, 1971.
2. Attebery, H. R., Finegold, S. M. A miniature anaerobic jar for tissue transport or for cultivation of anaerobes. *Am J Clin Pathol 53:*383, 1970.
3. Bartlett, J. G., Sullivan-Sigler, N., Louie, T. J., and Gorbach, S. Anaerobes survive in clinical specimens despite delayed processing. *J Clin Microbiol 3:*133, 1976.
4. Chow, A. W., Cunningham, P. J., and Guze, L. B. Survival of anaerobic and aerobic bacteria in a nonsupportive gassed transport system. *J Clin Microbiol 3:*128, 1976.
5. Finegold, S. M., and Rosenblatt, J. E. Practical aspects of anaerobic sepsis. *Medicine 52:*311, 1973.
6. Gorbach, S. L., and Bartlett, J. D. Anaerobic infections. *New Engl J Med 290:*1177–1184, 1237–1245, 1289–1294, 1974.
7. Holdeman, L. V., and Moore, W. E. C. *Anaerobic Laboratory Manual.* Blacksburg, Va.: Anaerobe Laboratory, Virginia Polytechnic Institute & State University, 1972.
8. Lennette, E. H., Spaulding, E. H., and Truant, J. P. *Manual of Clinical Microbiology.* Washington, D. C.: American Society for Microbiology, 1974.
9. MacLennan, J. D. Histotoxic clostridial infections in man. *Bacteriol Rev 26:*177–275, 1962.
10. Porschen, R. K., and Stalons, D. R. An evaluation of simplified dichotomous schemata for the identification of anaerobic bacteria from clinical material. *J Clin Microbiol 3:*161–171, 1976.
11. Rosenblatt, J. E. Isolation and identification of anaerobic bacteria. *Hum Pathol 7:*177, 1976.
12. Spaulding, E. H., Vargo, V., Michaelson, T. C., et al. A comparison of two procedures for isolating anaerobic bacteria from clinical specimens, pp. 37–46, in A. Balow, R. M. DeHaan, L. B. Guze, and V. R. Dowell, Jr. (Eds.), *Anaerobic Bacteria.* Springfield, Ill.: Charles C Thomas, 1974.
13. Sutter, V. L., Vargo, V. L., and Finegold, S. M. *Wadsworth Anaerobic Bacteriology Manual.* Los Angeles: Department of Continuing Education in Health Sciences University Extension, and School of Medicine, U.C.L.A., 1975.
14. Vargo, V. L., Korzeniowski, M., and Spaulding, E. H. Tryptic soy bile-kanamycin test for the identification of *Bacteroides fragilis. Appl Microbiol 27:*480–483, 1974.

10

Mycobacteria: Significance of Speciation and Sensitivity Tests

EMANUEL WOLINSKY

WHY SPECIATE MYCOBACTERIA?

Tuberculosis is transmitted from person to person and the isolation of *Mycobacterium tuberculosis* or *Mycobacterium bovis* from human material is always an indication for action on the part of the physician. Other mycobacteria, however, may be found as saprophytes in the environment and the recovery of acid-fast bacilli other than mammalian tubercle bacilli may represent nothing more than contamination of the specimen or temporary colonization of the respiratory tract, for which treatment is not indicated. Furthermore, the treatment and the prognosis of nontuberculous mycobacterial disease are quite different from those of ordinary tuberculosis. Because of these considerations, I believe that all initial mycobacterial isolates should be speciated as quickly as possible.

The reasons may be divided into those that concern the patient, the doctor, and the public health.

The diagnosis of tuberculosis brings fear and apprehension to the patient. There is often a sense of shame and embarrassment for his family. Prolonged hospitalization no longer is necessary for most patients but they still must face the prospect of long term treatment with potentially toxic drugs, some of which are expensive. Other members of the family will need to be questioned, skin tested, and x-rayed. There is a possibility of loss of nursing home accommodations for the older patient. The patient and his family can be spared most of these inconveniences if the organism can be identified as a mycobacterium other than *M. tuberculosis*.

The physician should know the species identification of the mycobacterium isolated by the laboratory in order to prescribe the proper treatment. First, he must differentiate between mycobacterial disease, temporary mycobacterial colonization of the respiratory tract, and casual isolation as a result of environmental contamination of the mouth, throat, or stomach. The rate of isolation of environmental strains in a given laboratory will depend on many factors, including harshness of the sputum digestion procedure, use of unsterile water for sputum induction or gastric lavage

(6), the amount of dust in the sputum processing room, and the mycobacterial flora of the dust in that geographic region. Single isolations of small numbers of environmental strains may be disregarded. On the other hand, the recovery of *M. kansasii* should suggest the possibility of association with disease, especially if the organism is isolated repeatedly. A summary of the mycobacterial species encountered in human material and their pathogenic potential is presented in Table 10.1. In addition, strains of environmental species *M. gordonae, M. gastri, M. terrae* complex, and *M. flavescens* may be recovered. These are usually not associated with disease. A summary of the association of mycobacteria with human disease and with the environment will be found in Table 10.2.

Speciation is especially important to evaluate the significance of the recovery of yellow acid-fast colonies from sputum cultures, when the pigment is not light dependent. *M. gordonae* and *M. flavescens* usually can be disregarded, while *M. scrofulaceum, M. xenopi,* and *M. szulgai* should be taken more seriously. Repeated cultures and clinical evaluation will be necessary for proper interpretation.

Recovery of any mycobacterium from a closed space is usually significant. The most common cause of mycobacterial lymphadenitis in children is not tuberculosis, but infection with one of the nontuberculous potential pathogens, especially *M. scrofulaceum* (16). The prognosis is excellent even without drug therapy. Lymphatic tuberculosis, on the other hand, is a serious disease which should be treated vigorously and for a long time (3). The prognosis for disseminated infection with nontuberculous mycobacteria is poor because it usually occurs in a setting of immunosuppres-

TABLE 10.1
Pathogenic or potentially pathogenic mycobacteria recovered from human secretions or tissues

Mycobacterial Species	Human Disease	Known Reservoirs
	A. Classical Pathogens	
M. tuberculosis	Tuberculosis	Man
M. bovis	Tuberculosis	Man, cattle, other mammals
M. africanum	Tuberculosis	Man
M. leprae	Leprosy	Man
M. avium (chick virulent)	Tuberculosis	Birds (also cattle and swine)
M. ulcerans	Deep skin ulcers	?
	B. Potential Pathogens	
M. avium-intracellulare	Pulmonary, lymphatic, disseminated	Soil, dust, swine, cattle
M. kansasii	Same, also skeletal	Water, ? animals
M. scrofulaceum	Lymphatic (also pulmonary and disseminated)	Soil, water
M. marinum	Superficial skin	Fish, water
M. xenopi	Pulmonary	Water
M. szulgai	Pulmonary soft tissue	?
M. simiae	Pulmonary	?
M. fortuitum-chelonei	Soft tissue (also pulmonary)	Soil, dust, water

sion. The slim chance for successful treatment depends upon the correct identification of the strain and the institution of appropriate drug therapy, together with correction of cellular immune functions.

The recommended treatment schedules for nontuberculous mycobacterial infections are summarized in Table 10.3. The relatively drug sensitive species, such as *M. kansasii*, may be successfully treated by a regimen of 2 or 3 drugs (8). Even with 4 and 5 drug regimens, however, disease caused by the resistant group responds poorly (21). *M. marinum* infections are usually self-limited and do not require treatment, but the combination of rifampin and ethambutol may be used successfully (20).

It is obvious that the necessity for speciation will depend upon how often nontuberculous mycobacteria are cultured from the sputum and the frequency with which nontuberculous mycobacterial disease is encountered in a given area. Both rates are variable throughout the world. Infection is seen more frequently in warm and moist regions at low altitude. Skin testing of U. S. Naval recruits with sensitins derived from *M. scrofulaceum* and *M. avium* complex has revealed a reactor rate of 30–40% (15). A higher rate was found in those men who lived in the southeastern United States. The results of dual testing of children with similar antigens and with tuberculin purified protein derivative (PPD) has been documented in several publications. In one study there were at least 4 times as many

TABLE 10.2
Association of mycobacteria with human disease and with the environment

Mycobacterial species often associated with disease: *M. tuberculosis, bovis, kansasii, avium-intracellulare, xenopi, szulgai, simiae, marinum, ulcerans, scrofulaceum, fortuitum-chelonei*

Species rarely (if ever) associated with disease: *M. gordonae, flavescens, gastri, terrae, phlei, smegmatis*

Species not commonly found in the environment; *M. kansasii, szulgai, ulcerans, simiae*

TABLE 10.3
Suggested treatment regimens for nontuberculous mycobacterial disease, based on reported in vitro susceptibility

Mycobacterium	Drugs*
A. Resistant Species	
M. avium-intracellulare, M. scrofulaceum, and M. fortuitum	INH or RMP + EMB + CS or ETA + SM or KM
M. simiae	EMB + ETA (+CS) + SM
B. Relatively Sensitive Species	
M. kansasii	RMP + INH + EMB or SM
M. marinum	RMP + EMB
M. szulgai	RMP + EMB + ETA or SM
M. xenopi	INH + RMP + SM (also susceptible to ETA, CS, VM)

* Cycloserine (CS), ethambutol (EMB), ethionamide (ETA), isoniazid (INH), kanamycin (KM), rifampin (RMP), streptomycin (SM), viomycin (VM).

reactions to PPD-B (made from a strain of *M. avium* complex) as to tuberculin PPD (12). Delayed hypersensitivity to these antigens presumably indicated infection with a mycobacterium other than a mammalian tubercle bacillus. It is difficult to determine the prevalence of nontuberculous mycobacterial disease. The most common manifestation of disease probably is mycobacterial lymphadenitis in children (16). A more significant clinical manifestation is chronic pulmonary disease resembling tuberculosis; its incidence varies from less than 1% to 16% of the new cases presenting as pulmonary "tuberculosis" (2). At Cleveland Metropolitan General Hospital this rate has averaged 2–4% for the last 15 years, but it was as high as 8.2% in 1973 (19).

In patients who present with tuberculosis-like disease, is it reasonable to accept misdiagnosis of up to 16% because of incomplete mycobacteriologic identification?

Public health considerations also are important. Tuberculosis usually is transmitted by person to person contact through contaminated droplet nuclei which are produced by the aerosolization of sputum. The other mycobacterial diseases are acquired by other mechanisms involving contact with the environment or possibly animal tissues and animal products (4). Therefore, patients with other mycobacterial disease need not be isolated even though their sputum continues to harbor large numbers of organisms, except from other individuals with chronic pulmonary disease and with states of immunosuppression. Proper speciation will avoid the time consuming and expensive search for index cases and contacts, and the subsequent institution of prolonged prophylactic treatment of the contacts, which is now considered proper health practice for newly discovered cases of tuberculosis.

There is one situation in which speciation, although desirable, probably is not necessary. That is the single isolation of 1 or 2 obviously scotochromogenic or rapidly growing acid-fast colonies from the sputum. Several additional sputum specimens should be requested if they are not already in process.

The laboratory tests recommended for speciation are outlined in Table 10.4. Most laboratories that are capable of culturing specimens for mycobacteria should be able to do the first 5 tests. Further identification may be reserved for appropriate reference centers (13).

DO WE NEED TO DO SENSITIVITY TESTS?

Sensitivity tests should be performed on all initial positive mycobacterial cultures to determine in vitro susceptibility to at least the four major antituberculosis drugs, isoniazid, rifampin, ethambutol, and streptomycin. The reasons involve both the public health and the individual patient and his contacts. For the public health it is important to monitor the prevalence of drug resistant strains of *M. tuberculosis* in the community. The rates of primary resistance to at least one drug determined by the Veterans Administration were approximately 4–10%, and the prevalence did not appear to be increasing (9). We have had a similar experience at the

TABLE 10.4
Tests recommended for speciation

1. Cell and colonial morphology
2. Pigment and light dependence
3. Growth at 25°, 32°, 37°, 45°
4. Speed of growth
5. Niacin
6. Nitrate reduction
7. Tween hydrolysis in 10 days
8. Semiquantitative catalase
9. 68°C catalase
10. Urease (disc test)
11. Arylsulfatase
12. Growth inhibition by 5% NaCl
13. Susceptibility to TCH (thiophene carboxylic acid hydrazide)

Cleveland Metropolitan General Hospital. In certain other areas of the world, however, the prevalence of such strains is much higher (11). It is especially important to be aware of the possibility of primary infection with drug-resistant strains in children. Experience at this hospital indicated a frequency of 9%. Steiner has reported the alarming rate of 24% from his community in Brooklyn, New York; 4 children had been infected with triply resistant strains (17). This experience prompted him to recommend that at least 3 drugs be used to initiate treatment of primary tuberculosis in children, and that a 4-drug regimen be given to those patients who have life-threatening infections (18).

For the initial episode of tuberculosis we should aim for successful drug treatment in 100% of patients. Fox (5) has emphasized the difficulty of performance and interpretation of drug sensitivity tests and he points to the studies of the Hong Kong Tuberculosis Treatment Services-British Medical Research Council for evidence that one can get along very well without these tests (10). These studies involved 3 groups of patients, the first treated with isoniazid, streptomycin, and PAS for 3–6 months, followed by isoniazid and PAS for a total of 1–2 years of treatment, regardless of the drug sensitivity result at the start of treatment; the second treated in the same manner except that other drugs were added appropriately after drug resistance had been recognized; and the third treated with appropriate drugs initially based upon the results of a rapid drug sensitivity test. Drug-resistant strains were found in 25–33% of the patients admitted to the study; 10–14% were resistant to 2 or 3 drugs, and the remainder were resistant to a single drug. The overall favorable response at 3 years was 87–90% for all 3 groups. Further analysis of group 1, however, revealed that in these 174 patients an unsatisfactory result was found in only 3% of those who entered the study with drug-sensitive strains as compared to 31% of those whose admission strains were drug resistant. A further breakdown showed that the proportion of patients with unsatisfactory results was 23%, 30%, and 71%, for those whose organisms were resistant to 1 drug, to 2 drugs, or to 3 drugs, respectively (10). Although the relatively small number of patients with unsatisfactory results was not enough to influence the overall figures, it cannot be denied that failure rates of that magnitude should not be acceptable in this country. It may be argued that all new cases of tuberculosis should be treated with isoniazid, rifampin, and ethambutol which would provide effective therapy for those patients whose organisms were resistant to isoniazid alone. Recent information from the Veterans Administration study documented the facts that of 24 primary isoniazid resistant strains, 14 were resistant to at least 1 other drug, and there were 11 primarily rifampin resistant strains of which 7 were resistant to at least 1 other drug (9).* Thus, it appears to me that to disregard the results of pretreatment drug susceptibility tests would lead to something less than optimum therapy for too many individuals.

Furthermore, methods for performing drug susceptibility tests have been simplified and standardized (7). Radiometric procedures have been described which can provide results within several hours (1). A committee of the American Thoracic Society has recommended that reference laboratories with demonstrated proficiency be designated to which cultures may be sent for susceptibility testing from smaller laboratories (13).

Results of pretreatment sensitivity tests are particularly significant for the patient who is being treated for the second or third time, and for children. In the case of the retreatment group, it is essential that this information be known so that the appropriate multi-drug regimen may be prescribed (14). The drug sensitivity pattern of the tubercle bacilli isolated from the index case should be determined as soon as possible so that the child with, for example, tuberculous meningitis may be treated properly. It would not be sensible to prescribe and maintain isoniazid prophylaxis for a year when the adult to whom a child had been exposed was excreting tubercle bacilli that were resistant to isoniazid.

One other benefit from pretreatment drug susceptibility tests is to flag the resistant cultures as possible mycobacteria other than tubercle bacilli which require further study.

LITERATURE CITED

1. Benitez, P., Medoff, G., and Kobayashi, G. S. Rapid radiometric method of testing susceptibility of mycobacteria and slow-growing fungi to antimicrobial agents. *Antimicrob Agents Chemother 6:*29–33, 1974.
2. Brown, M., Buechner, H. A., Bailey, W. C., and Ziskind, M. M. Atypical mycobacterial pulmonary disease at the New Orleans V.A. Hospital and metropolitan New Orleans (abstract). *Am Rev Respir Dis 103:*885–886, 1971.
3. Bwibo, N. O. Tuberculosis of the cervical lymph nodes. Clinical studies of children in East Africa. *Clin Pediatr 9:*733–735, 1970.
4. Chapman, J. S. The ecology of the atypical mycobacteria. *Arch Environ Health 22:*41–46, 1971.
5. Fox, W. Changing concepts in the chemotherapy of pulmonary tuberculosis. The John Barnwell Lecture. *Am Rev Respir Dis 97:*767–790, 1968.
6. Goslee, S., and Wolinsky, E. Water as a source of potentially pathogenic mycobacteria. *Am Rev Respir Dis* **113:** 287–292, 1976.
7. Griffith, M. E., Matajack, M. L., Bissett, M. L., and Wood, R. M. Cooperative field test of drug-impregnated discs for susceptibility testing of mycobacteria. *Am Rev Respir Dis 103:*423–426, 1971.
8. Harris, G. D., Johnson, W. G., Jr., and Nicholson, D. P. Response to chemotherapy of

*Primary resistance to rifampin also has been documented in a recent report from Massachusetts (K. D. Stottmeier: Emergence of rifampin resistant *Mycobacterium tuberculosis* in Massachusetts. *J Infec Dis 133:*88–90, 1976).

pulmonary infection due to *Mycobacterium kansasii*. *Am Rev Respir Dis 112:*31–36, 1975.

9. Hobby, G. L., Johnson, P. M., and Boytar-Papirnyik, V. Primary drug resistance. A continuing study of drug resistance in tuberculosis in a veteran population within the U. S. X. Sept. 1970 to Sept. 1973. *Am Rev Respir Dis 110:*95–98, 1974.
10. Hong Kong Tuberculosis Treatment Services/British Medical Research Council Investigation. A study in Hong Kong to evaluate the role of pretreatment susceptibility tests in the selection of regimens of chemotherapy for pulmonary tuberculosis—second report. *Tubercle 55:*169–192, 1974.
11. Horne, N. W. Drug-resistant tuberculosis. A review of the world situation. *Tubercle 50*(Suppl): 1–12, 1969.
12. Kendig, E. L. Unclassified mycobacteria. Incidence of infection and cause of a false positive tuberculin reaction. *New Engl J Med 268:*1001–1002, 1963.
13. Kubica, G. P., Gross, W. M., Hankins, J. E., Sommers, H. M., Vestal, A. L., and Wayne L. G. Laboratory services for mycobacterial diseases. *Am Rev Respir Dis 112:*773–787, 1975.
14. Lester, W. Treatment of drug resistant tuberculosis. *Disease-A-Month,* pp. 3–43, April 1971.
15. Palmer, C. E., and Edwards, L. B. Tuberculin test in retrospect and prospect. *Arch Environ Health 15:*792–808, 1967.
16. Salyer, K. E., Votteler, T. P., and Dorman, G. W. Cervical adenitis in children due to atypical mycobacteria. *Plast Reconstructr Surg 47:*47–53, 1971.
17. Steiner, P., Rao, M., Goldberg, R., and Steiner, M. Primary drug resistance in children. Drug susceptibility of strains of *M. tuberculosis* isolated from children during the years 1969 to 1972 at the Kings County Hospital Medical Center of Brooklyn. *Am Rev Respir. Dis 110:*98–100, 1974.
18. Steiner, P., Rao, M., Victoria, M., and Steiner, M. Primary isoniazid-resistant tuberculosis in children. Clinical features, strain resistance, treatment, and outcome in 26 children treated at Kings County Medical Center of Brooklyn between the years 1961 and 1972. *Am Rev Respir Dis 110:*306–311, 1974.
19. Wolinsky, E. Nontuberculosis mycobacterial infections of man. *Med Clin North Am 58:*639–648, 1974.
20. Wolinsky, E., Gomez, F., Zimpfer, F. Sporotrichoid *Mycobacterium marinum* infection treated with rifampin-ethambutol. *Am Rev Respir Dis 105:*964–967, 1972.
21. Yeager, H., Jr., and Raleigh, J. W. Pulmonary disease due to *Mycobacterium intracellulare*. *Am Rev Respir Dis 108:*547–552, 1973.

11

Findings of Significance in the Systemic Mycoses

CHARLOTTE C. CAMPBELL

For the compromised host who develops a bacterial infection, there is a large "armamentarium" of antibiotics in reserve for therapy. If the infection is due to a fungus or a yeast the armamentarium is reduced to two antimycotics and in many instances only one, a nephrotoxic drug. Therefore, I propose to review and to illustrate some of the tissue forms of mycotic agents as they appear in clinical specimens by direct (light) microscopic examinations of unstained and stained preparations; and to recommend that these examinations for yeasts, filamentous and other fungal forms become standard procedures in the microbiology laboratory in which there is no experienced medical mycologist. Finally, I shall try to support the rationale for reemphasizing these simple procedures for those whose primary concern is with antibiotic sensitivity tests for bacterial species.

INCIDENCE OF MYCOSES

Because mycotic infections and diseases—even those that are fully confirmed by recognized cultural, histologic and serologic procedures (2, 7, 9, 18)—are not required to be registered in the Morbidity and Mortality Reports issued by the Center for Disease Control (the official census of infectious diseases in the United States), the incidence of mycotic disease in either compromised or uncompromised hosts is unknown. The cost both in terms of dollars for hospitalization and in productive time lost is equally unknown.

The first effort to redress this omission, using modern statistical methods and tools, was undertaken in 1970 by Hammerman, Powell and Tosh (10) who approximated 8,000 cases at a hospitalization cost of slightly over 9 million dollars for six of the most prominent mycoses hospitalized in acute-care hospitals. By the design of the study, however, these figures merely confirmed Ajello's (1) comprehensive review of the world literature of case reports and series of case reports which led him to predict that medicine was recognizing only the top of the massive "medical mycological iceberg." Reports by the author have pointed out that nearly all of medicine's most significant and sophisticated advances in other fields merely increased the vulnerability of the human host to infection with yeasts and fungi, a factor which is further compounded by the restless jet transporting of populations from one geographic area to another and the massive soil excavations common to our contemporary lives (4, 5).

Unfortunately, the size of the mycotic disease problem in the United States remains obscure and can be estimated only by such subjective assessments after review of the literature from 1945 to 1975. These are the 30 years from the time it first became known that certain fungi produced pulmonary infections that varied in severity from mild to fatal in millions of otherwise healthy people (6, 16, 21, 22) to the present, when any species of yeast or fungus is a potential and possibly lethal hazard to an increasing population of compromised hosts. According to a more recent partial literature review by Huppert, Harper, Sun, and Delanerolle (11), *Candida albicans* alone accounts for a significant rate of mortality as well as morbidity in patients with burn wounds, shock trauma, postoperative thoracic surgery, heart surgery, renal transplants and cancer, especially in the leukemia-lymphoma group (11). Much of the mortality is due to infection that is not recognized until autopsy or recognized too late for either of the two antimycotics to be effective before death (11). *Candida albicans,* the one fungus which is as indigenous to the normal gastrointestinal flora as *Escherichia coli* either still is not as readily recognized as the bacterium, is regarded as inconsequential if it is, or simply is not looked for.

ANTIMYCOTICS VS. ANTIBACTERIAL ANTIBIOTICS

A second legacy to the compromised host from the failure to *count* the mycoses and thereby overlook their numbers and "cost" is that he now has nearly a generation of surgeons, physicians and clinical pathologists—as well as clinical microbiologists—who have received only the most minimal, if any, training about the mycoses and their causative agents. To these physicians it is not only acceptable but "routine" medicine to literally exhaust the large armamentarium of bacterial antibiotics to which even bacterial species grow ever more resistant before thinking about infections due to any other microbes that can be seen through the light microscope.

This is unfortunate even for the "uncompromised" host who develops primary or secondary "complicating" infection or disease from yeasts or fungi. Even to him the use (and misuse) of antibacterial antibiotics in "exclusion diagnosis" for earlier infections has already become a *compromising factor* by leaving him virtually a "sitting duck" to invasion by the ubiquitous fungi in his environment, endogenous and exogenous. Since the advent of the first penicillin it has been recognized that mycotic agents are resistant to antibacterial antibiotics. They were early incorporated into culture media merely to isolate the slower growing yeasts and fungi from the competition of the infinitely more numerous and more rapidly growing bacterial species, in vitro as well as in vivo (24).

Consider, then, the compromised host who may be receiving one or more of the other compromising drugs (which are the same for the mycotic agents as for bacteria) in addition to the antibacterial antibiotics and who is "invaded" by a fungus or a yeast. For him there is not only the lack of recognition of the ubiquitousness of fungi among contemporary clinicians blinded by the profusion of antibacterial antibiotics and the failure to update their own thinking about the fungi whose "distribution" has changed over the past 20 years owing to factors quite unrelated to medical progress (5). There is also the grim fact that due to this neglect of the mycoses—beginning with education—there are only four *antimycotics* in 1975 that have been approved by the Food and Drug Administration (3).

Amphotericin B. This is the fungal "broad-spectrum" in that it is effective in infection and disease caused by either yeasts or fungi. It is also notoriously nephrotoxic. It must be administered intravenously—very slowly—over long periods of time (months) with repeated courses the rule in chronic, disseminated disease (3). Yet it is also the one and only drug of last resort for mycoses caused by filamentous and dimorphic fungi as well as the yeasts which develop resistance or invade tissues that are not reached by:

5-Fluorocytocine C. This antimycotic is effective against the yeasts *Cryptococcus neoformans* and *Candida albicans* (which also develops hyphae under appropriate environmental conditions which are not understood, in vivo and in vitro), and other yeasts. It is ineffective against filamentous and other dimorphic fungi. It is administered orally and has relatively few side effects (3). Some strains of *C. neoformans* especially develop resistance (19) during the long course the drug must be administered in cryptococcal meningitis (the most commonly recognized form of disease due to this agent). This then, leaves amphotericin B as the drug of last resort in some cases of this devastating disease, which has a mortality rate approaching 100% (10).

Mycostatin. While effective in "infections" of *C. albicans* limited to the mucocutaneous tissues of the human gastrointestinal tract following oral administration, this antimycotic is not absorbed through these tissues (3). It can not be injected parenterally and thus is unavailable as a therapeutic agent in systemic candidiasis (including colonization of deeper tissues) which also constitutes a significant mortality as well as morbidity rate in the compromised host (11). Applied topically, the drug is also effective in vaginal candidiasis whose incidence is legion; but because it does not produce overt clinical symptoms in males, as does the gonococcus which is also sexually transmitted, its high incidence has not been equally noted despite the untold distress it has caused to countless women; and to their babies infected during passage through the birth canal, and who stepwise produce "outbreaks" of infant diarrhea in hospital nurseries (23).

Griseofulvin. This antimycotic is effective against only those fungi which produce infection in hair, skin, and nails and are incapable of surviving in living tissues—the dermatophytes (2, 7, 9, 18). It too is administered per os, is possibly the safest and most effective drug of the four (depending upon the type of keratinaceous tissue infected). Paradoxically, however, it is of no value in treating mycotic infection or disease due to any other yeast or fungus, including yeast infections of the skin.

This, then, is the "armamentarium" of antimycotics available for treatment of the mycoses other than lymphocutaneous sporotrichosis for which potassium iodide is the treatment of choice (3). It is orally administered. Because it is misleading, I have deliberately omitted the numerous other drugs that have been used in the past and are sometimes still noted in the texts (2, 7, 9, 18). Drugs such as 2-hydroxystilbamidine for blastomycosis (3), and the sulfonamides for paracoccidioidomycosis have not been accepted either because of their toxicity or that they amounted to little more than placebos. Drugs such as clotrimazole (3) or miconazole, an even newer drug, and an oral or less nephrotoxic preparation of amphotericin B are still in a developmental stage with acceptable trials in human mycoses still in the future. Vaccines for prevention and immunotherapy are in even earlier stages of development (5) as is combined therapy with an antimycotic and certain of the bacterial antibiotics (15).

Which of the two available drugs to use to initiate therapy for a systemic mycosis is rarely, if ever, based on the more inhibitory in an antimycotic sensitivity test. This obtains even when the etiologic agent is a yeast against which both drugs are inhibitory. Clinicians understandably prefer to use 5-fluorocytocine in these cases. It is less nephrotoxic than amphotericin B and it is administered orally. Unfortunately, it has not proved to be effective in any of the mycoses not caused by yeasts. Amphotericin B must be used despite its nephrotoxicity and difficulties of slow intravenous injection (3).

ADVANTAGES OF DIRECT EXAMINATION

Therapy obviously will not be initiated with either drug if a clinician is unaware that a fungus is the cause of his patient's deteriorating course. Since many clinicians no longer have a very high index of suspicion for the mycotic diseases, including many who practice in known "endemic" areas, specific requests for fungal examinations have correspondingly decreased.

Thus, if a mycotic agent is to be looked for, the microbiologist must frequently do so on his own initiative. Culture for such agents on selective fungal media does not always satisfy this requirement, particularly in the compromised patient. Most yeasts require 48–72 hr to isolate (11); the most rapidly growing filamentous fungi 72–120 hr, i.e., *Aspergillus* spp. and the several genera of Phycomycetes which also produce a significant mortality in compromised patients (17); and some of the dimorphic pathogens as long as 3 weeks. This is time that a severely compromised patient might not have and is all the more tragic when his clinical specimens are often teeming with the fungus that could have been readily observed in direct microscopic examinations had these been done. Fungi disseminate rapidly in these patients since they are usually immunologically incompetent and their competitive bacterial flora has been reduced by antibiotics. Moreover, some of the fungi can be just as specifically identified, both as to genus and species, in direct microscopic examinations of clinical specimens as in subsequent cultures. For those that cannot be, the species is generally irrelevant as to which of the two drugs can be used. If the agent is a true yeast such as a species of *Candida* or *Cryptococcus* (or any yeast *that is not a yeast form of a dimorphic fungus*), the safer and orally administered drug can be used. Otherwise, the only drug available is amphotericin B.

From the standpoint of therapy for the patient, therefore, there is nothing to be gained by speciating the fungi beyond the three broad general categories by which they parasitize human tissues: true yeasts (two of the most serious for compromised patients, *Candida albicans* and *Cryptococcus neoformans,* are rapidly identifiable in clinical specimens); filaments (hyphae); and the parasitic forms of the dimorphic fungi—yeasts that are distinctly different from true yeasts (*Histoplasma capsulatum, Blastomyces dermatitidis, Paracoccidioides brasiliensis, Sporothrix schenckii*), the "spherules" of *Coccidioides immitis* and the sclerotic bodies of the *Fonseceae* spp.

Few fungal forms in clinical specimens are difficult to differentiate from bacteria simply by virtue of their larger size—and their shape. It is urged that microbiologists learn to recognize these relatively few forms, reinstate the more extensive direct microscopic examinations of clinical specimens, and report the presence of any of the fungal forms to the attending clinician at once, without waiting to confirm the observation by culture. These have always been a first routine operating procedure for the experienced medical mycologist (2, 7, 9, 18). I refer to the unstained wet preparation, India ink preparation, Giemsa as well as Gram stain (since some fungi are not stained by Gram), and germ tube tests. Because of the slow growing nature of the fungi in vitro these simple tests are the best route to stemming the rising incidence of morbidity and mortality of the systemic mycoses in compromised patients (1, 4, 5, 10, 11, 17, 23). They should become routine operating procedures in all clinical microbiology laboratories as they once were, preantibiotic medicine.

This emphasis on the direct examination of the clinical specimen is not intended to exclude cultural study of these same specimens, whether fungal forms are or are not observed. However, the more thorough direct examination frequently makes it possible to establish an early diagnosis and prognosis as well as to institute therapy for an unsuspected mycosis long before the organism can be isolated. As clinical microbiologists we are obliged to keep our priorities rightly aligned. The patient must always come first. Where the mycoses are concerned we must also remember that there are no antimycotics which are more inhibitory against one *Candida* sp. than another, none even more inhibitory to the genus *Candida* than to the genus *Cryptococcus,* and none more effective in treating histoplasmosis than coccidioidomycosis or aspergillosis, for example. Thus efforts to speciate the fungi in clinical diagnostic laboratories should not preempt the time that would be better spent in more thorough and extensive direct examinations for the fungal forms in patient's clinical specimens—especially from compromised hosts.

ENDOGENOUS VS. EXOGENOUS SOURCES

As noted above, *C. albicans* is a normal inhabitant of the human gastrointestinal flora but in reduced numbers relative to the normal component of bacterial species. All other yeasts and fungi exist saprophytically or as commensals in man's environment.

In the compromised host, however, this clear-cut definition breaks down with respect to exogenous sources. As in tuberculosis, for example, an incompletely healed histoplasmoma, coccidioidoma, or aspergilloma from an earlier possibly unrecognized infection with one of the normally exogenous fungi may serve as an endogenous source for recrudescence of the respective organisms. Since *C. neoformans* is widespread in nature because of the large pigeon population and is rarely recognized as the pulmonary infection preceding meningitis (1, 2, 7, 9, 18), it is speculated that recrudescence accounts for many of the cases of cryptococcal meningitis observed in compromised hosts. Recrudescence may also account for extension to the meninges by fungi (17) in which meningitis is relatively rare in "uncompromised" hosts.

In brief, it is unsafe to rely on the former classifications of exogenous and endogenous sources (2, 7, 9, 18). During the bacterial antibiotic—jet transportation—massive soil excavation past two decades (4, 5) man has also become the carrier of yeasts and fungi that were formerly exogenous to the human host.

It is equally unsafe to rely on recent exposure during travel to a recognized endemic area such as the relatively limited but well defined one for coccidioidomycosis (5). Two recent examples in central Illinois, a highly endemic area for histoplasmosis, illustrate this point. Meningitis due to *Coccidioides immitis* developed in a renal transplant patient who had not visited the endemic area since WWII. Close examination virtually excluded "fomitic" transmission (2, 7, 9, 18). Diagnosis was established by discovering "spherules" in the sediment of the patient's centrifuged spinal fluid. The second case was in a pregnant woman (which is also a compromised state) who had recently visited Phoenix, Arizona, prior to the development of central nervous system manifestations for which she visited a neurologist. Spherules of *C. immitis* were found in the placenta following premature delivery 2 days before the woman's death from coccidioidal meningitis. She had not responded to therapy with antibacterial antibiotics. However, the patient's family revealed that she had made many previous trips to Arizona to visit her family. Earlier x-rays disclosed many calcified lesions and one or two cavities which could have been due to *H. capsulatum*, *C. immitis*, or indeed *Mycobacterium tuberculosis*. Her skin test reactions were unknown (5). The date of her primary exposure to *C. immitis* thus remains a moot point.

For similar reasons, a former classification of mycotic diseases as subcutaneous or systemic also does not hold in the compromised host. Sporotrichosis, for example, formerly manifesting itself almost exclusively by subcutaneous lesions from an initial physical prick by a thorny plant (rose bush, spaghnum moss, etc.) harboring the fungus (*Sporothrix schenckii*) is found increasingly as a primary pulmonary infection in alcoholics (also a genre of compromised hosts receiving renewed attention in our contemporary lives along with other drug addictions) (18). While this may be due to improved diagnostic procedures, the widespread use of bacterial antibiotics, reducing this agent's usual competition in the human host can not be discounted as a reason for the "changes" in the presentation of cases of this widespread fungus. Certainly "alcoholism" is not a new disease. At the expense of redundancy, the bacterial antibiotics *are* a *compromising factor* where infections with yeasts and fungi are concerned (2, 7, 9, 12, 18).

SEROLOGIC DIAGNOSIS OF THE MYCOSES

As my bibliography firmly attests I have been a professional life long proponent of serologic diagnostic tests for the respiratory mycoses in uncompromised hosts. Yet one of my earliest reports (6) included a series of cases of widely disseminated histoplasmosis whose internal organs were teeming with the yeasts of *H. capsulatum* at autopsy. Their serologic reactions were low or negative for as long as 6 months prior to death. These patients were victims of Hodgkin's disease and/or leukemia-lymphomatous diseases which have long been recognized as underlying diseases to rampant infectious diseases of whatever origin, including the mycoses.

In the compromised host, therefore, the standard serologic diagnostic tests may also be unreliable indicators of the extent of disease by a mycotic agent even in those increasingly rare instances in which one is suspected antemortem. Additionally, many of the antigens for performing such tests are not locally available and sera must be sent to state laboratories (some of which do not carry these antigens or have the skilled personnel needed to carry out these tests) or to the Center for Disease Control, Atlanta, Ga. The CDC is understaffed for the large volume of work expected. Reports are thus sometimes slow as well as unhelpful diagnostically or prognostically and not suitable in the "stat" situations which generally exist in the advanced state of disease by the time fungal "invasion" is considered.

Further, research on the mycotic antigens is urgently needed as indeed is research in the entire area of the immunology of the mycoses. Possibly this is another legacy of the low prioritizing of the mycotic diseases. Less is known about the antigens of the common fungal "opportunists"—the Jekyll-Hydes (8)—than the long recognized mycotic pathogens, although the importance of the fungi as "opportunists" was addressed in a special conference as early as 1961 (12). Yet only recently was an amino acid liquid synthetic medium developed in which the yeast or mycelial form of *C. albicans* could be exclusively grown for antigen production (13). Completely synthetic media meeting the specific requirements of certain other pathogens have also only been recently developed (14).

There doubtless are many reasons why medicine finds itself unprepared for the onslaught of mycotic disease in the increasing numbers of compromised hosts. That there is an increase in these there can be no doubt. However, they must certainly include the late and low index of suspicion of mycotic disease by surgeons and physicians educated during the bacterial antibiotic decades. They must also include the inadequate, both in training and in numbers, cadre of clinical diagnostic medical mycologists, mycological investigators and teachers (both M.D.s and Ph.D.s) educated during the last decade.

FUNGAL FORMS IN CLINICAL SPECIMENS

For this reason, I believe the most helpful contribution I can make to medical microbiologists, clinical pathologists and medical microbiological technologists is to review the yeasts, filamentous and other fungal forms that can be readily identified, at least as fungal forms, in the direct examination of clinical specimens, and to urge that these simple procedures be reintroduced as routine operating ones on all specimens—but especially those from the severely compromised host receiving immunosuppressants, corticosteroids, cytotoxic drugs, etc., in addition to bacterial antibiotics.

These forms are described and illustrated below. They should serve at least as an "alert" that a yeast or a fungus is present which, regardless of genus or species, must be regarded as "pathogenic" for the compromised host. The types of specimen(s) to be examined by the wet and fixed Gram- and Giemsa-stained preparations are also described.

Yeasts. These may vary widely in shape or size but one or more of the yeasts will doubtless have unmistakable buds. This is the mode of reproduction.

A. When *only* yeast forms are observed, a second wet preparation with India ink should be done. This will reveal capsules if they are present, which the Gram or Giemsa stain will not; thereby causing you to miss the alert to one of the most lethal yeasts—*C. neoformans*—to the compromised host. Do not expect all yeasts to be equally encapsulated. One with a "suspicious" capsule is enough to alert! *Rarely* a strain of *C. neoformans* is observed which exhibits pseudohyphae or clamp connections (20).

B. If the yeast forms only are unusually small (2–5 μ) and except for the buds difficult to differentiate from large cocci, *Torulopsosis glabrata* is a potential candidate; so are *Histoplasma capsulatum* and *Sporothrix schenckii* which are more difficult to see but can be if they are sufficiently numerous. Some in this group may be intracellular.

C. If the yeast forms are unusually large (5–50 μ) with a "hyaline" ring surrounding the interior cell wall, *Blastomyces dermatitidis* is a potential candidate.

D. If the yeast forms are similar to those in (C), but have multiple buds sometimes ringing the entire external surface of some of the yeasts, *Paracoccidioides brasiliensis* is a potential candidate.

E. If the yeast forms are accompanied by septate hyphal ("peudohyphae") forms of essentially the same diameter as the yeasts, this is an alert to a *Candida* spp.; though not necessarily *C. albicans,* and not necessarily the genus *Candida*. The hyphal forms may be short (2–6 septa) or long (crossing several microscopic fields, especially in histologic sections or "pressed" unstained wet preparations of resected tissues or skin). The hyphae may be branching or unbranching.

(Note that (E) is the only category in this group of significant potential and known pathogens which has more than one morphological form in clinical specimens: the tissue or in vivo form).

Hyphal Forms. These may be short, long (extending over several fields), septate or nonseptate, branching or nonbranching, pigmented or colorless; depending upon the agent. However, such structures in the absence of budding yeasts are indicative of such potential fungal pathogens (especially for the compromised host with severe burns) as the rapidly growing *Aspergillus* spp. and Phycomycetes (*Mucors, Rhizopus, Absidia,* etc). The latter are nonseptate (coenocytic). Nearly all of these organisms are ubiquitous in our environment.

Granules. These structures are in vivo colonies, aggregates of microbial species formed in living tissues. They are produced by certain bacterial species (e.g., *Staphylococcus, Actinomyces,* and *Nocardia*) as well as true fungi (e.g., maduromycosis agents). It is especially important to note the microscopic composition of such granules. In those composed of bacterial species the patients sometimes respond to therapy with antibacterial antibiotics, whereas patients with granules composed of true fungi do not. This is also useful information from the microbiological stand point. If the granules are composed of bacillary-like elements one can proceed to the acid-fast stain which differentiates acid-fast *Nocardia* spp. from *Actinomyces* spp. Some *Actinomyces* spp. are facultative *anaerobes* and require an anaerobic environment as well as media without bacterial antibiotics for isolation. Both genera are Gram-positive.

Fungal Forms (neither hyphal nor yeasts).

A. *Sclerotic Bodies:* Dark, thick-walled cells, 6–12 μm in diameter, occurring singly or in clusters. The cells divide by splitting, *not* by budding. Occasionally "septa" can be seen within cells. The potential candidates in this instance are the agents of Chromomycosis (*Fonsecaea pedrosoi, F. compacta,* and *Phialophora verrucosa.*)

B. *"Spherules":* The tissue form of *Coccidioides immitis* which may vary in size from 5 μ (the smallest) to 80–100 μ, the mature, thick walled sporangium filled with endospores which ruptures to release the endospores each of which repeats the maturation process. However, there are *no* budding forms and *no* hyphal forms except rarely the latter in tissue at the edge of a pulmonary cavity.

The notation of these various tissue forms of yeasts and fungi has acquired renewed significance in the present era of the compromised host; and all clinical diagnostic microbiologists should make every effort to become familiar with and to recognize these tissue forms in clinical specimens. Excellent illustrations are also included in the texts (2, 7, 9, 18).

The wet preparation consists of placing a small drop of the clinical specimen on a scrupulously clean slide and covering the drop with an equally clean cloverslip. For India ink preparations the small drop is mixed with a small drop of India ink before applying the coverslip. Both types of preparations are examined while wet. The 10× lens is used to scan the entire preparation. The 40× lens for closer examination of "suggestive" forms—both under reduced light. These preparations remain wet for 24 hr or more if coverslips are first ringed with petrolatum or a petrolatum-parafilm mixture.

Specimens of urine, cerebrospinal fluid, synovial fluid (septic arthritis due to yeasts or fungi in corticosteroid-treated patients) bronchial secretions, pleural fluid, small volumes (2–3 ml), saline in which swabs of the vagina or surgical or burn wounds, etc., are collected, are centrifuged; and drops of the sediment used for the direct examinations (and subsequent culture). Specimens of pus, sputum, bioptic brushings, bone marrow (heparinized), etc., may be examined directly or diluted with saline if necessary. Resected tissues may be finely minced with sterile scissors in a small volume of saline (0.5–2.0 ml, depending upon the amount of tissue) for making "pressed" preparations. The saline may also be centrifuged and the sediments examined. Giemsa stains in addition to the Gram stain should be routine for all specimens, but are especially important for resected tissues, bone marrow and blood specimens. The "pressed" preparation is as the name implies. A tiny snip of the tissue is pressed as flat as possible between the slide and the coverslip.

The Giemsa stain should also become routine in the hematology laboratory for all differential blood smears from compromised hosts. It is also recommended that the India ink preparation become a standard routine in virus diagnostic laboratories, especially on patients with symptoms of meningitis or encephalitis.

SUMMARY AND DISCUSSION

It is impossible to cover the laboratory findings of significance in a field as vast and as varied as the systemic mycoses in one brief report. I have used the opportunity, instead, to try to heighten general awareness of (1) the mycotic diseases themselves especially in an increasing number of compromised hosts; (2) the dearth of antimycotics available for treatment of patients whose diagnosis is established as mycotic antemortem; and (3) reviewed some very simple procedures which hopefully will increase the number of cases identified as mycotic long before death so that the two available antimycotics can at least be used more effectively.

I have also stressed the increasing neglect of this field over the past one and one half decades. I believe this stems in no small measure from failure to record the incidence of mycotic disease—much less its cost in dollars and productive time lost. This *has* been done with other, principally person to person transmitted, infectious diseases in the CDC Morbidity and Mortality Reports.

Finally, I trust I have made it possible to read the several very comprehensive, excellent and well illustrated texts on the mycoses and medical mycology with greater understanding and confidence; to differentiate what is important for the medical microbiologist and clinical pathologist not trained in medical mycology to know from what should be left to the trained and experienced medical mycologist (both M.D. and Ph.D.). The ranks of both have grown perilously thin during the decades of antibacterial agents medicine.

Figure 11.1. Yeasts

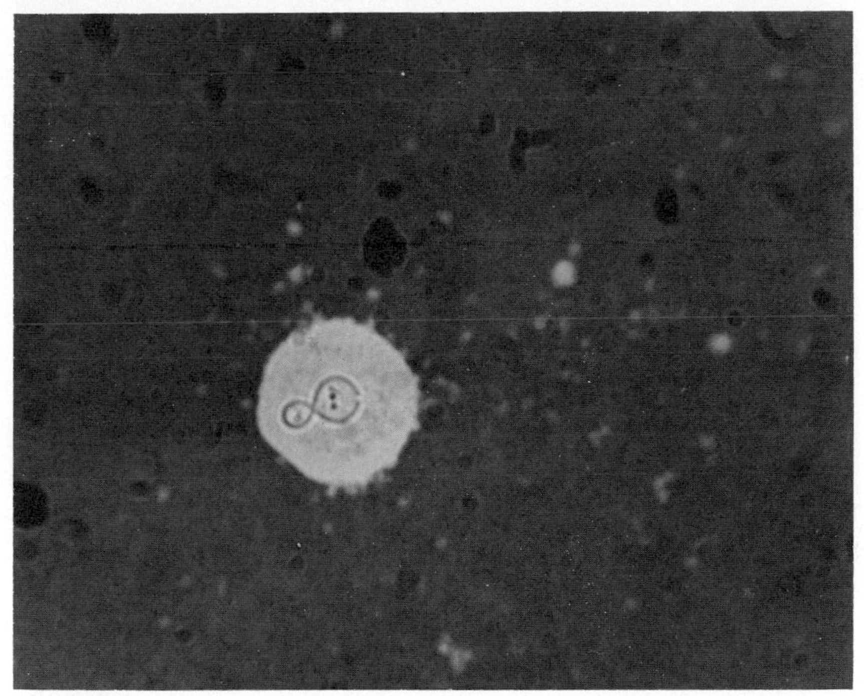

A. *Cryptococcus neoformans,* India ink preparation (×1000). Sediment of centrifuged cerebrospinal fluid. Organisms usually few. Capsules, however, are usually larger than in initial primary isolates.

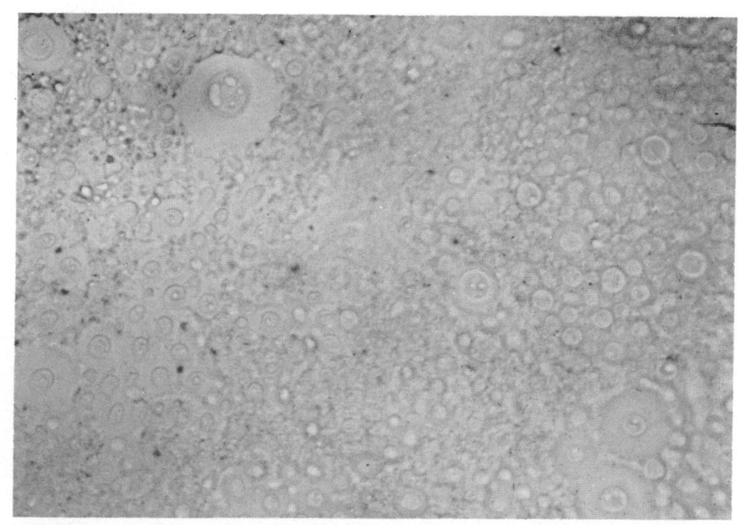

B. *C. neoformans* in lung, unstained, pressed preparation (×400). If background tissue is dense capsules may be seen without India ink.

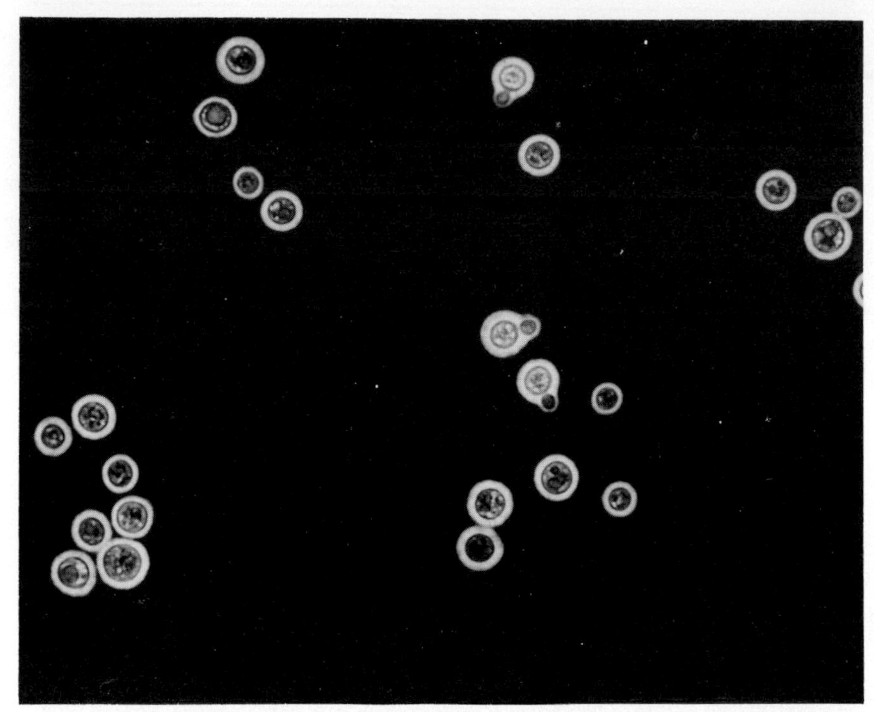

C. *C. neoformans* in culture; India ink preparation (×400) of organism isolated from 1a after two weeks' incubation of 37 C and several subsequent transfers on Sabourand agar. Mycosel is inhibitory to this organism (2, 7, 9, 18). Note capsules are not present on all yeasts. This is typical even for strains with predominantly large capsules (2, 7, 9, 18).

Figure 11.2. Yeasts. Generally intracellular (mononuclear cells or macrophages)

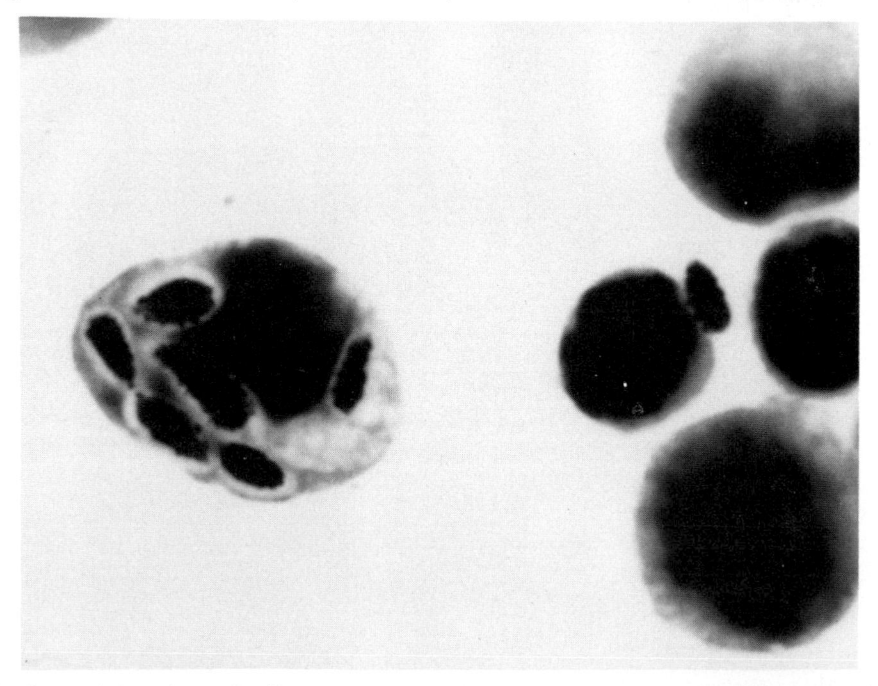

A. *Sporothrix schenckii.* Cigar-shaped yeasts in a mononuclear cell (Giemsa stain ×1250).

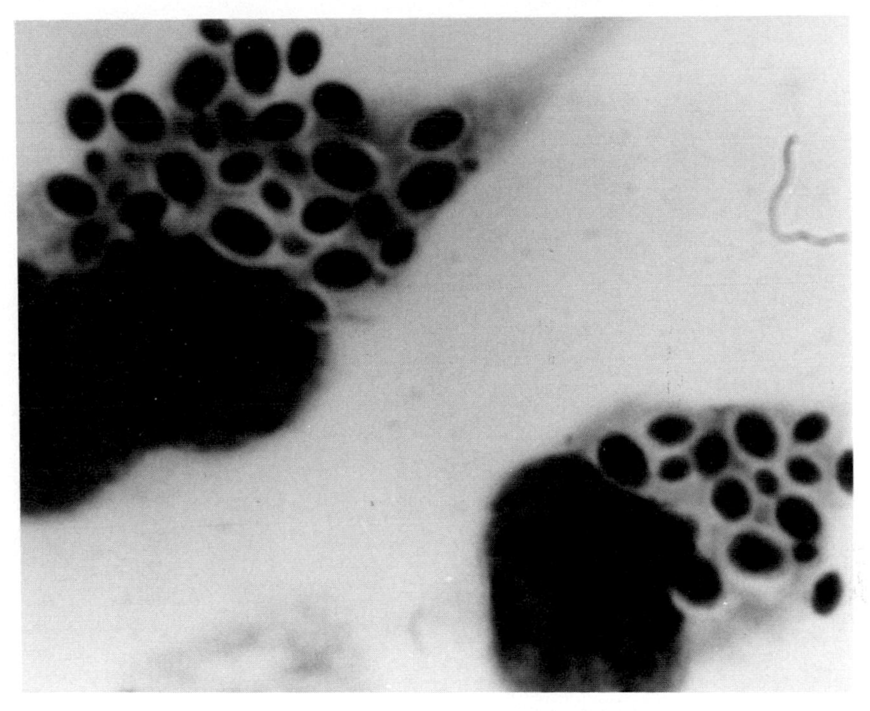

B. *Torulopsis glabrata*. Mononuclear cell (Giemsa stain ×1250).

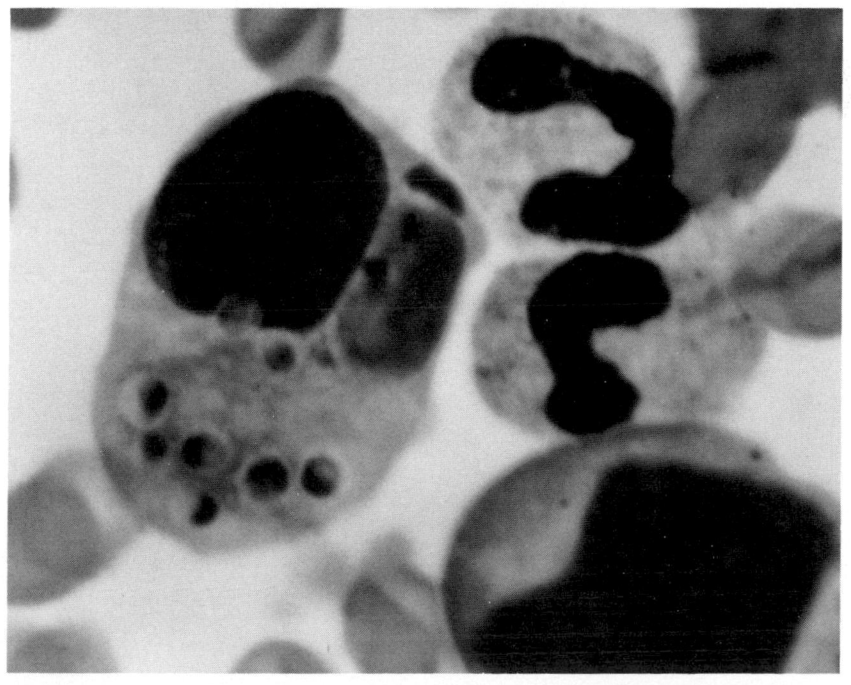

C. *Histoplasma capsulatum*. Bone marrow smear. (Giemsa stain ×1250).

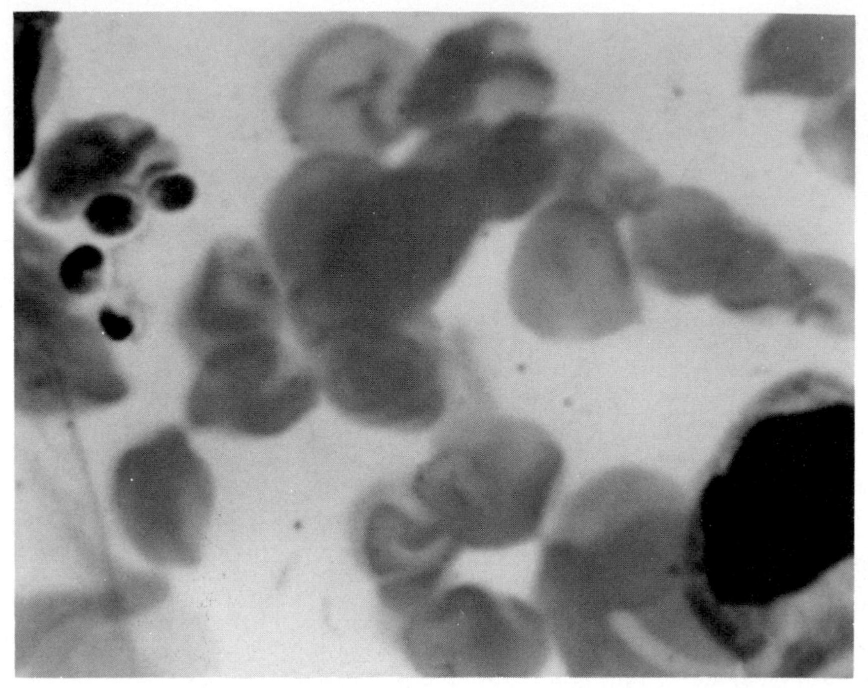

D. *Histoplasma capsulatum*. Bone marrow smear. Note extracellular yeasts (Giemsa stain ×1250).

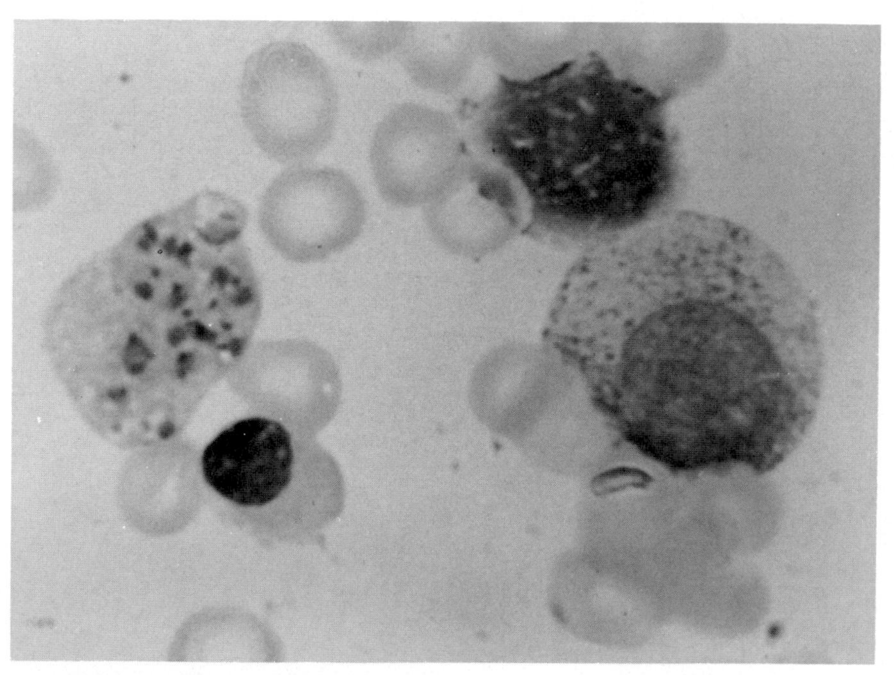

E. *Leishmania donovani*. Not a fungus but can be confused with any of the above to the inexperienced eye (Giemsa stain ×1000).

Figure 11.3. Yeasts, generally larger and not intracellular

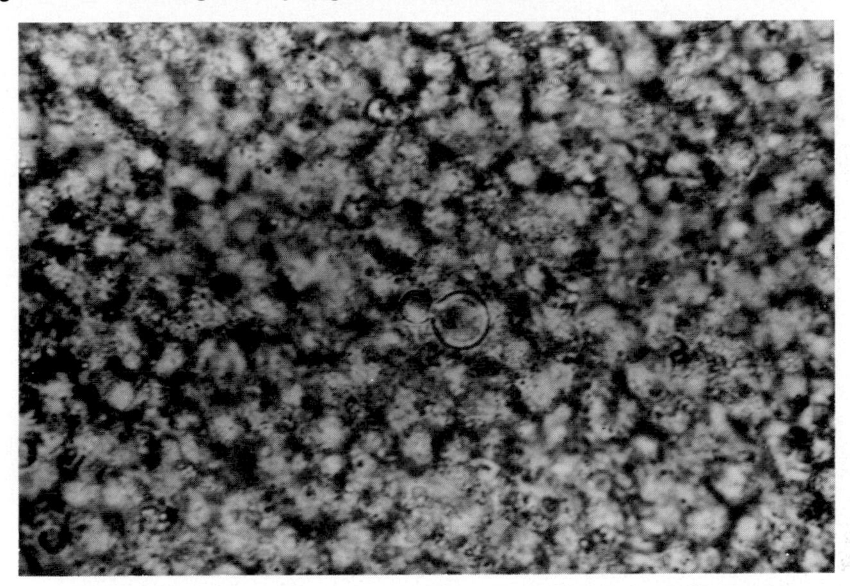

A. *Blastomyces dermatitidis*. Wet preparation of pus (unstained ×400).

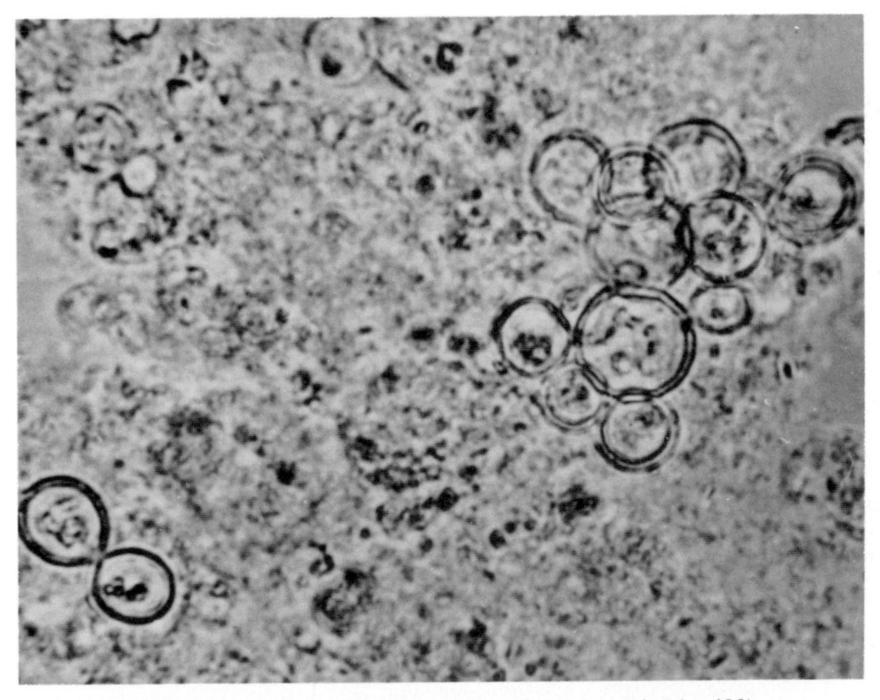

B. *Paracoccidioides brasiliensis*. Wet preparation of pus (unstained ×400).

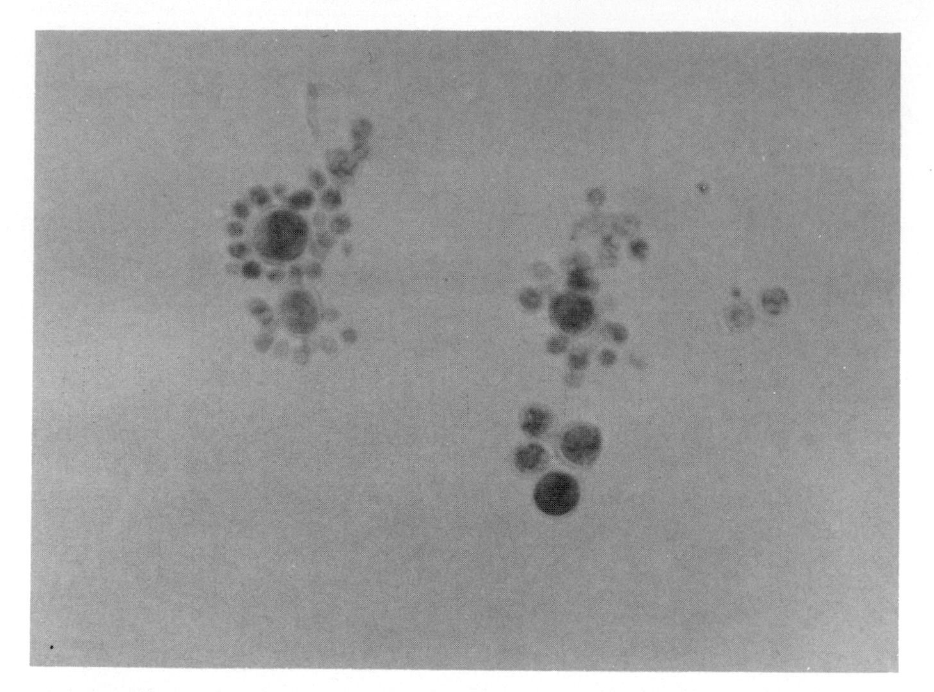

C. *P. brasiliensis* in culture at 37°C_0, revealing the "pilot wheel", multiple budding character of the organism (unstained ×400). This form is also observed in histopathology sections stained with the Grocott-Gomori-methenamine silver (GMS), or one of the modifications of the periodic acid schiff (PAS) stains (e.g., Gridley).

Figure 11.4. Yeasts and "pseudohyphae"

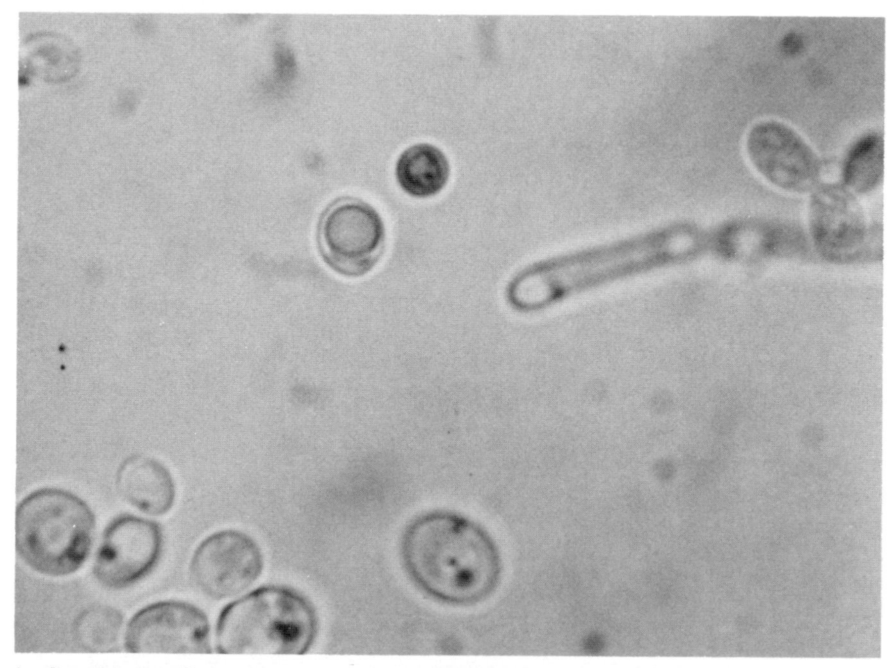

A. *Candida* sp. Wet preparation of urine sediment (unstained ×1000).

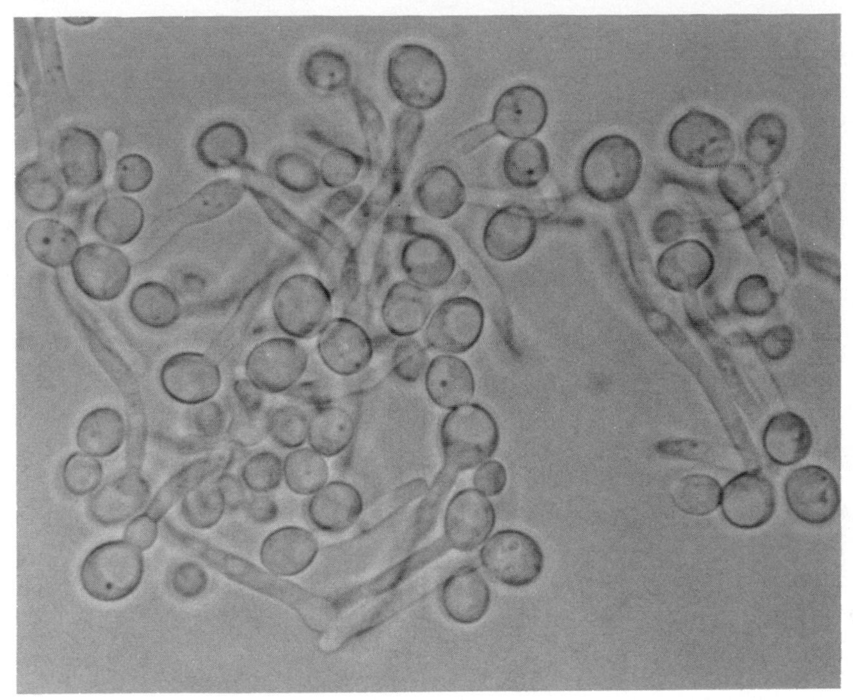

B. *Candida* sp. "confirmed" as *C. albicans* by germ tube test. Incubated small quantity clinical specimen in serum (2, 7, 9, 18) for less than 3 hr (unstained ×1000).

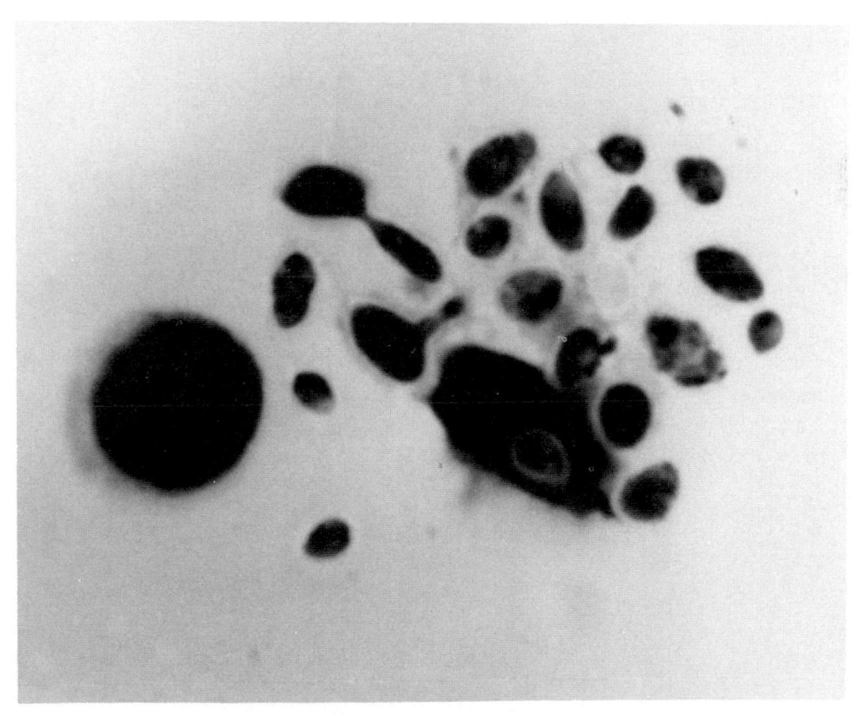

C. *Candida albicans.* Germ tubes formed in vivo in mononuclear cell (Giemsa stain ×1550). Rare except in extensively disseminated, usually fatal candidiasis.

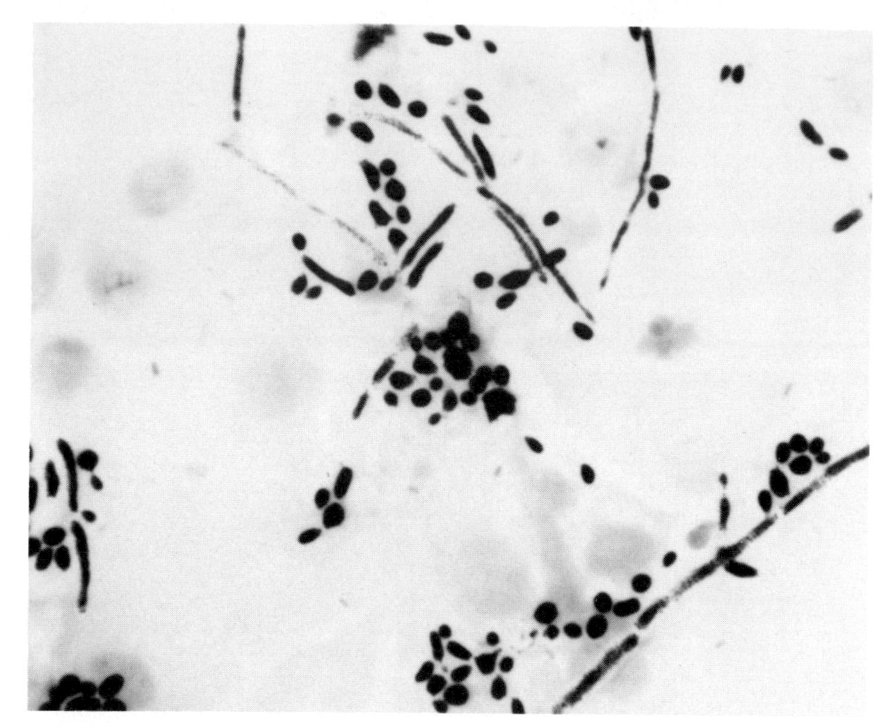

D. *Candida* sp. Kidney impression smear (Gram stain ×450).

Figure 11.5. Hyphal forms without yeasts

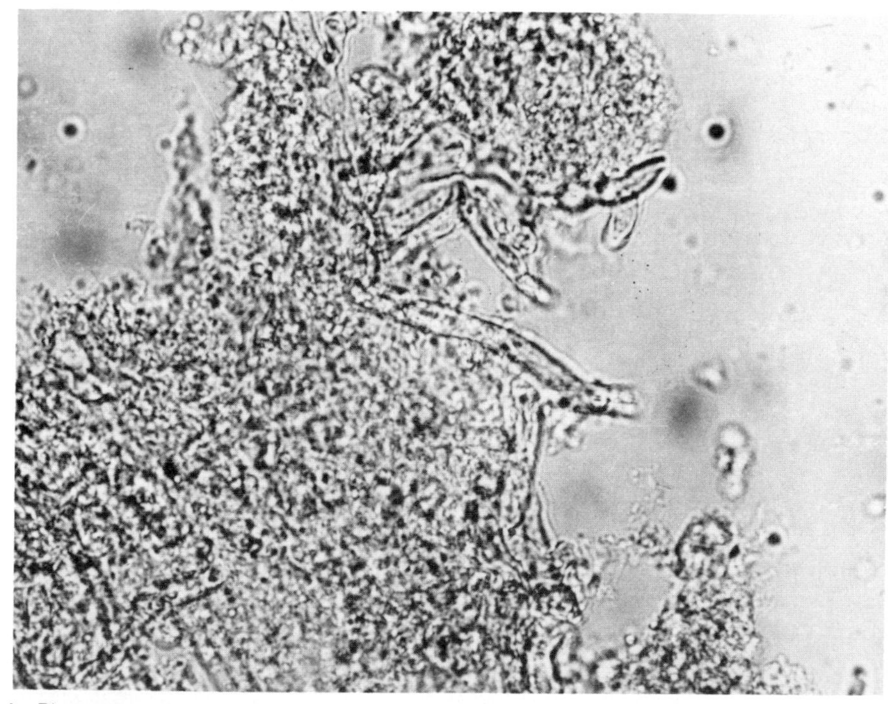

A. Phycomycete hypae. Wet preparation of scrapings from gastric mucosa (unstained ×450).

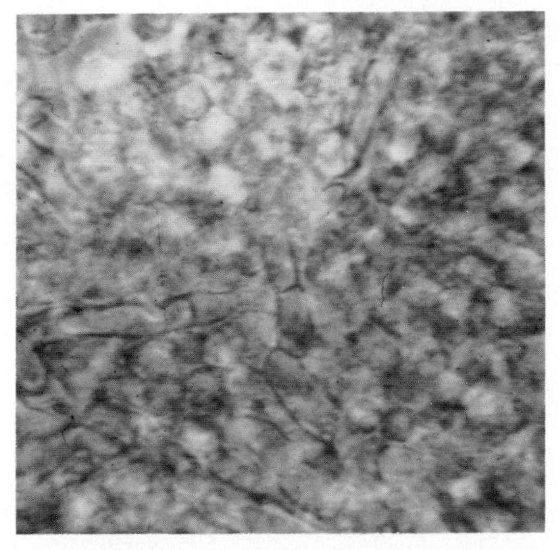

B. Phycomycete hyphae. Wet preparation of nasal crust (unstained ×200).

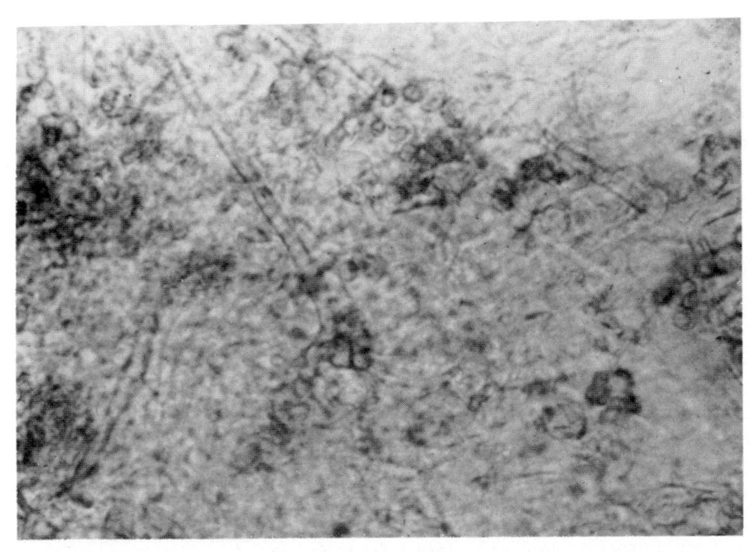

C. Aspergillus hyphae. Wet preparation of resected fungus ball in lung cavity (unstained ×400). Note fragmented spores. Density is due to natural pigment of the invading species.

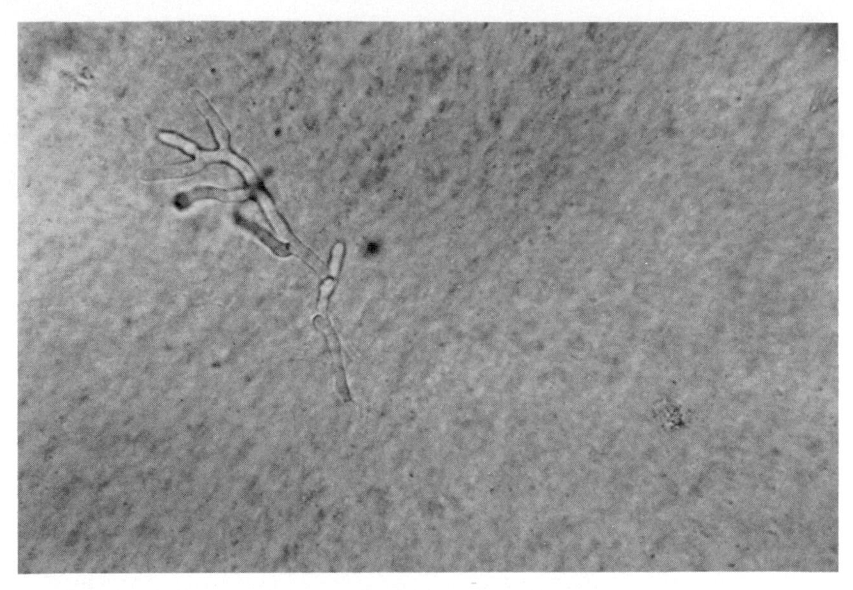

D. *Aspergillus* sp. Sputum, wet preparation (unstained ×400).

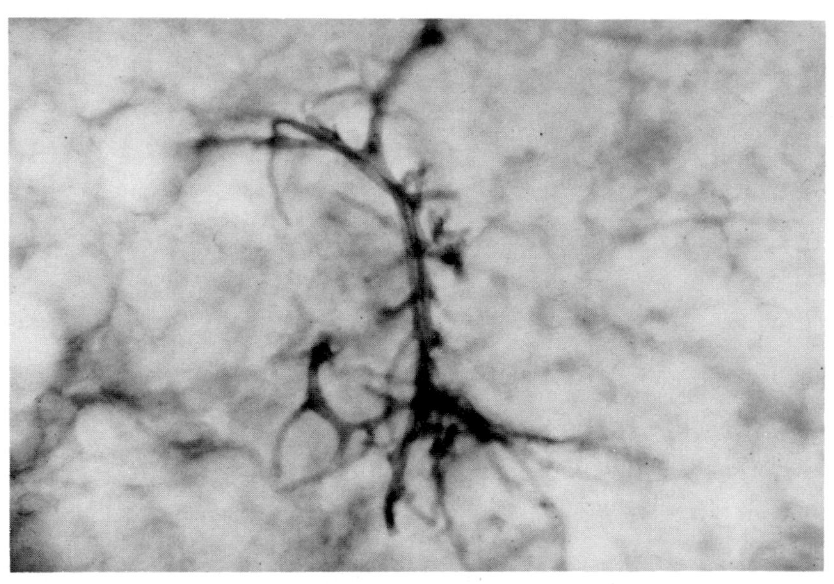

E. *Aspergillus* sp. Lactophenol (2, 7, 9, 18) stained wet preparation of bronchial aspirate (×1000).

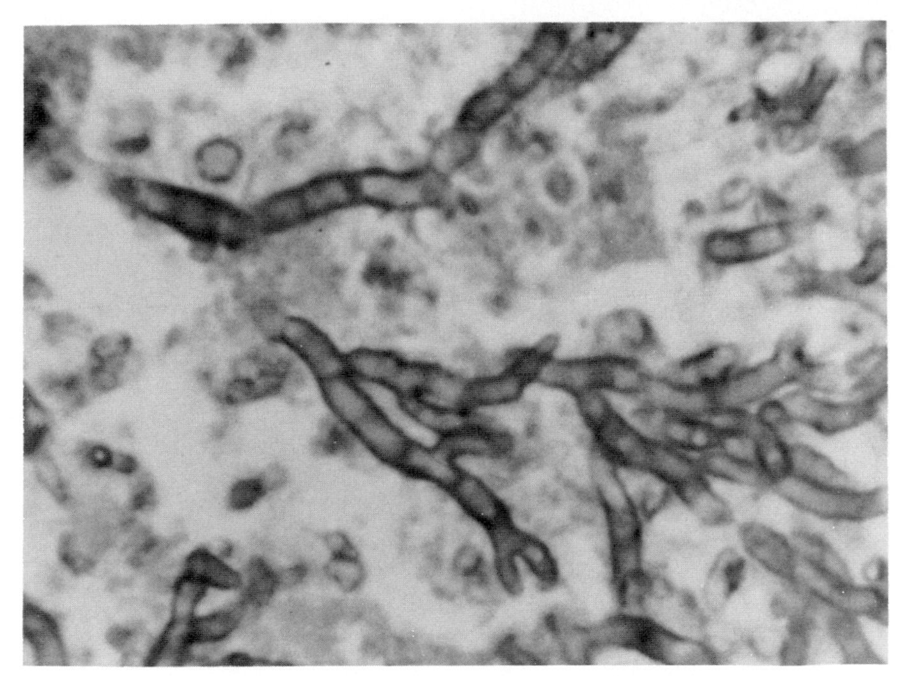

F. *Aspergillus* sp. Lung biopsy (H&E stain ×1200).

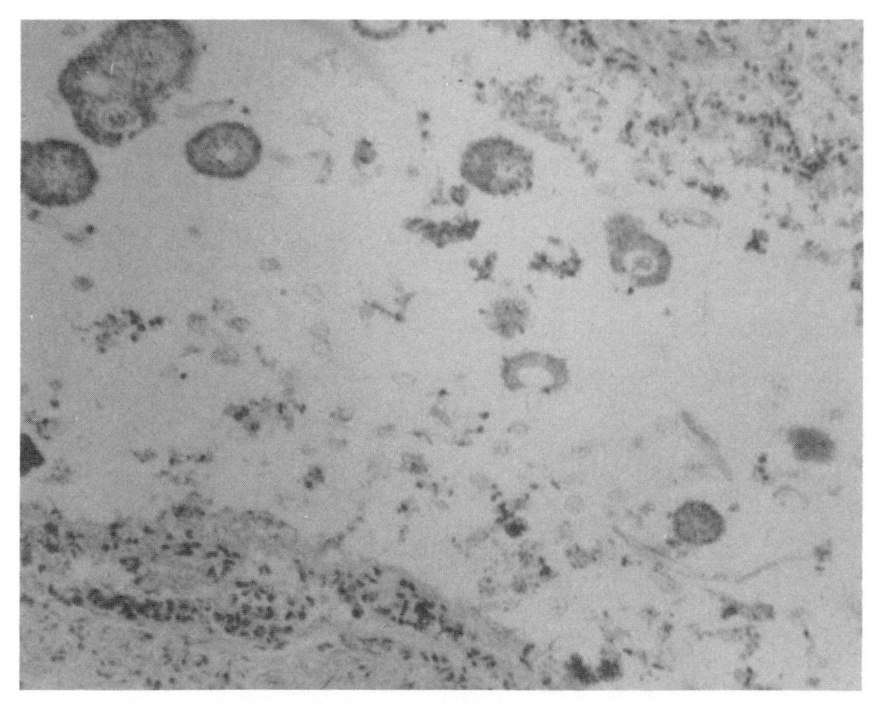

G. *Aspergillus* sp. Lumen of blood vessel occluded with fully matured "heads" at autopsy of undiagnosed case antemortem (H&E ×1000).

Figure 11.6. Other fungal forms

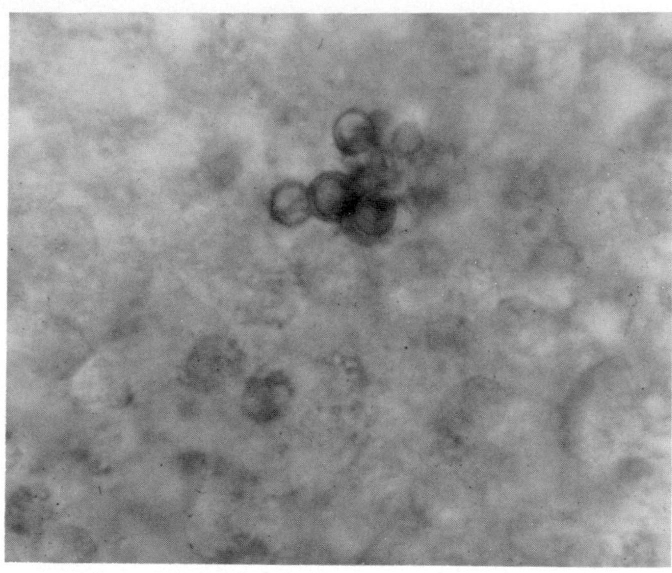

A. Sclerotic bodies of agents of chromomycosis. Cells are dematiaceous (dark), and split rather than bud. Wet preparation of pus (unstained ×400).

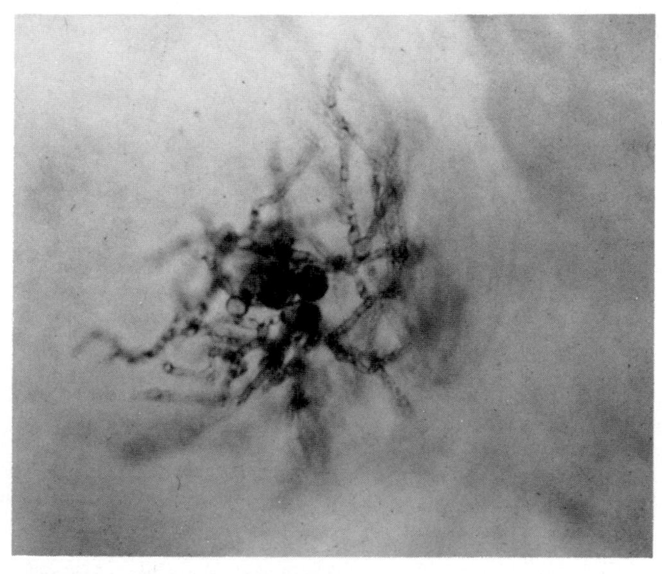

B. Ibid. Demonstrated filamentous formations from sclerotic bodies (unstained ×400). Rare.

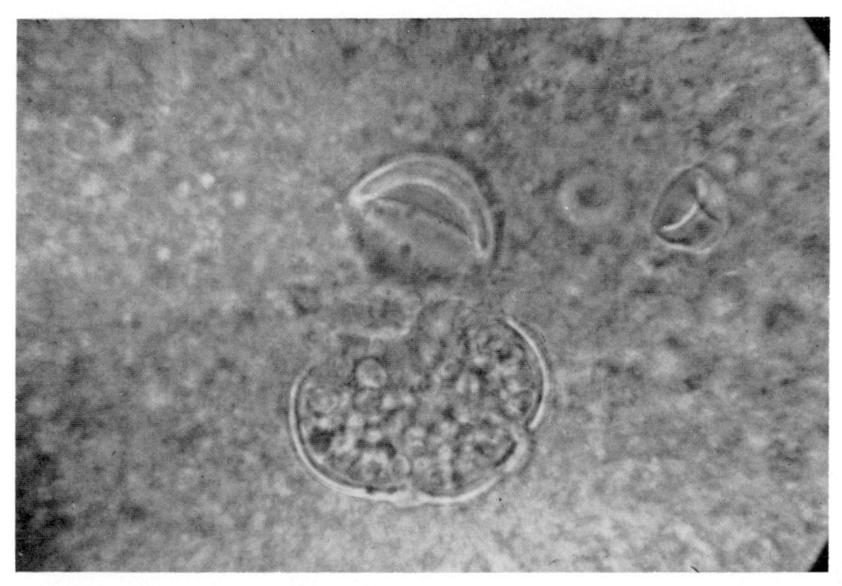

C. Spherules of *Coccidioides immitis.* Wet preparation of pus (unstained ×1000). Note mature spherule at point of rupture, and empty ruptured spherule walls.

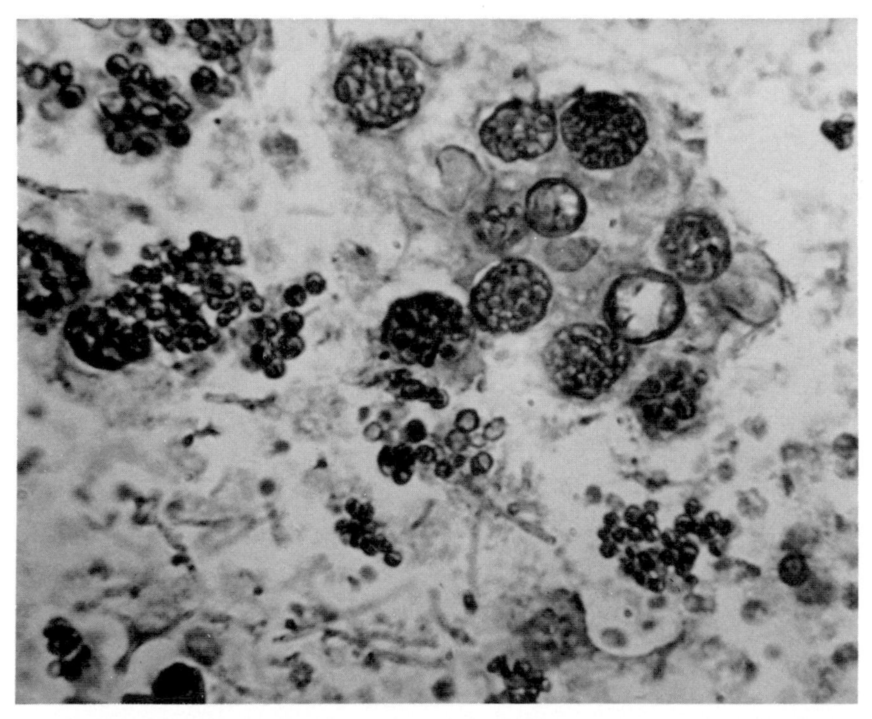

D. Spherules of *C. immitis.* Histopathological section (Brown & Brenn ×1000) of edge of cavity. Note hyphal filaments. Rare.

Figure 11.7. Granules

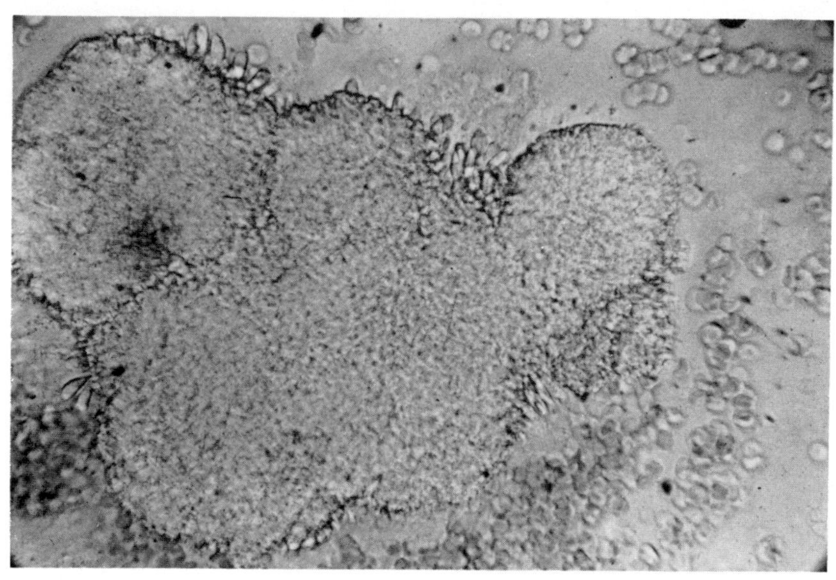

A. Maduromycotic: Wet preparation of pus and scrapings from lesion (unstained ×400). Note broad hyphal elements at edge of granule, indicating *true* fungus etiology.

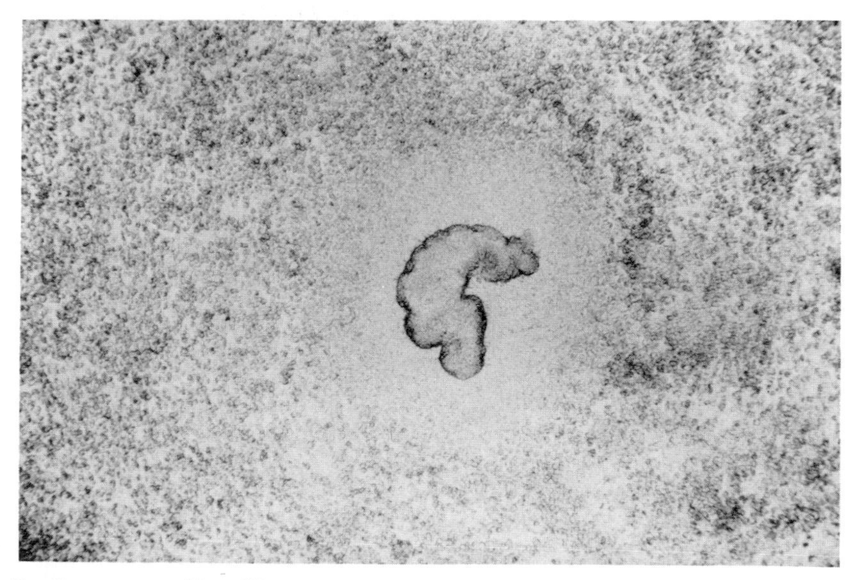

B. Bacillary composition. Wet preparation of pus (unstained ×100).

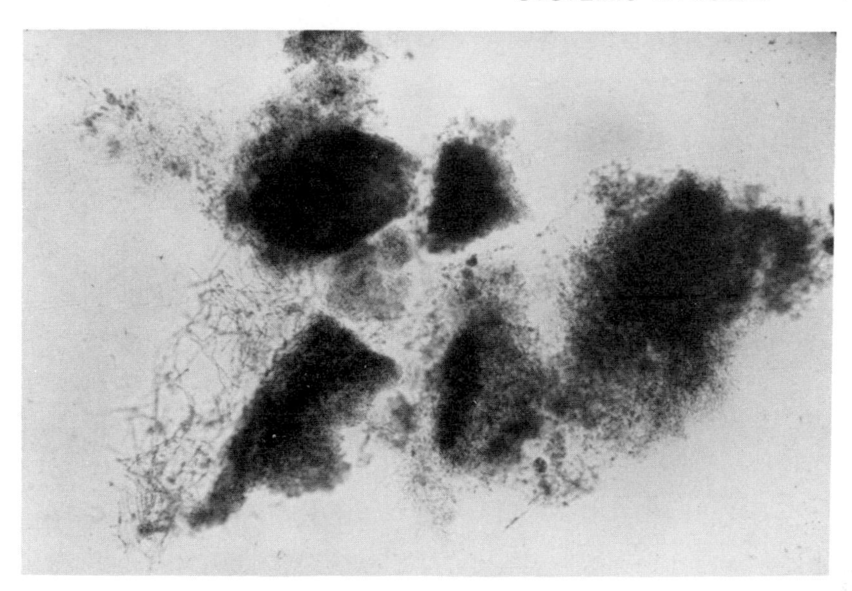

C. Bacillary composition: Wet preparation of pus (unstained ×1000). Note branching elements emerging from granules.

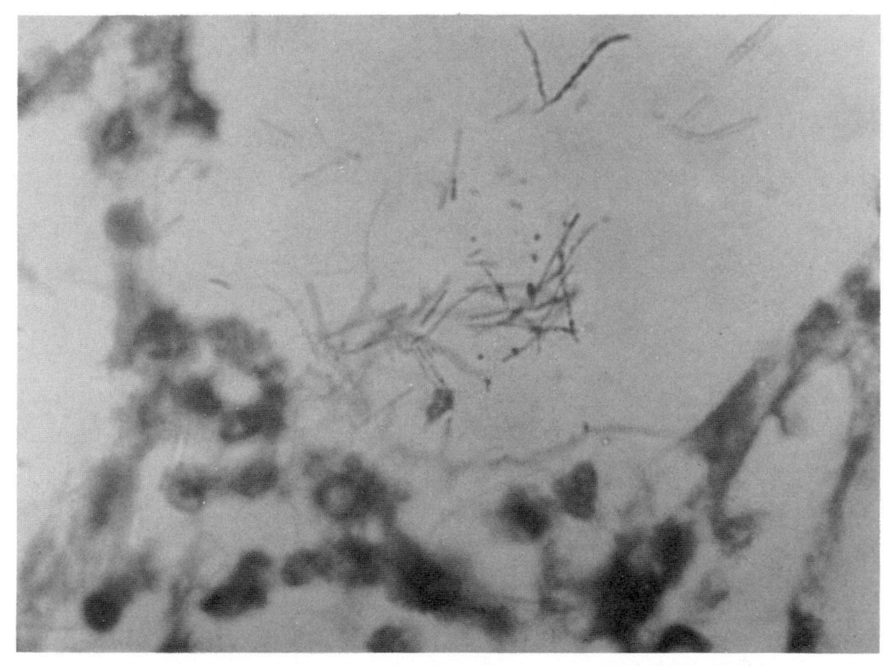

D. Bacillary elements. Acid fast stained preparation (×1000). Acid fast *Nocardia* spp. are more strongly acid-fast in clinical specimens than in culture. Confusion may exist with *Mycobacterium* spp. *Nocardia* sp., especially the acid-fast *N. asteroides,* does not require enriched medium for isolation as do *Mycobacterium* spp. Bacterial antibiotics should not be used.

LITERATURE CITED

1. Ajello, L. The medical mycological iceberg. *HSMHA Health Rept 86:*437–448, 1971.
2. Baker, R. D. (Ed.) *Human Infection with Fungi, Actinomycetes and Algae*. New York, Springer-Verlag, 1971.
3. Bennett, J. E. Chemotherapy of systemic mycoses. *New Engl J Med 290:*30–32, 320–323, 1974.
4. Campbell, C. C. The pilot wheel: a change in course. Proc. 1st Int. Conf. on Paracoccidioidomycosis. *Pan Am Health Org Sci Publ. 254:*306–312, 1972.
5. Campbell, C. C. Respiratory mycotic infection. *Prev Med 3:*517–528, 1974.
6. Campbell, C. C., and Binkley, G. E. Serologic diagnosis with respect to histoplasmosis, coccidioidomycosis and blastomycosis and the problem of cross reactions. *J Lab Clin Med 42:*896, 1953.
7. Conant, N. F. et al. *Manual of Clinical Mycology,* Ed. 3. Philadelphia: W. B. Saunders Co., 1971.
8. Emmons, C. W. The Jekyll-Hydes of mycology. *Mycologia 52*(5):669–680, 1960.
9. Emmons, C. W., Binford, C. H., and Utz, J. P. *Medical Mycology,* Ed. 2. Philadelphia: Lea & Febiger, 1970.
10. Hammerman, K. J., Powell, K. E., and Tosh, F. E. The incidence of systemic mycoses. *Sabouraudia 12:*33–45, 1974.
11. Huppert, M., Harper, G., Sun, S. H., and Delanerolle, V. Rapid methods for identification of yeasts. *J Clin Microbiol 2*(1):21–34, 1975.
12. International Symposium on Opportunistic Fungus Infections. *Lab Invest 11*(Part 2):1017–1241, 1962.
13. Lee, K. L., Buckley, H. R., and Campbell, C. C. An amino acid liquid synthetic medium for the development of mycelial and yeast forms of *Candida albicans*. *Sabouraudia 13*(2):148–153, 1975.
14. Lee, K. L., Reca, M. E., and Campbell, C. C. Amino acid synthetic media for fungal pathogens based on aminopeptidase specificities: *Histoplasma capsulatum, Blastomyces dermatitidis, Paracoccidioides brasiliensis* and *Cryptococcus neoformans*. *Sabouraudia 13*(2):142–147, 1975.
15. Medoff, G., Kobayashi, G. S., Kwan, C. N., et al. Potentiation of rifampicin and 5-fluorocytosine as antifungal antibiotics by amphotericin B. *Proc Natl Acad Sci USA 69:*196–199, 1972.
16. Palmer, C. E. Nontuberculous pulmonary calcification and sensitivity to histoplasmin. *Public Health Rep 60:*513, 1945.
17. Rifkind, D., Marchior, T. L., Schneck, S. A., and Hill, R. B., Jr. Systemic fungal infections complicating renal transplantation and immunosuppressive therapy. *Am J Med 43:*28–38, 1967.
18. Rippon, J. W. *Medical Mycology, the Pathogenic Fungi and the Pathogenic Actinomycetes*. Philadelphia: W. B. Saunders Co., 1974.
19. Shadomy, S. Further *in vitro* studies with 5-fluorocytocine. *Infect Immun 2*(2):484–488, 1970.
20. Shadomy, J. Clamp connections in two strains of *Cryptococcus neoformans,* pp. 45–48, in D. Ahearne (Ed.), *Recent Trends in Yeast Research*. School of Arts and Sciences Georgia State University, 1970.
21. Smith C. E., Beard, R. R., Whiting, E. G., and Rosenberger, H. G. Varieties of coccidioidal infection in relation to the epidemiology and control of the disease. *Am J Public Health 36:*1394, 1946.
22. Smith, C. E., Whiting, E. G., Baker, E. E., Rosenberger, H. G., Beard, R. R., and Saito, M. T. 1948. The use of coccidioidin. *Am Rev Tuberc 57:*330, 1948.
23. Taschjdian, C. L., Kozinn, D. J., and Toni, E. F. Opportunistic yeast infections with special reference to candidiasis. *Ann NY Acad Sci 174*(2):606–622, 1970.
24. Thompson, L. Note on a selective medium for fungi. *Proc Staff Meet Mayo Clin 20*(14):241, 1945.

12

Clinical Significance of Mycoplasma and Bacterial L-Forms

SARABELLE MADOFF
WILLY N. PACHAS

Mycoplasma are an extremely interesting group of organisms with a multiplicity of pathogenic propensities and unique biologic properties. They represent a challenge to the student of microbiology since they defy many of the usual criteria of taxonomic classification, and their position in the evolution of the microbial world remains to be explained. The increasing interest in the mycoplasma by microbiologists, molecular biologists, electron microscopists, and clinicians also stems from the fact that they are the smallest free-living microorganisms thus far described that are capable of causing disease in many animal species and in man. The distinctive feature of the mycoplasma is that they possess all the elements found in bacteria with the exception of the cell wall. The classical feature of mycoplasma is the well-known appearance on agar medium of the colonies formed by small granules and "large bodies." Granular growth penetrates the medium at the center of the colony; at the periphery, surface growth and large bodies give the colony the typical "fried egg" appearance.

In terms of their distribution in nature, mycoplasma have been found both as commensals and pathogens in man and in animals, and probably in plants and insects. In the human host, mycoplasma are found in the oropharynx and in the genitourinary tract. *Mycoplasma hominis*, *Mycoplasma orale* I, II, and III, and *Mycoplasma salivarium* usually inhabit the oropharynx, whereas *Mycoplasma hominis*, *Mycoplasma fermentans*, and the T-strains (*Ureaplasma urealyticum*) are found in the genitourinary tract. *Mycoplasma pneumoniae*, the only mycoplasma pathogenic to man, is isolated from nasopharyngeal secretions and tracheobronchial exudates. Mycoplasma are fastidious organisms, difficult to grow in the laboratory. The requirements for growth include the presence of enrichment factors such as animal serum and yeast extract. An important growth factor for most mycoplasma is cholesterol. However, the saprophytic strain *Acholeplasma laidlawii* does not require cholesterol, and some *Acholeplasma* strains can be grown in chemically defined media. In recent years, a large number of publications have thoroughly described the biologic characteristics of the mycoplasma (10, 15, 19, 22).

MYCOPLASMA AS PATHOGENS

The first indication that mycoplasma could play a pathogenic role in man arose from Dienes' finding of mycoplasma in a culture from an infected Bartholin's cyst (7). Further work by Dienes gave some inkling as to the possible etiologic significance of these organisms in patients with abacterial hemorrhagic cystitis and Reiter's syndrome. This early work created the incentive to study mycoplasma in relation to a number of pathologic processes of human origin. In 1942, Eaton and collaborators were able to transmit the agent of primary atypical pneumonia (PAP) into cotton rats. After years of work, the viral etiology of PAP became dubious and indirect methods of identification suggested the presence of mycoplasma. In 1962, Chanock, Hayflick, and Barile succeeded to grow *Mycoplasma pneumoniae* on artificial media and to clearly establish its role in human disease (2). Following this discovery, a burst of enthusiasm in the field led to important contributions in pathogenesis and immunologic mechanisms of *M. pneumoniae* disease. Many efforts to establish a role for mycoplasma in other areas of human pathology have not been equally rewarding in that all of Koch's postulates could not be met. These aspects have been thoroughly reviewed in several publications (10, 17, 19, 22).

Research efforts in veterinary medicine with pathogenic mycoplasma have been more fruitful. A legion of animal diseases is clearly linked to mycoplasma infections. The severity of these illnesses is reflected in the great economic reversals resulting from epidemics of pleuropneumonia in cattle, agalactia in sheep and goats, and upper respiratory infections in poultry and in swine (10, 19, 22).

Mycoplasma pneumoniae Infection

In working up a patient with possible primary atypical pneumonia, the indications for special cultures or serological tests for *M. pneumoniae* should be viewed on the basis of the clinical manifestations and other auxillary information such as chest radiographs and blood studies, including cold agglutinins. Clinically, *M. pneumoniae* is a disease of young people often manifested by severe tracheobronchitis or pneumonitis. Occasionally, concomitant otitis media or myringitis can occur. Low grade fever or fever of moderate degree is usually present. The illness has a self-limiting course in most patients with an average duration of 1 to 2 weeks. With antibiotic treatment, the course of the illness can be shortened to between 3 and 7 days. The most disturbing symptom is persistent hacking cough. The pneumonitis is usually mild and does not cause severe respiratory failure as often seen in bacterial or viral pneumonias. Involvement of the neighboring structures to the lung such as pleura is uncommon, but has been reported. Pericarditis is probably very unusual. Peripheral manifestations of *M. pneumoniae* infections such as Stevens-Johnson syndrome and skin ulcers, probably related to the presence of cold agglutinins, have been observed. The chest x-rays in cases of pneumonitis show soft infiltrates near the parahilar areas. Enlargement of mediastinal nodes is not seen. The laboratory studies show mild elevation of white blood cell counts or normal WBC. Often there is a tendency to lymphocytosis. The presence of neutrophilia should make the physician suspicious of bacterial pneumonia.

The hallmark of *M. pneumoniae* infections is the appearance of cold agglutinins shortly after the onset of clinical disease, and their increase in the convalescent period. Cold agglutinins are autoantibodies of the IgM type directed against cell surface components of the red blood cells (i antigen). Occasionally high titers of cold agglutinins have been the cause of complicating mild cases of hemolytic anemia. Apparently, this phenomenon occurs in exposed areas of the skin where the low temperature may favor the agglutination of these cold antibodies. The histology of this infection in the upper respiratory tract and in the lungs has not been studied in humans. However, experimental evidence has shown colonization of the surface epithelium with mycoplasma and impairment of the function of the ciliated processes. The pneumonitis itself may represent a localized allergic response to *M. pneumoniae* infection.

From the above discussion it seems obvious that in general the clinical data, the radiological studies, and the cold agglutinin determination may suffice to establish the diagnosis of primary atypical pneumonia. In occasional cases, however, other studies may be needed. A significant number of patients develop antibodies to the strain of streptococcus called MG (4) although the specificity of this reaction has not been clearly established. If the cold agglutinin test is strongly positive, isolation and identification methods for *M. pneumoniae* are probably not practical. Nevertheless, it is highly desirable to proceed with cultural techniques for isolation, because cold agglutinins take 2–3 weeks to peak and because this test is not always specific for PAP. Although the cultural methods still require several days to a few weeks to yield growth of *M. pneumoniae,* a positive culture establishes a definitive diagnosis for PAP (2, 4, 5, 10, 19).

Serologic techniques may give earlier evidence of *M. pneumoniae* infection (4, 10, 18, 19). With the metabolic inhibition test, results can sometimes be obtained in 3–5 days. However, this test is available only in certain research laboratories. Immunofluorescent techniques may also be helpful for early detection of PAP, but the cost of these studies is still excessive. The complement fixation (CF) test is more widely used and more readily available for diagnostic purposes than the metabolic inhibition test. Most state laboratories are set up to perform CF testing for *M. pneumoniae*. It has been shown that a positive CF test for *M. pneumoniae* usually correlates with positive isolations, and CF titers may remain elevated for long periods of time. In addition, very high titers of cold agglutinins seem to indicate a high probability of isolating *M. pneumoniae* in culture. It should be emphasized that in order to make proper utilization of these serologic techniques, acute and convalescent sera must be provided. It is advisable that in the search for mycoplasma etiology a combination of tests including cultural and serologic methods be used (Table 12.1).

Mycoplasma in the Genitourinary Tract

The significance of the presence of mycoplasma in the genital and urinary tracts has been a controversial matter over the years. It is of historical interest that the first mycoplasma in humans were cultured from infectious material obtained from a Bartholin's cyst (7). Certain strains of mycoplasma have been associated with nongonococcal urethritis (NGU). In particular, the T-mycoplasma (*Ureaplasma urealyticum*) have been implicated as the cause of this disease (10, 19, 20, 21). However, their isolation from asymptomatic individuals raises questions as to their etiologic role. T-mycoplasma have also been implicated as a primary cause of premature births, spontaneous abortions, and reproductive failure (1, 10, 12). Similar objections can be raised as to their role as pathogens in these conditions. Definitive studies to demonstrate the relationship of T-mycoplasma with these clinical entities as well as appropriate experimental models are difficult to realize. The mechanism of pathogenicity of these mycoplasma may be complex and may require either special conditions in the host or activation of pathogenicity through mutational mechanisms. The existence of various serotypes of T-strains may offer the opportunity to test these various possibilities of pathogenicity.

Interest in this area of disease should be encouraged. In a number of clinical entities such as bartholinitis, infection of the Fallopian tubes, "sterile" deep-seated abscesses, ovarian cysts, and focal glomerulonephritis, T-strains and other mycoplasma have been the only microorganisms present. Obviously, much work is needed to fully understand the mechanisms by which these organisms may cause pathologic processes in humans.

Mycoplasma in Rheumatic Diseases

The etiology of some rheumatic diseases including rheumatoid arthritis has been linked to a variety of infectious agents. At one time or another, bacteria, viruses, mycoplasma, and chlamydia have been implicated as possible etiologic agents. Disappointingly, some of the isolations of myco-

TABLE 12.1
Diagnostic tests for *Mycoplasma pneumoniae* infection

Serum Cold Agglutinins	Complement Fixation	Metabolic Inhibition	Culture
Significant titers Acute stage >1:32 Convalescent rising titer	4-fold increase in antibody titer to *M. pneumoniae* between acute and convalescent sera. Titer may remain elevated for several months.	Antibody to *M. pneumoniae* often present in young adults. Rising titer indicates infection in 90% of patients with PAP	Isolation of *M. pneumoniae* in about 40–50% of patients with PAP. Isolation may be possible for several months following the acute episode
Positive in 50–100% of patients with PAP	Positive in 50–60% of patients with PAP		

plasma failed to be confirmed, or critical evaluations showed the presence of contaminants in culture material. It is of interest to point out, however, that in many animal species (cattle, sheep, goats, swine, poultry, and rodents) mycoplasma have been found to be the etiologic agent of arthritis (10, 19). In this respect the similarity of chronic arthritis of swine caused by *Mycoplasma hyorhinis* and human rheumatoid arthritis is very striking. The role of mycoplasma in the rheumatic diseases merits further exploration.

Mycoplasma in Miscellaneous Conditions

Mycoplasma isolates have been reported in a variety of disorders including hemorrhagic cystitis, postpartum septicemia, leukemia, focal glomerulonephritis, nontropical sprue, brain abscesses, meningitis, and infections of the upper respiratory tract (4, 10, 15, 17, 19). The significance of these isolates is difficult to establish. In many instances, the basic disease process had already been established and the mycoplasma probably played the role of an opportunistic invader. In other cases, mycoplasma strains appear to have had a pathogenic function, such as in the cases of salpingitis and brain abscesses where no other causative organisms were found. The corollary of these interesting observations is that to develop a more complete understanding of the interaction of the mycoplasma with the host identification of the mycoplasma and serologic studies should be pursued. The application of sophisticated immunological techniques such as macrophage migration-inhibition and lymphocyte transformation may provide useful markers to clarify the relationship between the organisms and the disease process.

From the above, it is obvious that the mycoplasma research laboratory has the primary responsibility of conducting the proper studies that will increase the knowledge of mycoplasma in human disease. IN PRACTICE IT WOULD APPEAR THAT A SEARCH FOR MYCOPLASMA FOR THE ACTUAL CARE OF PATIENTS IS WARRANTED ONLY IN SOME PATIENTS SUSPECTED OF HAVING PRIMARY ATYPICAL PNEUMONIA. On the other hand, without the cooperation and interest of the clinician and the clinical laboratory personnel, progress in other important areas will remain at a standstill.

TECHNIQUES OF ISOLATION AND IDENTIFICATION OF MYCOPLASMA

The identification of mycoplasma requires skillful techniques, considerable experience, and the availability of special culture media. The colonies grown in agar media can usually be seen with the help of a magnifying lens. Some strains, however, can be recognized only with microscopic techniques. The Dienes technique which utilizes light microscopy is the most reliable way to identify mycoplasma colonies (14, 15). This technique is simple and reproducible and distinguishes true mycoplasma colonies from artifacts. Other microscopic techniques include the use of phase microscopy, epi-immunofluorescence with fluorescent antibody, and electron microscopy. Phase microscopy is a good tool in experienced hands but it does not distinguish some artifacts that resemble mycoplasma colonies. The technique of epifluorescence allows for specificity and strain identification, but this technique is costly and only applicable at present in research laboratories. Electron microscopic studies have been invaluable in the study of the ultrastructure of mycoplasma. This technique has also provided significant information in the area of plant and insect pathology where the ultrastructural studies gave the first clue as to the possible mycoplasma etiology of certain diseases.

The mycoplasma laboratory utilizes a variety of culture media to isolate and identify mycoplasma. In brief, the basic solid media should be fairly soft and consist of a good nutrient agar base enriched with 10–20% animal serum and yeast extract or yeast autolysate. Penicillin (1000 units/ml) is incorporated into the medium as an antibacterial agent (4, 5, 14, 15).

For the specific isolation of *Mycoplasma pneumoniae* enrichment with 20% uninactivated horse serum and 2.5% of *fresh* yeast extract is required (Hayflick's medium) (2, 9). Thallium acetate is used as an antibacterial agent in addition to penicillin. Some laboratories also incorporate methylene blue and amphotericin B into the medium with the purpose of eliminating other mycoplasma species and fungi, respectively (4, 5).

The T-mycoplasma have special requirements for optimum growth (20, 21). These include soft agar, a low pH (5.5–6), the presence of urea and 20% horse serum. Thallium acetate is omitted because of its inhibitory effect on T-strain growth. Anaerobiasis by the Fortner method is helpful for primary isolation of the T-strains (14, 15).

In our experience a blood agar plate is a useful medium for isolation of mycoplasma. Many strains fail to grow on primary isolation on enriched media, particularly tissue culture strains and occasional isolates from the oropharynx and from the genitourinary tract (15). Mycoplasma growth can also be obtained in liquid media. For this purpose, PPLO broth enriched with similar nutrients as the solid media can be used. Subcultures can be made from agar to agar by the push-block method and by agar block inoculation.

The recognition of mycoplasma growth can be unequivocally made using the Dienes technique. The special stain is readily absorbed by the viable organisms and allows their examination in detail. The staining solution is made up by dissolving 2.5 g of methylene blue, 1.25 g of azur II, 10 g of maltose, 0.25 g of Na_2CO_3 and 0.2 g of benzoic acid in 100 ml of water. The procedure involves the selection of an appropriate mycoplasma growth area on the surface of the agar medium. This step is made with a magnifying hand lens or using the low power objective of the microscope. The selected area on the agar is then cut out with a dissecting knife, lifted out, and placed face up on a glass slide. The agar block is then covered with a Dienes-stained coverslip cut to size with a diamond point pencil. The coverslip is prepared by painting one surface with the Dienes stain by means of a cotton applicator and allowing it to dry. Many coverslips can be prepared in a few minutes and stored for future use. The preparation is now sealed around the edges with a heated mixture of melted paraffin softened with a small amount of vaseline. This is done with a Pasteur pipette. Any excess paraffin on the surface of the preparation should be removed with the dissecting knife or with a razor blade. An experienced technician can easily make such a preparation in about 1 min. Microscopically, the mycoplasma colonies stain a light blue at the periphery and a deep blue at the center. With oil immersion magnification, the granules and the large bodies that comprise the colonies can be studied. The method is equally applicable in the study of L-forms. Photographic impressions can also be made from these preparations (15, 20) (see Fig. 12.1, *M. hominis* and T-strains).

The simplest method of strain identification is the growth inhibition test. For this purpose an agar block containing mycoplasma growth is deposited in a test tube containing 2 ml of 0.85% NaCl. The tube is agitated with a Vortex mixer for a few minutes to detach the organisms. The suspension is then streaked on appropriate media with a Pasteur pipette and allowed to soak in until all moisture is absorbed. Antisera-impregnated disks are then placed on the inoculated agar surface. The specific strain is identified by an area of inhibition of growth surrounding the disk.

For the proper cultivation of mycoplasma from clinical material, it is of the utmost importance that a suitable specimen be sent to the laboratory. In cases of suspected primary atypical pneumonia, the physician should contact the laboratory ahead of time so that the appropriate media will be prepared. The specimen should be fresh sputum, if possible, or a properly taken throat swab or throat washings as required for viral isolations. At any rate, it is desirable that the specimens be brought to the laboratory with promptness to prevent loss of viability of the organisms. Some experts believe that the best method of isolating *M. pneumoniae* is by planting the culture directly at bedside. The same principles apply for the isolation of mycoplasma from other clinical material such as urines, gynecological specimens, urethral scrapings, etc. On special occasions, when immediate processing of the culture is not feasible, overnight refrigeration or freezing at −20 C has been found useful.

L-FORMS

The concept of L-forms has been a very confusing one to many people over the years. Their resemblance to mycoplasma has been the major factor in the confusion. It is now clearly established that many bacteria—and perhaps all bacteria—may produce L-form growth when properly stimulated. The end result is the loss of cell wall in part or in toto and the generation of organisms with new biologic properties that resemble the mycoplasma in many respects (3, 6, 8, 10, 11, 15, 16, 19).

Partial loss of cell wall by the bacteria yields the highly revertible B-type L colonies, whereas a complete loss of the cell wall produces the "stable" or difficult-to-revert A-type L colonies. When fully developed, a typical L-form colony bears a strong resemblance to a mycoplasma colony and may even adopt a "fried egg" appearance. When L colonies are studied with the Dienes technique, the L-form granules appear to be somewhat

Figure 12.1 (top) and 12.2 (bottom)

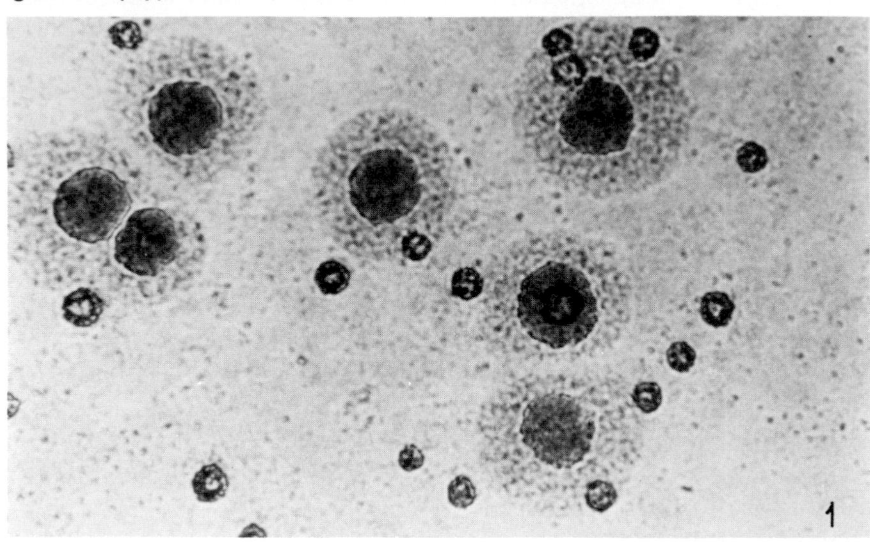

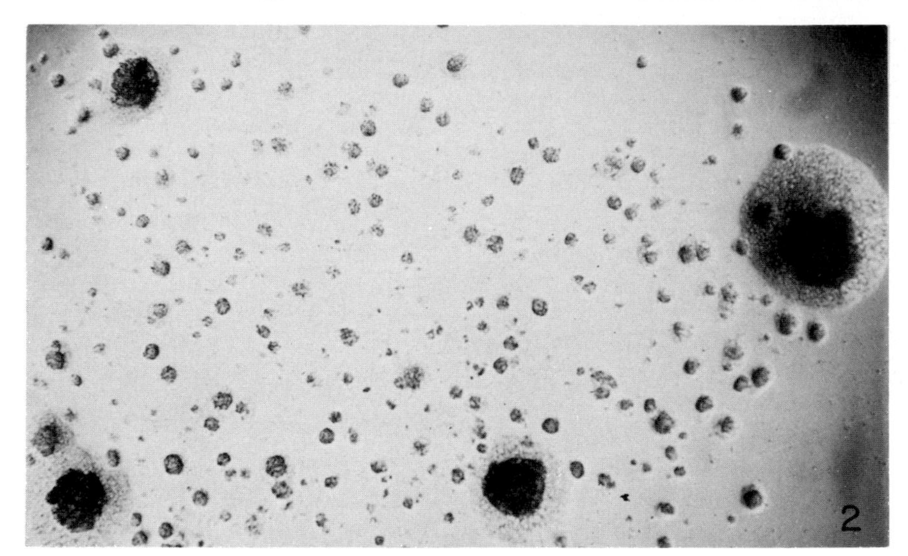

1. Colonies of *Mycoplasma hominis* stained by the Dienes technique and T-mycoplasma colonies stained by urease color test (×350). Courtesy of Maurice C. Shepard (20).
2. A and B type L-colonies of Proteus (×100, unstained). (Dienes collection.)

more variable in size than those of mycoplasma. The "large bodies" which are usually found at the periphery of the colony are more abundant than in mycoplasma and may actually represent reproductive forms. Ultrastructural studies have clearly illustrated the complete absence of bacterial cell wall in the A-type L colonies. However, it should be emphasized that certain Enterobacteriaceae may produce both A- and B-type L colonies, or either one or the other under special experimental conditions (Fig. 12.2). Streptococci, staphylococci, diphtheroids, and other Gram-positive organisms tend to produce exclusively A-type colonies. L-forms may occur spontaneously in nature as in the case of *Streptobacillus moniliformis*. In most instances, however, the L-forms are products of the laboratory. The usual procedure is to induce L-form growth by means of an agent that inhibits cell wall synthesis. Penicillin and similar antibiotics are the most commonly utilized for this purpose. Stripping the cell wall with lysozyme with the production of protoplasts may occasionally produce L-form growth when the protoplasts are transferred to appropriate agar media. An important requirement for L-form production by some bacterial species is the presence of hyperosmolar media. This is particularly noted with Gram-positive organisms. Many of the Gram-negative bacteria, on the other hand, do not require a hyperosmolar environment for L-form production (6, 11, 15).

Studies of pathogenicity with L-forms have failed to produce conclusive data. Most experimental work suggests that L-forms are innocuous, and the recovery of pathogenicity occurs only with the reappearance of the cell wall. One important objection to these studies, however, is that pathogenicity was not studied with organisms specific for the animal host. Future studies must be done using the L-forms of organisms pathogenic for the animals. The adaptation of L-forms to growth in tissue culture with production of cytopathic effects may be an effective way to induce pathogenicity in animals, as suggested by recent studies (13).

In addition to the L-forms, variant and atypical bacterial forms have been mentioned as being possibly related to recurrent infections of the urinary tract and in subacute bacterial endocarditis (3, 16, 19). These aberrant forms are characterized by having defective cell wall and they may require special media for their survival. The organisms have been described as initially pleomorphic and fragile and growing in agar as atypical colonies. Later, apparently as they build a more complete wall, they slowly return to the normal state. The significance of these bacterial variants is unknown and the evidence for their pathogenicity remains controversial. Protoplasts and spheroplasts, on the other hand, generally are laboratory products produced by exposure of the organisms to lysozyme or other similar enzymes. As such, they have been useful tools for the study of cell wall synthesis and membrane organelles (mesosomes).

In our present state of knowledge, the bacterial L-forms are considered to be the products of manipulations of bacteria in the laboratory. The concept of an "L phase" of bacteria has been proposed by many workers. This implies a cyclic change from bacteria to L-form as part of a normal life process that, with the exception of *Streptobacillus moniliformis*, has not been shown in other organisms. The occurrence of L-forms in vivo cannot be categorically denied, and indeed it may be a provocative consideration in cases of recurrent and persisting infections. The possibility of cell-wall dissolution by cellular enzymes, immune mechanisms, and environmental factors such as hypertonicity may well reproduce the experimental circumstances offered in the laboratory. However, the cultivation of true L-forms from clinical material has never been documented. A great amount of work needs to be done with experimental models to establish the definitive role of bacterial L-forms in human and animal diseases. At the present time it does not appear to be warranted to recommend a search for L-forms in clinical material except on an investigative basis.

LITERATURE CITED

1. Braun, P., Lee, Y. H., Klein, J. O., Marcy, S. M., Klein, T. A., Charles, D., Levy, P., and Kass, E. H. Birth weight and genital mycoplasmas in pregnancy. *N Engl J Med 284:*167–171, 1971.
2. Chanock, R. M., Hayflick, L., and Barile, M. F. Growth on artificial medium of an agent associated with atypical pneumonia and its identification as a pleuropneumonia-like organism. *Proc Natl Acad Sci USA 48:*41, 1962.
3. Clasener, H. Pathogenicity of the L-phase of bacteria. *Ann Rev Microbiol 26:*55–84, 1972.
4. Crawford, Y. E. Mycoplasmas of human derivation, pp. 3–70, in C. Panos (Ed.), *A Microbial Enigma: Mycoplasma and Bacterial L-forms.* Cleveland: World Publishing Co., 1967.
5. Crawford, Y. E. Mycoplasma, pp. 251–262, in J. E. Blair, E. H. Lennette and J. P. Truant (Eds.), *Manual of Clinical Microbiology.* Baltimore: Williams & Wilkins Co., 1970.
6. Dienes, L. Morphology and reproductive processes of bacteria with defective cell wall, p. 74–93, in L. B. Guze (Ed.), *Microbial Protoplasts, Spheroplasts and L-forms.* Baltimore: Williams & Wilkins Co., 1968.
7. Dienes, L., and Edsall, G. Observations of the L-organism of Klieneberger. *Proc Soc Exptl Biol Med 36:*740–744, 1937.
8. Guze, L. B. (Ed.) *Microbial Protoplasts, Spheroplasts and L-forms.* Baltimore: William & Wilkins Co., 1968.
9. Hayflick, L. Cell cultures and mycoplasmas. *Tex Rep Biol Med 23:*285–303, 1965.
10. Hayflick, L. (Ed.) *The Mycoplasmatales and the L-phase of Bacteria.* New York: Appleton-Century-Crofts, 1969.
11. Hijmans, W., vanBoven, C. P. A., and Clasener, H. A. L. Fundamental biology of the L-phase of bacteria, pp. 67–143, in L. Hayflick (Ed.), *The Mycoplasmatales and the L-phase of Bacteria.* New York: Appleton-Century-Crofts, 1969.
12. Kundsin, R. B. Mycoplasma infections of the female genital tract, pp. 291–306, in M. L. Taymor and T. H. Green, Jr. (Eds.), *Progress in Gynecology, Vol. VI.* New York: Grune & Stratton, 1975.
13. Leon, O., and Panos, C. The adaptation of an osmotically fragile L-form of *Streptococcus pyogenes* to physiological osmotic conditions and its ability to destroy human heart cells in tissue culture. *Infect Immun 13:*252–262, 1976.
14. Madoff, S. Isolation and identification of PPLO. *Ann NY Acad Sci 79:*383–392, 1960.
15. Madoff, S., and Pachas, W. N. Mycoplasma and the L-forms of bacteria, pp. 195–217, in C. D. Graber (Ed.), *Rapid Diagnostic Methods in Medical Microbiology.* Baltimore: Williams & Wilkins Co., 1970.
16. McGee, Z. A., and Wittler, R. C. 1969. The role of L-phase and other wall-defective microbial variants in disease, pp. 697–720, in L. Hayflick (Ed.), *The Mycoplasmatales and the L-phase of Bacteria.* New York: Appleton-Century-Crofts, 1969.
17. Morton, H. E. Mycoplasmas from man with the undetermined specific relationships to their human host, pp. 147–171, in J. T. Sharp (Ed.), *The Role of Mycoplasmas and L-forms of Bacteria in Disease.* Springfield, Ill.: Charles C Thomas, 1970.
18. Purcell, R. H., Chanock, R. M., and Taylor-Robinson, D. 1969. Serology of the mycoplasmas of man, pp. 221–265, in L. Hayflick (Ed.), *Mycoplasmatales and the L-phase of Bacteria.* New York: Appleton-Century-Crofts, 1969.
19. Sharp, J. T. (Ed.) *The Role of Mycoplasmas and L-forms of Bacteria in Disease.* Springfield, Ill.: Charles C Thomas, 1970.
20. Shepard, M. C., and Howard, D. R. Identification of "T" mycoplasmas in primary agar cultures by means of a direct test for urease. *Ann NY Acad Sci 174:*809–819, 1970.
21. Shepard, M. C., and Lunceford, C. D. A differential agar medium (A 7) for identification of *Ureaplasma urealyticum* (human T mycoplasmas) in primary cultures of clinical material. *J Clin Microbiol 3:*613–625, 1976.
22. Smith, P. F. (Ed.) *The Biology of Mycoplasmas.* New York; Academic Press, 1971.

13

Significance of Virus Isolation and Identification and Virus Serology in the Care of Patients

R. GORDON DOUGLAS, JR.
CAROLINE B. HALL

Diagnostic virology laboratories presently suffer from a reputation earned some years previously that they are of little help in clinical medicine. How often have we heard physicians say, "Viral specimens can be obtained for interest, but they will not help the patient." This feeling is justified in many cases, since the diagnostic virology services available to the physician are often lengthy isolation procedures with a low yield of positive results, or serologic procedures, which, because of the requirement for convalescent sera, provide diagnostic information only in retrospect. As a result, a common remark by a house staff member is that he used to obtain viral specimens, but that he never received a positive result or any result, until months later when he had forgotten the patient. Another frequent complaint in this day of "cost-benefit awareness" is that viral identification is too expensive, especially when the clinical value is low.

We can argue, however, that the physician is missing a large number of accurate diagnoses if he ignores viral etiology. Respiratory viral infections alone account for close to 10% of all admissions for children under 10 years of age to Strong Memorial Hospital in Rochester, New York (4). In addition, viral infections of the central nervous system and infections with the herpes virus group in compromised hosts are important causes of hospital admission. Nosocomial infections with respiratory viruses, especially influenza, parainfluenza and respiratory syncytial, cause appreciable morbidity and prolonged hospitalization as well as occasional mortality (3, 10, 11). In addition, it should be noted that viral infections are even more common in nonhospitalized patients. Thus, viral infections are important causes of illness in terms of frequency and morbidity. Accurate diagnosis is important to avoid the expense and risk of unnecessary diagnostic tests looking for other conditions, and to prevent unwarranted and potentially toxic antimicrobial therapy.

In this chapter, we will not present detailed methods for isolation of common viruses for these methods are well presented in several standard references sources (5–7). Rather, we will present the rationale and goals for a virus diagnostic laboratory, and some views on how these goals may best be accomplished.

GOALS OF DIAGNOSTIC VIRAL LABORATORY

What then should be the goals of a viral diagnostic laboratory? Foremost, the viral laboratory should aid the primary physician in the care of his patients, much like the diagnostic bacteriology laboratory. This goal may be best attained by keeping in mind what information the physician would most need to help him in the care of his patients. Perhaps the two major questions that the physician asks the laboratory are: (1) Is this patient infected with a virus? (2) And, if so, what is the relationship between the viral infection and the patient's illness?

IMPORTANCE OF VIRAL ISOLATION

One of the major problems of viral diagnostic laboratories is providing physicians with results within a time period that is clinically useful. In most cases, the physician needs the information within the first few days after submitting the specimens, if it is to help in the management of his patient. For example, if the presence of a virus in a specimen can be quickly ascertained, antibiotic therapy may be avoided or discontinued before untoward effects occur.

In view of these goals, the optimal methods are viral isolation techniques or detection of viral antigens rather than serology. This emphasis is analogous to that placed on bacterial isolations rather than serology by most clinical microbiology laboratories. Diagnosis by viral isolation may be accomplished for many viruses within a suitably short time period if the tentative identification is based on the cytopathogenic effect (CPE) in selective cell cultures. The use of several cell lines often allows tentative identification of the virus by observing which cell lines will support its growth, much as the diagnostic bacteriology laboratory identifies bacteria by the use of selective media.

Many laboratories require rigorous identification and serotyping of viruses before releasing results. Although such identification is highly desirable from the epidemiologic point of view, it is not important for the clinician. A viral laboratory should be able to supply whatever information is available as soon as possible. For example, the report of an "hemadsorbing agent in monkey kidney cells," or "probably herpes simplex CPE" only a few days after inoculation of cultures is preferable to a report of "parainfluenza virus type 3" 2 or 3 weeks later. Such reports are analogous in many respects to the report of "a lactose fermenting gram negative rod" in blood cultures.

Respiratory syncytial virus (RSV) is the most important respiratory viral pathogen of young children. During the yearly epidemics of RSV infection in our community, RSV can be recovered from 89% of children hospitalized with lower respiratory tract disease (3). As shown in Figure

13.1, CPE characteristic of RSV was identifiable in HEp-2 cell cultures in 72% of cases by the 4th day after inoculation of nasal washes (3). This information has been helpful in the management of these infants, allowing antibiotic therapy to be stopped before troublesome side effects or bacterial superinfection occur.

Similarly, when influenza virus activity is suspected in the community, rhesus monkey kidney cell cultures may be hemadsorbed on the 3rd day after inoculation, rather than the more commonly employed 5th day. We have been able to detect hemadsorption on the 3rd or 4th day in 80% of cultures subsequently shown to be positive. A representative set of data is shown in Table 13.1, comparing frequency of detection of A/Victoria/75 H3N2 in rhesus monkey kidney cell cultures by hemadsorption on each of five different days.

NEWER METHODS

More recently, techniques have been developed that allow definite viral diagnosis within the first day after collection of specimens. Hopefully, this will be the new direction of most viral diagnostic laboratories. Techniques that currently appear to be most applicable are direct and indirect immu-

Figure 13.1

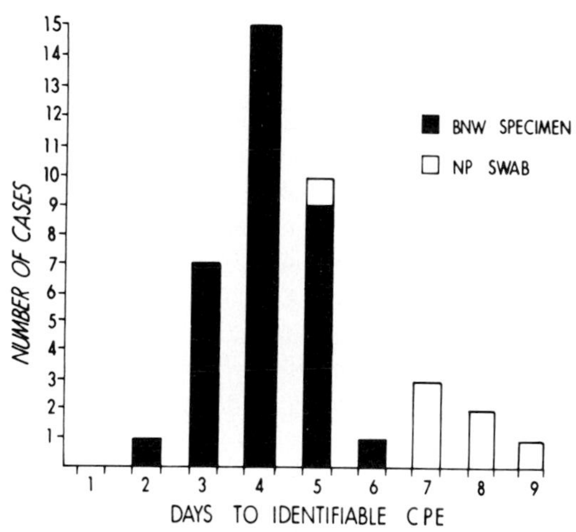

Comparison of number of days from inoculation to identifiable CPE by the bedside nasal wash (BNW) and nasopharyngeal (NP) swab methods.

TABLE 13.1
Numbers and cumulative percent of nasopharyngeal specimens containing influenza A/victoria/H3N2 virus exhibiting hemadsorption in rhesus monkey kidney cell cultures

Total No. of Specimens	No. of Specimens Positive on Indicated Day of Testing				
	3	4	5	6	10
60	32 53%	19 85%	4 92%	3 97%	2 100%

nofluorescence for the detection of viral antigen in specimens from patients or in recently inoculated cell cultures. Such techniques have been shown to be useful for RSV, influenza, and parainfluenza viruses (2, 8). Hopefully, radiometric assays for a variety of viral antigens will also become available. Electron microscopic examination of patient specimens, often combined with immunofluorescent techniques, are also now being explored for rapid diagnosis (1, 9).

SEROLOGIC TECHNIQUES

Classical serologic techniques have little place in a clinical viral diagnostic laboratory. Serologic techniques are certainly sensitive and accurate for making diagnoses of such viral illnesses as mumps, measles, adenovirus, influenza A and B, herpes simplex, cytomegalovirus, and many others. However, the delay required in obtaining a convalescent serum makes serology of little help clinically, and relegates it for the most part to an epidemiologic and research tool. On occasion, serologic testing is of help in supporting a diagnosis, or it may provide intellectual satisfaction in arriving at a correct diagnosis in retrospect, but such information is rarely of help in patient management.

COST

Few viral diagnostic laboratories show financial profit. Yet many physicians are hesitant to obtain specimens for viral diagnosis because of the costly bill incurred by the patient. Thus, confining cost must be a major goal. In part, this may be accomplished by knowing precisely what information the physician is requesting from each specimen. A physician may be only interested in knowing if a swab of the oral pharynx contains herpes simplex virus, or if a lesion of the pharynx is due to herpes simplex virus versus coxsackie virus. In these cases only one or two specific cell lines may be used, and the cultures kept for a shorter period of time than when searching for such viruses as cytomegalovirus. Specimens submitted for a "viral screen" are rarely, if ever, justified. Tailoring the laboratory work to answer specific questions helps keep costs minimal. Hopefully, this can also be reflected in differential charges to the patient, dependent on the amount of work performed.

CHOICE OF SPECIMEN AND ASSAY SYSTEM

Obtaining a basic amount of clinical information about the patient on the form accompanying submitted specimens also aids the laboratory in performing accurate diagnosis at minimal cost. This information should include the age of the patient and the pertinent diagnosis or acute symptoms, as well as such underlying conditions as immunocompromised states. Receiving a specimen from a patient with "kidney disease" is of little help in the laboratory. The patient might have acute hemorrhagic cystitis or be a renal transplant recipient with acute pneumonia. In the former patient, an adenovirus should be sought, whereas in the second patient a variety of respiratory pathogens for immunocompromised hosts should be sought. In Table 13.2, we have listed the viruses most frequently associated with the more common viral syndromes. Knowledge of the suspected pathogens leads to selection of the appropriate cell cultures.

Age also influences the number and types of viruses to be considered. Specimens from infants with viral pneumonia may contain respiratory syncytial virus, parainfluenza virus, adenovirus, or influenza virus, whereas those from an adult are most likely to contain influenza virus.

Collection of the correct specimen is also essential to the efficiency of the laboratory. The proper specimens for recovery of most of the common viruses are listed in Table 13.3. It may be necessary for laboratory personnel to telephone physicians asking for additional or properly collected specimens. For instance, the chance of detecting a virus from a patient with aseptic meningitis may be increased by obtaining specimens other than cerebrospinal fluid. Similarly, in paramyxovirus infection, a nasal wash or combined nose and throat swab specimen is preferable to a throat swab specimen (3). A urine for the identification of cytomegalovirus, if not freshly obtained, may be worthless. Many physicians are not familiar with the preferred specimen for each virus or clinical entities, nor with the preferred methods of storage. Laboratory personnel should be able to offer this information by telephone in addition to having printed sheets with this information available for clinicians.

Laboratory personnel are obligated to select the appropriate cell culture system following guidelines such as those listed in Table 13.4. Where more than one virus is sought, several cell lines may be necessary. However, appropriate use of the information in Tables 13.2–13.4 can keep to a minimum the number of cell lines and, therefore, reduce costs. Other factors, such as awareness of community outbreaks of RSV, influenza or enterovirus infections, offer similar guidance in the appropriate selection of methods and cell lines.

VIRAL IDENTIFICATION

Some comments concerning this subject have been made previously in reference to the rapidity of reporting results. In addition, final specific characterization of a viral isolate also may be costly and unnecessary. When a virus is recovered from a specimen, the physician usually will wish to know to which group or family of viruses the isolate belongs. Knowing the specific type of virus, such as an adenovirus type 1 or type 2, generally is of little additional value to the physician in the management of his patient. Typing can add extraordinary cost to the processing of a specimen, particularly in terms of personnel time, and reagents required. Typing might better be reserved for those cases in which the physician specifically requests further characterization of the isolate.

The following exceptions might be made when typing may add valuable clinical or epidemiological information. Viral agents that are apt to cause epidemics in the community should be typed when the first such isolates are obtained. As an example, determining the type of the first parainfluenza virus isolates that appear in the fall may allow clinically helpful

TABLE 13.2
Viruses associated with clinical syndromes

Syndrome	Associated viruses	
	Common	Less common
Respiratory tract:		
Pneumonia		
Child	Respiratory syncytial virus (RSV) Parainfluenza 3, 1, 2 Influenza A	Adenoviruses Influenza B
Adult	Influenza A	Coxsackie A Coxsackie B
Bronchiolitis	RSV Parainfluenza 3, 1, 2	Influenza A, B Adenoviruses
Croup	Parainfluenza A 1, 3, 2 RSV	Influenza A, B Adenoviruses
URI	Rhinovirus RSV Coronavirus Parainfluenza 1, 3 Adenovirus 1, 2, 3, 5, 14, 21	Influenza A, B Parainfluenza 2 Coxsackie A, B Echo
Pharyngitis	Herpes simplex Influenza A, B Parainfluenza 1, 2, 3 Rhinovirus EB Virus	Adenovirus
Central nervous system:		
Meningitis	Mumps Coxsackie A, B ECHO	Herpes simplex type 2
Encephalitis	Mumps Herpes simplex type 1	Arbovirus
Paralytic disease	Coxsackie A, B (Polio)	Echo
Myocarditis and pericarditis	Coxsackie B	Coxsackie A Influenza Parainfluenza Mumps
Exanthemata (nonspecific with fever)	Coxsackie A 9, 16	Echo 1–6, 14, 18, 19 Coxsackie A (2, 4, 23) Coxsackie B (1–5) Adenovirus
Nonspecific febrile illness	Coxsackie A, B ECHO Influenza A, B	Arbovirus
Gastroenteritis	Orbivirus Parvovirus Reovirus	? ECHO 11, 14
Urinary tract:		
Acute hemorrhagic cystitis		Adenovirus 2, 11 ECHO 9 Adenovirus 7
Orchitis and epididymitis	Mumps	Coxsackie B
Parotitis	Mumps	Parainfluenza 1, 3 Coxsackie A
Herpangina	Coxsackie A(1–6, 8, 10, 16, 22)	ECHO 9, 17 Coxsackie B
Hand, foot, mouth syndrome	Coxsackie A	
Lymphonodular pharyngitis	Coxsackie A(10)	
Herpes genitalis	Herpesvirus type 2	

TABLE 13.3
Clinical Specimen to be obtained for viral isolation*

Virus Group Suspected	Nasal Wash (PBS wash in VIB (2) or HBSS (3) with BA (4) or gel (5))	Combined Nose and Throat Swab (VIB, HBSS with BA or gel)	Stool (all media)	Rectal Swab (VIB, HBSS)	CSF (Sterile, all media)	Blood (all media)	Effusions (sterile, all media)	Urine (sterile, all media)	Skin eye lesions (VIB, HBSS, BA, or gel)	Autopsy Specimens Sterile, (all media)	
Enterovirus											*Viruses in group:*
Stable, refrig.	+	+	+	+	+	×	×			×	Coxsackie
Rhinovirus											ECHO, polio
Labile, obtain fresh	+	×									"H" and "M" strain
No freezing											
Myxoviruses											
Paramyxoviruses											Influenza A, B, parain-
Labile, obtain fresh	+	×				×	×	×		×	fluenza, mumps, RSV
No freezing											
Adenoviruses											
Stable, refrig.	+	+	+	+	+	×	×	×	×	×	
Herpesviruses											HSV
Labile, obtain fresh	+	+			(×)	×		+	+	+	Varicella-zoster, CMV
No freezing											Vaccinia (E B)
Rubella											
Labile	+	+	×	×	×	×		+	×	+	
Coronavirus	+	+									
Orbivirus			+								
Parvovirus			+								

* 1 = PBS—phosphate buffered saline, 2 = VIB—veal infusion broth, 3 = HBSS—Hanks balanced salt solution, 4 = BA—bovine albumin (0.5%), 5 = gelatin, + = best specimens, × = other appropriate specimens.

TABLE 13.4
Optimal system for isolation of common viruses

Virus Group	Virus Type	Suckling Mouse	Embryonated Egg	Cell Cultures				
				Primary monkey kidney	Primary human embryonic kidney	Primary rabbit kidney	Semicontinuous[a] human diploid	Contin- uous[b] hu man hete oploid
Picornavirus	Rhinovirus	0[c]	0	+	++	0	++	±
	Coxsackievirus A1–24	++	0	0 to ++	0 to ++	0	0 to ++	0 to +
	Coxsackievirus B1–6	++	0	++	++	0	+	++
	ECHO virus 1–32	0 to ++	0	++	++	0	++	0 to +
Myxovirus	Influenza A, B, C	0	++	0 to ++	+	0	0	0
Paramyxovirus	Parainfluenza 1, 2, 3, 4	0	0	++	+	0	±	±
	Respiratory syncytial	0	0	+	+	0	+	++
Adenovirus	Adenovirus types 1–31	0	0	± to +	++	0	+	+
Herpesvirus	Herpes simplex	0	++	0	++	++	++	0
	Varicella zoster	0	0	0	0	0	++	0
	Cytomegalovirus	0	0	0	0	0	++	0
	Epstein Barr virus	0	0	0	0	0	0	0
Rubella		0	0	+	0	0	0	0
Coronavirus		0	0	0	0	0	±[d]	0
Orbivirus[e]		0	0	0	0	0	0	0
Parvovirus[e]		0	0	0	0	0	0	0

[a] WI-38 cells.
[b] HEp-2 cells for RSV, KB, or Hela cells for adenovirus, picornavirus.
[c] 0 = Insensitive, ± = slightly sensitive, + = moderately sensitive, ++ = very sensitive.
[d] Optimal isolation system is human tracheal organ cultures.
[e] Not detected in cell cultures.
Adapted from R. G. Douglas, Jr.: The Laboratory Diagnosis of Viral and Mycoplasmal Infection, pp. 11- in V. Knight (Ed.), *Viral and Mycoplasmal Infections of the Respiratory Tract.* Philadelphia: Lea & Febig 1973.

predictions. Type 1 isolates would indicate an outbreak of croup would be likely to follow, whereas type 3 isolates often occur sporadically and are associated with lower respiratory tract infections affecting younger children. Initial influenza isolates should be typed for similar reasons. The complications, prophylaxis and treatment of influenza A infections are very different from those for influenza B. The specific typing of influenza A isolates is also important for defining the antigenic change or drift that has occurred from previous years, allowing predictions concerning the extent of morbidity that will be associated with the current strain. This is also clinically important in determining the adequacy of immunization by the currently available vaccines and in advising use of such prophylaxis as amantadine.

For the routine diagnosis, however, the laboratory may identify many viruses with reasonable accuracy by their characteristic CPE in selected cell cultures. The appearance and speed of growth of herpes simplex virus is so characteristic that the physician may often be called within 24–48 hr and told with reasonable accuracy that herpes simplex virus is present. Similarly, RSV and adenovirus cause such characteristic CPE as to be identified tentatively on that basis alone. In contrast to a research laboratory, a clinical diagnostic laboratory has the obligation of balancing accuracy with clinical usefulness.

SERVING COMMUNITY PHYSICIANS

The viral diagnostic laboratory should serve not only hospitalized but ambulatory patients. Too often the great majority of specimens submitted are from inpatients, although the prevalence and importance of viral illnesses are much greater in outpatients. However, specimens are infrequently submitted from private offices because of the difficulty in obtaining the correct transport media and in returning it promptly to the laboratory. This problem may be partially solved by providing private physicians with media to keep in their offices. Printed sheets, such as Tables 13.2 and 13.3, accompanying this media with information concerning correct specimens to be obtained for the various viruses and correct storage can be helpful.

Another function of a viral diagnostic laboratory for the community is local surveillance for viral diseases. This may be accomplished in several ways depending on the size of the community and its facilities. In a medical center, the diagnostic viral laboratory may serve to process specimens obtained through an ongoing program for surveillance of infectious diseases. In communities where such a program is not developed, the viral diagnostic laboratory should serve as a barometer of the activity of viral agents in the community. This may best be accomplished by asking one or two groups of practicing physicians to routinely submit specimens from patients with acute illnesses. These specimens should be processed without charge. In return, the laboratory's determination of the current trends in the community will allow it to both offer the clinician more information and to run more efficiently. Delineating the seasonal patterns of the viruses allows the laboratory reasonable predictions of the work load, cell lines, materials and personnel necessary. As an example, in Rochester, RSV has for the last few years, caused an epidemic of infection at twelve month intervals with a December debut. Thus, by the first of December, a special cell line (HEp-2) for RSV is obtained for the expected increase in respiratory specimens which usually lasts for two to three months.

VIRAL DIAGNOSTIC LABORATORY AS SOURCE OF PHYSICIAN EDUCATION

Physicians are generally less well versed in virology than in bacteriology. The clinician often relies upon the diagnostic viral laboratory to guide him in his search for a virus in his patient. First, he may not be sure which are the correct specimens to obtain from his patient. Secondly, he may expect that the laboratory will determine which viruses should be sought, if he simply tells him that the patient is a "child with pneumonia" or an "adult with a rash." Hence, the laboratory personnel must possess sufficient knowledge to give the physician these guidelines, and summary sheets containing much of this information, such as Tables 13.2 and 13.3, are often helpful.

Laboratory personnel are also asked to interpret the results for the physician. The physician may often ask not only if the patient is infected with a virus, but also what is its significance? This second question may only partially be answered by the laboratory. The virus group and its source is information that will help him in such a judgment. For example, an adenovirus recovered from the stool may be less significant than one isolated from tracheal secretions of the upper respiratory tract. A commonly related question is, "How often is a particular virus isolated from a normal person"? Much of this information is available in the literature, and the laboratory might benefit from having these references. Further interpretation, however, of the significance of a viral isolate is often out of the range of that which the laboratory personnel can or should offer. Further questions should be referred to a physician trained in clinical virology.

SUMMARY

Viral infections constitute man's most frequent afflictions. While some result in little morbidity, others result in hospitalization, permanent disability, or death. Accurate viral diagnosis is essential for therapeutic reasons only in a few instances. However, it is important to avoid the unnecessary expense and risk of other diagnostic procedures, and to terminate the use of antibiotics, many of which have toxic side effects or may lead to bacterial superinfection. Accurate viral diagnosis, to be performed as efficiently and cheaply as possible, requires knowledge of virology by physicians and some knowledge of clinical medicine by virology laboratory technicians. To accomplish these goals, emphasis must be placed on viral isolation with final typing and serology playing secondary roles. We have presented data to show that most RSV isolates can be recognized in 4 days, and influenza virus isolates in 3 days. In addition, our experience in viral diagnosis indicates that herpes simplex virus can commonly be detected in 1–2 days, and enterovirus isolates in 4–5 days. These time intervals are sufficiently short to accomplish the above stated goals. To be sure, more rapid methods are desirable, and it appears that wider application of immunofluorescence and radiometric methods will yield the desired shortening of the diagnostic interval.

Thus, in this time of rapid technical advances destined to bare the virion's inner chemical soul, the most effective viral diagnostic laboratory is the one that remains pragmatic, but not stagnant. It is the laboratory that most closely tries to match its goals to those of the clinician in the care of his patient with major or minor illnesses. It must overcome the heresy of its heritage by emphasizing the rapidity and usefulness of its diagnostic techniques.

Appendix: How to Begin a Viral Diagnostic Laboratory

It must be obvious that a fully functioning laboratory utilizing all of the systems shown in Table 13.4 requires space, equipment and trained personnel. In smaller hospitals, such facilities may not be available, and in larger hospitals one may need to begin with less than optimal facilities.

Our recommendations are to begin with a single technical person who should have an appropriate background in microbiology and who should receive additional training by taking a course in viral diagnosis at the Center for Disease Control in Atlanta or other appropriate institutions. The only facilities required would be a small room separate from the bacteriology laboratory, a desk-top hood, a microscope, an incubator set at +34 C capable of holding a rotating drum, and some −20 C and −70 C freezer space. Initially, cell lines should include only WI-38 and rhesus monkey kidney cells, both of which, and maintenance media, can be purchased commercially. Expansion of cell lines, equipment, space and personnel to the needs of the hospital can then take place at a pace consistent with utilization and interests of the institution.

LITERATURE CITED

1. Doane, F. W., Anderson, N., Zbitnew, A., and Rhodes, A. J. Application of electron microscopy to the diagnosis of virus infections. *Can Med Assoc J 100:*1043–1049, 1969.
2. Gardner, P. S., and McQuillin, J. *Rapid Virus Diagnosis Application of Immunofluorescence.* London: Butterworth and Co., 1974.
3. Hall, C. B., and Douglas, R. G., Jr. Clinically useful method for the isolation of respiratory syncytial virus. *J Infect Dis 131:*1–5, 1975.
4. Hall, C. B., and Douglas, R. G., Jr. Respiratory syncytial virus and influenza: Practical community surveillance. *Am J Dis Child 130:*615–620, 1976.
5. Hsiung, G. D. *Diagnostic Virology.* New Haven, Conn.: Yale University Press, 1973.
6. Lennette, E. H., and Schmidt, N. J. (Eds.) *Diagnostic Procedures for Viral and Rickettsial Infections,* Ed. 4. New York. American Public Health Association, Inc., 1969.
7. Lennette, E. H., Spaulding, E. H., and Truant, J. P. (Eds.) *Manual of Clinical Microbiology,* Ed. 2. Washington, D.C.: American Society for Microbiology, 1974.
8. Liu, C. Rapid diagnosis of human influenza infection from nasal smears by means of fluorescein-labeled antibody. *Proc Soc Exp Biol Med 92:*883–887, 1956.
9. Long, G. W., Noble, J. Jr., Murphy, F. A., Herrmann, K. L., and Lourie, B. Experience with electron microscopy in the differential diagnosis of smallpox. *Appl Microbiol 20:*497–504, 1970.
10. Louria, D. B., Blumenfeld, H. L., Ellis, J. T., Kilbourne, E. D., and Rogers, D. W. Studies in the pandemic of 1957–1958. II. Pulmonary complications of influenza. *J Clin Invest 38:*213–265, 1959.
11. Mufson, M. A., Mocega, H. E., and Krause, H. E. Acquisition of parainfluenza 3 virus infection by hospitalized children. I. Frequencies, rates and temporal data. *J Infect Dis 128:*141–147, 1973.

14

The Antibiotic Sensitivity Test. Variability. Interpretation. Rapid Versus Overnight. Agar Versus Broth

J. C. SHERRIS

When penicillin and streptomycin were first introduced into clinical practice, in vitro susceptibility tests, if they were done at all, were usually made by broth dilution procedures. Results were often reported both as the minimal inhibitory concentration of the antibiotic for the agent and as the relationship of the susceptibility of the organism to that of a control strain such as the Oxford Staphylococcus. As more antimicrobics were introduced and more resistant organisms were selected, these procedures became increasingly cumbersome and costly.

The development of diffusion testing procedures showed early promise of providing a satisfactory and inexpensive system for simultaneous testing of significant isolates with several antimicrobics, and commercially prepared dried filter paper discs containing various contents of the various antibiotics became available to the clinical laboratory. However, the apparent simplicity of the procedure was deceptive and a number of reports appeared of poor reproducibility between laboratories (12, 18, 26, 29, 35). This was not too surprising because many different procedures, media, and interpretive criteria were in use, and the actual disc content sometimes varied widely from labeled potency (8, 16). Poor reproducibility was not limited to diffusion tests and was also demonstrated between results of dilution procedures in different laboratories (9).

These problems led to an International Collaborative Study (ICS) to define further the variables influencing the results of susceptibility tests and to develop proposals for reference procedures (12). They also led to the adoption by the U. S. Food and Drug Administration (FDA) of a slightly modified "Kirby-Bauer" (4) diffusion procedure which was recommended for clinical laboratory use (1, 13, 14).

While progress in standardization of diffusion tests was being made, technical developments were simplifying dilution tests for routine use in the clinical laboratory. Inoculum replication devices, such as that of Steers et al. (34), were developed and permitted many different strains to be spot-inoculated simultaneously on to antimicrobic-containing agar plates. This procedure has been routinely used at the Mayo Clinic for many years (G. M. Needham, personal communication), is applicable to fairly large volume work, but is obviously dependent on an adequate level of technical precision and accuracy in preparing antibiotic dilutions of appropriate concentration and incorporating them in known volumes of agar. More recently, routine broth dilution procedures of increased practicality have been developed based on semiautomated versions (25, 38) of the microtitration procedure of Takatsy (36). A number of devices are now marketed for preparing dilutions of antimicrobic in microtitration plates, for inoculating the tests and for facilitating endpoint reading. Operating experience has shown that it is best to pre-prepare plates containing dilution of antimicrobics in broth, and store them at −20 C or below for up to a week before use (or sometimes longer depending on the stability of the antimicrobics used). As a result, new equipment has been developed for bulk distribution of selected antimicrobic dilutions in broth rather than for preparing each series of dilutions in the individual plates, and this should improve reproducibility. There is also developing interest now in the commercial preparation and distribution of pre-prepared frozen trays which require only inoculation.

Even more recently, there has been increasing interest in the possibility of more fully automated susceptibility test procedures based on the use of liquid media, various methods of measuring microbial growth, and with read-outs in less than 6 hr. One such piece of equipment has already been evaluated and is available on the market (37).

Unfortunately, we still have no nationally or internationally recognized reference dilution procedures, although the ICS proposed two in 1971 which have been widely used (12). Absence of such reference points is reflected in variations in results of different procedures and has complicated the approach to early-read methods. Some of the factors contributing to variations between different procedures and their influence on the selection of routine procedures for testing rapidly growing aerobic or facultative pathogens will constitute the rest of this chapter. Other aspects of susceptibility testing have recently been reviewed elsewhere and the reader is referred to these sources for a more comprehensive coverage (2, 21, 31).

FACTORS INFLUENCING QUANTITATIVE RESULTS OF SUSCEPTIBILITY TESTS

The quantitative results of broth and agar dilution procedures, and of diffusion tests, are influenced by differences in test conditions and reagents. The minimum inhibitory concentration (MIC) is thus a method dependent variable. The major factors influencing test results are shown in Table 14.1 and are briefly discussed below.

Differences of inoculum size influence the results of all types of tests to an extent which varies with different organism-antimicrobic combinations and is particularly marked when bacteria produce antibiotic inactivating enzymes. This is illustrated in Table 14.2 which shows the effect of inoculum variation on the ampicillin and cephalothin susceptibility of a number of different species. It will be noted that the greatest inoculum effect was seen with penicillinase-producing staphylococci and ampicillin.

Medium variation can also have marked impact on results. The influence of *p*-aminobenzoic acid concentration on sulfonamide susceptibility has long been known. Concentrations of thymidine above 0.01 μg/ml will reduce the antibacterial effect of both sulfonamides and trimethoprim (15). More subtle are the effects of differences in ionic content of medium. For example, divalent ions will chelate with tetracyclines and reduce their activity as illustrated in Table 14.3. Differences in both free calcium and free magnesium ions will effect the permeability of *Pseudomonas* cell membranes to gentamicin and tobramycin and result in MIC differences of up to 16-fold or of equivalent variations with diffusion tests (Table 14.4). Some effects of duration of incubation are shown in Table 14.5 and these become increasingly marked as intervals between readings are extended.

TABLE 14.1
Some factors influencing result of susceptibility tests

A. *All Methods*
- Inoculum size
- Medium constitution
- pH
- Atmosphere of incubation
- Incubation time
- Incubation temperature
- End point criteria

B. *Diffusion Tests*
- Disc content
- Medium depth
- Prediffusion time
- Boundaries of interpretative categories

TABLE 14.2
Increase in MIC with 100-fold increase of inoculum size in agar dilution technique*

Bacterial Strains	Antibiotic	-Fold Rise in MIC					
		0	2	4	8	16	>16
Enterococci (12)	Ampicillin	11†	1				
	Cephalothin	9	3				
Staphylococcus aureus (18)	Ampicillin	3	5		2	1	7
	Cephalothin	7	11				
Enterobacter-Klebsiella (10)	Ampicillin	2	2	6			
	Cephalothin	1	8	1			
Escherichia coli (10)	Ampicillin	6	4				
	Cephalothin	1	7	2			
Haemophilus (12)	Ampicillin	12					
	Cephalothin	12					

* Inocula = ca. 10^8 and 10^6 organisms/ml in replicator cups.
† Number of strains.
From J. C. Sherris et al. (33), reproduced with permission of the New York Academy of Sciences.

TABLE 14.3
Effect of supplementation with magnesium and citrate on broth diltuion MICs of tetracycline against S. aureus

Supplement	Trypticase Soy MIC	Mueller-Hinton MIC
	μg/ml	*μg/ml*
0	5.0	1.25
5×10^{-3} M $MgCl_2$	10.0	2.5
5×10^{-3} M $Na_3C_6H_5O_7$	0.63	0.63
$MgCl_2 + Na_3C_6H_5O_7$	2.5	1.25

From V. C. Brenner and J. C. Sherris (10), reproduced with permission of the American Society for Microbiology.

TABLE 14.4
Gentamicin MICs and zone diameters of *Pseudomonas aeruginosa* tested on Mueller-Hinton agar containing different concentrations of magnesium

Medium Content of magnesium	Zone Diameter (10-μg Disc)	Agar Dilution MIC
mg/liter	*mm*	*μg/ml*
10	21.5	1
25	16.9	4
200	13.6	8

L. B. Reller et al. (28).

TABLE 14.5
Increase of ampicillin and cephalothin MICs for 18 bacterial strains* between 12 and 24 hours of incubation

Antibiotic	Test Method	-Fold Increase in MIC		
		0	2	4
Ampicillin	Broth dilution	5†	9	4
	Agar dilution	13	5	0
Cephalothin	Broth diltuion	8	9	1
	Agar dilution	14	4	0

* Strains tested were 8 *Stapylococcus aureus*, 4 Enterococci, 4 *Escherichia coli*, and 2 *Enterobacter*.
† Number of strains.
From J. C. Sherris et al. (33), reproduced with permission of the New York Academy of Sciences.

There may also be considerable differences between results of agar or broth dilution tests. An example is shown in Table 14.6, in which broth dilution cefamandole MICs of *Enterobacter* were consistently higher than agar dilution results. In this instance the discrepancy is due to a high mutation rate to resistance: a single mutant will yield full growth in a broth concentration which inhibits the rest of wild type populations. In an agar dilution procedure it will only represent a single colony if it is present in the inoculum.

These, and the other variables illustrated in Table 14.1, can interact and increase the interlaboratory variability between the results of different dilution test methods. This does not imply that any particular procedure is "wrong," it is simply an inevitable concomitant of the absence of agreed technology or reference procedures.

Even when procedural details are agreed, interlaboratory reproducibility may still be considerable. Table 14.7 compares the results of dilution susceptibility tests performed with the same strains, reagents, media, and protocol in different laboratories as part of the ICS study (12). There were still some considerable variations between results of some individual laboratories reflecting, presumably, differences in endpoint judgment or in minor uncontrolled technical factors. Such differences should have been, at least partially, eliminated by the use of stable control strains of established performance and this approach has been adopted for the FDA standardized diffusion test.

One cause of variability which requires further resolution is that between media of the same name produced by different manufacturers or, sometimes, between batches produced by the same manufacturer. This is illustrated in Tables 14.8 and 14.9 with the products of two manufacturers of Mueller-Hinton medium; a more extended study (J. C. Sherris and C. Thornsberry, unpublished observations) has shown greater variability when the products of more recent producers are included. There is, thus, a

TABLE 14.6
Influence of method and inocula on cefamandole dilution susceptibility test results with four strains of *Enterobacter*

Enterobacter	Agar Dilution MIC (μg/ml) Inoculum Size/Spot		Broth Dilution MIC (μg/ml) Inoculum Size/ml	
	10^4	10^5	10^4	10^5
E. cloacae #2	1	2	16	64 (8)
E. aerogenes #4	2	4	2	16 (4)
E. aerogenes #8	1	1	4	32 (1, 4, 8)
E. liquefaciens #11	4	16	16	64

Figures in parentheses show the antibiotic concentrations in "skipped" tubes which failed to grow in the broth dilution series.

From C. M. Findell and J. C. Sherris: *Antimicrobial Agents & Chemotherapy, 9:* 970, 1976; reproduced with permission of the American Society for Microbiology.

TABLE 14.7
Reproducibility of dilution susceptibility tests performed in 15 different laboratories with same reagents, organisms, and procedures

Test	Ratio of Results from Individual Laboratories to Mode of Results from All Laboratories										
	<0.06	0.06	0.13	0.25	0.05	Mode	2	4	8	16	>16
Broth dilution	2	1	8	31	107	242	74	29	9	5	1
Agar dilution	1	0	0	18	96	266	72	34	11	2	4

Approximately 95% of test results were within ±2 dilutions of the mode.
H. M. Ericsson and J. C. Sherris (12).

TABLE 14.8
Differences in zone diameters on six batches of Mueller-Hinton agar

Antibiotic	*Staphylococcus aureus*		*Escherichia coli*	
	*k**	lmd†	*k*	lmd
Penicillin	1.5	1.9	Not done	Not done
Methicillin	0.9	1.7	Not done	Not done
Cephalothin	1.3	1.8	1.3	2.4
Streptomycin	0.7	1.6	1.2	1.6
Kanamycin	0.9	0.9	0.9	1.9
Erythromycin	1.0	0.9	Not done	Not done
Tetracycline	1.2	4.0	1.2	5.6
Ampicillin	Not done	Not done	2.0	2.5
Polymyxin B	Not done	Not done	0.7	2.4

* The *k* value represents the minimum difference in zone diameters (in millimeters) between any two batches required for statistical significance.

† The largest average difference that occurred among the six batches.

From V. C. Brenner and J. C. Sherris (10), reproduced with permission of the American Society for Microbiology.

TABLE 14.9
Differences in zone diameters between batches of Mueller-Hinton agar from manufacturers A and B

Organism and Drug	Manufacturer		
	A (3 batches)	B (3 batches)	A and B (6 batches)
Staphylococcus aureus			
Tetracycline	3.2*	1.2	4.0
Escherichia coli			
Cephalothin	0.5	1.0	2.4
Tetracycline	2.8	1.0	5.6
Ampicillin	1.0	0.9	2.5
Polymyxin B	0.9	1.5	2.4

* The largest mean difference between zone diameters on different batches.

From V. C. Brenner and J. C. Sherris (10), reproduced with permission of the American Society for Microbiology.

rather pressing need for the establishment of agreed performance standards which can be applied to the products of different manufacturers.

SOME ASPECTS OF INTERPRETATION OF SUSCEPTIBILITY TESTS

Successful chemotherapy depends on many factors, including clinical pharmacological characteristics of the antimicrobic, dosage, tissue levels, access to the site of infection, interaction with host defenses, and the inherent susceptibility of the infecting organism. The latter can be measured in the laboratory with reasonable levels of reproducibility if standardized procedures are used.

The interpretation of susceptibility data on a particular strain causing an infection has been based on three main factors, in addition to those mentioned above.

1. Relationship of MIC of Organism to Levels Attained in Blood (*or Other Tissue Fluid*) *on a Particular Dose Schedule*. This has to be recognized as an incomplete model of the in vivo situation because of the relative artificiality of culture media and conditions, the method dependency of the MIC, the interaction (or lack of it) of host defenses, and the question (which probably varies for different antimicrobic organism combinations and clinical situations) of whether peak, valley, or mean levels are most significant and of the extent to which protein binding should be taken into account. Despite these uncertainties, blood level to MIC relationships have been widely regarded as useful guides in chemotherapy, although attempts to develop precise and generally applicable mathematical predictive relationships between blood level and MIC are certainly not justified in view of the complexities of the parameters involved.

2. Relationship of Susceptibility of Infecting Strain to Particular Antimicrobic to That of Other Members of Same Species. When each new antimicrobic was first introduced, the response of particular species was generally quite homogeneous, and susceptibility of the species could be expressed as a normal distriubtion curve covering a fairly narrow range of susceptibilities. As recombinants, mutants, or the rare pre-existing naturally resistant strains were selected out by chemotherapy, a second population of greater resistance appeared. In many cases this was quite well separated in degree of susceptibility from the wild type and thus a bimodal distribution of susceptibilities developed among many of the common pathogenic species such as those illustrated for *Staphylococcus aureus* in Figure 14.1. When this has occurred among species which had earlier been shown to respond well to chemotherapy with the particular antimicrobic (e.g., *S. aureus* to penicillin or *Salmonella typhi* to chloramphenicol), there has been a generally good correlation between test results allocating organisms to the resistant population and loss of reliable clinical response.

3. Clinical Experience. Clinical experience with particular species and infections influences the selection of antimicrobics among those to which the organism may appear to be susceptible by both the above two criteria. Thus, treatment of subacute bacterial endocarditis with bacteriostatic agents is ineffective because of lack of access of the infected site to immune mechanisms, and typhoid fever responds poorly to tetracycline despite the in vitro susceptibility of the species by criteria 1 and 2 above.

Because of the complexities and the semiquantitative nature of many diffusion test procedures, broad categorizations of susceptibility have been widely used for reporting test results with organisms being allocated to "sensitive," "resistant," and to one or more intermediate groups. In the case of the FDA and Kirby-Bauer (4) procedures, three groupings have been used, with the intermediate range serving as both an equivocal technical buffer zone to prevent or reduce the risk of a resistant organism being allocated to the sensitive category or vice-versa, and to indicate that the MIC may sometimes be attained in sites where the antimicrobic is

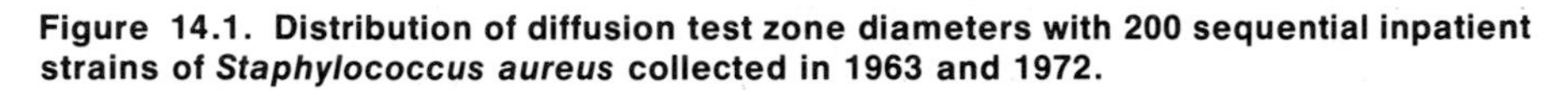

Figure 14.1. Distribution of diffusion test zone diameters with 200 sequential inpatient strains of *Staphylococcus aureus* collected in 1963 and 1972.

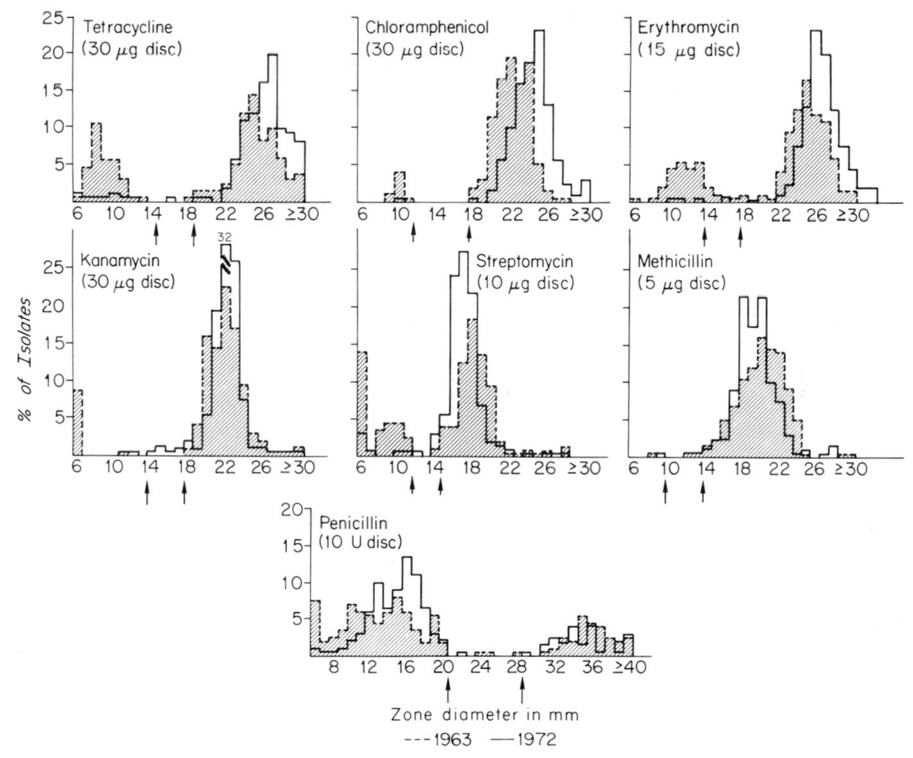

Zones between the *arrows* fall into the intermediate category. Those *to the right* are sensitive and those *to the left* resistant.

concentrated or in blood with unusually high dosage of nontoxic agents. These breakpoints are derived first from regression line studies relating diffusion test zone diameters to MICs, and to in vivo levels on "usual" dosage (see 1 above), and they are then tested against population distributions as described in 2 above (and refined if necessary). Their limits, thus, involve best judgement decisions whose reliability in terms of avoiding major errors is increased by the use of the intermediate range.

Some workers have objected to the complexities introduced by an intermediate range (6). However, the problems posed by its absence are well illustrated in a recent study comparing dilution test results with those of both the diffusion test and of a new automated apparatus which uses a single concentration of antimicrobic. The automated method is based on light scattering measurements and has an inevitably narrow intermediate range. In this study, MIC values for susceptible, resistant and intermediate categories were allocated to the dilution procedure by the collaborators in the study. As shown in Table 14.10, a significantly higher proportion of major ($S \rightarrow R$ or $R \rightarrow S$) discrepancies in relation to the dilution test occurred with the automated equipment than with the diffusion test, largely due to the absence of a sufficient intermediate or equivocal range

TABLE 14.10
Comparison of results of automated system and of diffusion tests when tested against dilution procedure

Technique	No. of Observations	Complete Agreement	Discrepancy*		
			V. Major	Major	Minor
		%	%	%	%
Automated vs. dilution	6166	90.0	3.1	1.2	5.7
Diffusion vs. dilution	6286	91.1	1.1	0.9	7.0

* V. major = dilution R : other test S; major = dilution S : other test R; minor = one test in intermediate range.
C. Thornsberry et al. (37).

with the automated procedure. Thus, small day to day variations can result in large interpretive differences. The lack of an intermediate range for gentamicin in the FDA diffusion test is also causing some problems in that strains of enterococci and some variants of *Escherichia coli* which have MICs of 8–16 μg may give zones in the sensitive range. The clinical microbiology laboratories at the University of Washington are now using an intermediate range of 13–16 mm for gentamicin to mitigate this problem and changes in the FDA recommendations appear to be indicated.

Recently, the concept of an intermediate range has been approached statistically. T. L. Gavan (personal communication) has proposed that the range should be selected to reduce to a specified and very low level the chance of a major error occurring with an organism whose true susceptibility is at a breakpoint. This would be calculated from the experimentally determined reproducibility of the method. Metzler and DeHaan (27) have proposed that breakpoints be derived by establishing maximum permissible error rates in comparisons of MIC breakpoints and zone size values. Their approach assumes the correctness of the MIC determination, an assumption which is only valid if a standardized procedure with several replicates is used for the MIC readngs. This approach would also require developing criteria for the selection of a representative population of organisms for testing.

Another method of data presentation is possible which would provide the clinician with all the microbiological parameters needed for interpreting in vitro susceptibility results (32). This depends on the collection of statistical data on the results of susceptibility tests of the various commonly encountered species which would be printed on charts relating them to MICs or MIC equivalents of zone diameters from regression line studies. The susceptibility of the strain under test would be indicated on the chart together with a bar showing ± 2 S.D. as determined from experiments with control strains. Recommended breakpoints could also be included. An example of such a report is shown in Figure 14.2 for strains of *E. coli* and *S. aureus*. Computer data handling could provide such charts automatically, and the clinician would know, (a) the approximate MIC of the strain (or the MIC if a dilution test were used) and the variability of the method, (b) the relationship of the susceptibility of the test strain to that of other

Figure 14.2. Possible strip chart reports of susceptibility test results.

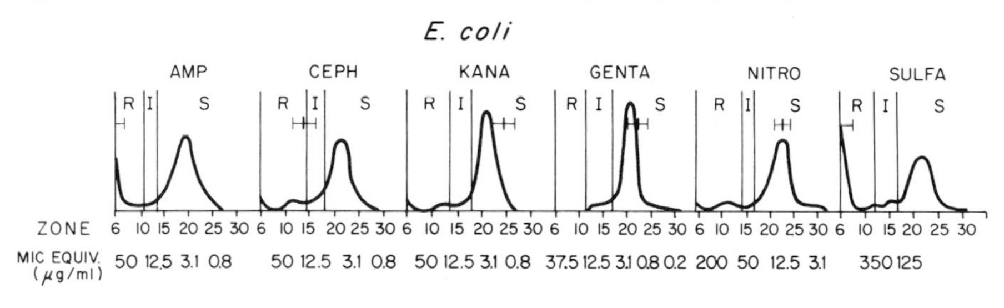

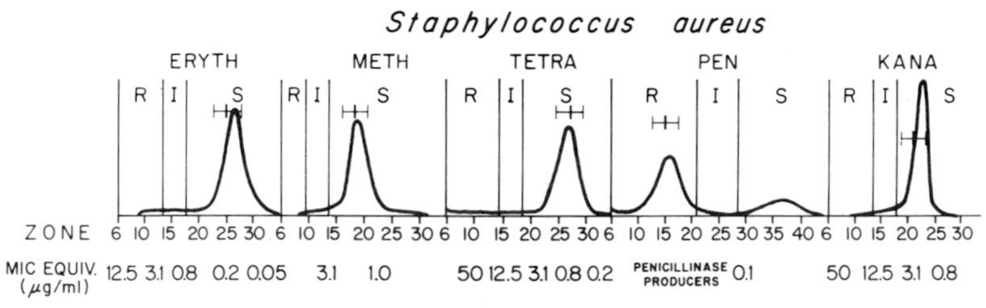

Distribution of susceptibilities of recent isolates of the same species. ⊢⊣Shows results with the strain under test ±2 S.D.

members of the species, and (c) the relationship of the susceptibility of the strain to recommended clinical categories. This would provide a rather complete and visual report and would appear to have value beyond the reporting methods now generally used.

ADVANTAGES AND DISADVANTAGES OF BROTH DILUTION, AGAR DILUTION, AND AGAR DIFFUSION PROCEDURES

Among traditional "overnight" susceptibility testing procedures, the choice involves agar or broth dilution methods or the agar diffusion procedure. Each has certain advantages and disadvantages for particular laboratories, based both on inherent properties of the test systems and on logistical considerations. Table 14.11 compares a number of these, some of which merit further comment.

The absence of agreed standard or reference dilution procedures constitutes a disadvantage because of the method dependency of results. Many workers use the proposed reference procedures of the ICS (12) or methods which have been shown to yield closely comparable results; nevertheless, there is need for formal agreement.

Agar diffusion procedures have the disadvantage of being only indirectly referable to MICs through experimentally derived regression equations, and the accuracy of the correlation is reduced when organisms of relatively slow growth rate are tested. In this country few laboratories have reported extrapolated MICs, and most have used qualitative categorizations of susceptibility which limit the information available, especially in problem cases.

The need to prepare accurate antibiotic stock solutions and dilutions in

TABLE 14.11
Relative advantages and disadvantages of agar dilution, broth dilution and agar diffusion susceptibility testing procedures

	Agar Dilution	Broth Dilution	Agar Diffusion
Agreed standardized procedure available	−	−	+
Yield MIC readings directly over wide range	+	+	−
No antimicrobic dilutions needed	−	−*	+
Applicable to sulfonamide (and TMP)	+	−	+
Special incubation conditions for testing (e.g., methicillin) most easily build into routine	+	−	−
Uninfluenced by agar	−	+	−
Uninfluenced by diffusion rates	+	+	−
Undesirably low Mg^{++} and Ca^{++} of many media	−	+	−
Results influenced by high mutation rates	−	+	−
Antimicrobics panel easily individualized	−	−	+
Applicable to urgent direct susceptibility tests	−	−	+

* Pre-prepared plates for the microtitration procedure available.

agar or broth involves greater technical expertise than the application of discs, and requires that a laboratory have more than one individual with the ability to perform and quality control these operations reliably. The development of pre-prepared frozen microtitration plates or plastic tubes containing lyophilized antibiotic in appropriate concentrations could reduce or eliminate this problem and make the use of broth dilution tests feasible for the smaller laboratory.

Sulfonamide susceptibility tests are unsatisfactory with routine broth dilution methods because of presence of sulfonamide inhibitors which result in much greater endpoint reading difficulty than is the case with agar dilution or diffusion tests. With the latter, endpoints are not sharp because organisms grow through several generations before being inhibited; nevertheless, an 80% inhibition endpoint is reasonably easy to read and results have been shown to be clinically useful in urinary tract infections (5).

Agar dilution tests have the advantage of facilitating incubation of plates containing particular antimicrobics under specialized conditions. For example, methicillin or oxacillin plates can be incubated at 30 C, which is optimal for detection of "methicillin-resistant" *S. aureus*. Similarly, sulfonamide containing plates can have thymidinase or lysed horse blood incorporated to reduce the concentrations of the inhibitory substrate.

Agar itself introduces some reactive components into the medium and is subject to some variability. It often contributes substantial amounts of divalent cations which influence the activity of the tetracyclines and of the aminoglycosides, especially against *Pseudomonas*. The divalent cation content of most agar media is actually closer to physiological values than is that of most broths. However, concentrations can be readily adjusted in broth and performance standards could be developed with those antibiotics specially effected by cation variation.

As indicated previously, broth dilution MICs may be substantially higher than agar dilution results with organisms with a relatively high mutation rate to resistance.

Finally, the agar diffusion test has the virtues of simplicity, easy individualization of antimicrobics tested, and easy applicability to urgent direct susceptibility tests to provide preliminary and tentative results.

At present, the FDA agar diffusion procedure is the best defined and established method, and well characterized strains for quality control are available. A considerable amount of data on interlaboratory reproducibility is being developed with this method through proficiency testing surveys. It, thus, probably remains the method of choice for the majority of smaller laboratories at present. In the future, however, it seems probable that some type of broth dilution procedure will become increasingly commonly used, because liquid media can contain fewer undefined components than agar media and lend themselves to automated or mechanized distribution, inoculation and read-out. There is, however, need for agreement on broth dilution procedures and for more data on interlaboratory reproducibility. Even with the development of highly practical broth dilution methods, it is probable that the agar diffusion method will continue to be used for such purposes as sulfonamide testing, testing supplementary antimicrobics and for direct tests on clinical material in cases of special urgency.

RELATIONSHIP OF EARLY-READ SUSCEPTIBILITY TEST RESULTS TO OVERNIGHT READINGS

Diffusion Tests

The factors determining the position of zones of inhibition have been extensively studied by Cooper and Gillespie (11), who defined a "critical time" occurring usually within the first few hours at which growth beyond a certain concentration of antimicrobic will continue despite higher concentrations that diffuse out later. This suggests that early readings will correspond to overnight readings; however, the principle is not universally applicable because growth of penicillin-susceptible organisms at the margins of zones may lyse in higher concentrations, and Rosenblatt and Schoenknecht (30) have shown that early growth of *E. coli* round aminoglycoside discs under anaerobic conditions will be inhibited if the plates are then transferred to aerobic conditions. Pragmatically, it has been shown by Barry et al. (3), by Kluge (19), and by Liberman and Robertson (22) that 5–8-hr readings give better than 85% or more correspondence of susceptibility category with overnight results and can yield useful tentative results. Boyle et al. (7) modified the FDA diffusion procedure by spraying with tetrazolium after 6–7-hr incubation. They found excellent correspondence in terms of zone diameters in comparison with 18-hr tests. Thus, readings of diffusion tests at approximately 6 hr appear to be reliable with their method, although they would obviously involve certain logistic problems if used routinely. In cases of clinical emergency, however, early read diffusion tests can provide valuable data.

Dilution Tests

Early readings of dilution tests bear a more complex relationship to overnight results because a "critical time" is not involved, and in many cases amounts of antimicrobic substantially below the overnight MIC will cause a prolonged inhibition of growth or even killing of the inoculum before the survivors grow out. As a consequence, MIC readings at 3–8 hr, whether measured visually or by automatic particle counting, are generally below those at 18 hr, sometimes with a difference of 32-fold or more (20), and this applies with many different organism antimicrobic combinations. The phenomenon is illustrated in the data shown in Table 14.12. There are a number of possible explanations for these differences including (a) enzymic inactivation of antimicrobic to subinhibitory concentrations by the inoculum, (b) spontaneous inactivation of antimicrobic, (c) adsorption of antimicrobic to the inocula, (d) the growing out of a small proportion of the inoculum which has higher resistance, or (e) more nebulously, initial inhibition followed by physiological reorganization in the organism permitting subsequent growth. Whatever the explanation, the phenomenon poses difficulties in developing early read systems which will yield results which are comparable to overnight procedures. Some automated procedures (37) have attempted to bypass the problem by pragmatically reducing the amount of antimicrobic in the test system to a level which gives the best possible correlation to a traditional overnight test. This has reduced the number of major qualitative discrepancies, but does not appear likely to fully resolve the difficulty.

We investigated the possibility that increasing the inoculum for an early read test might lead to better correspondence with overnight readings (20). Some results are presented in Table 14.13 representing tests with several species and antimicrobics. Comparing these with the results given in Table 14.12, it can be seen that much closer correspondence between early and late test readings was obtained when the inoculum for the early read test was increased by 100-fold. This appears to be an approach which merits further exploration with automated equipment and which may prove more

TABLE 14.12
Relationship of 3-hour MIC readings to overnight readings using ICS broth dilution procedure

	Ratio of 18-Hour to 3-Hour Readings						
	1	2	4	8	16	32	64
Both readings within range tested	10*	18	7	1	2	1	0
18-Hour MIC above range tested	—	6	2	1	1	0	0
3-Hour MIC below range tested	—	0	3	2	1	1	1

* Number of results.

From M. F. Lampe et al. (20), reproduced with permission of the American Society for Microbiology.

TABLE 14.13
Relationship of 3-hour MIC readings with inocula of 10^7 per ml to overnight readings with inocula of 10^5 per ml

0.03	0.125	0.25	0.5	1	2	4	8
–	2	5	20	32	9	2	1

From M. F. Lampe et al. (20), reproduced with permission of the American Society for Microbiology.

comparable to traditional methods than manipulation of antibiotic concentrations.

The question may be posed as to why the overnight MIC should be used as the aiming point for tests designed to be read in 3–5 hr, especially when it is well documented that concentrations of antimicrobic below the 18-hr MIC level may cause inhibition for several hours as well as striking morphologic changes with many organisms (17, 23, 24). One reason is that essentially all clinical laboratory correlations, whether determined experimentally or through accumulated clinical experience, have been made with overnight tests. Furthermore, in a number of instances, variants with specific resistance determinants may not be easily differentiated from sensitive strains in early read broth dilution tests with inocula of 10^5–10^6. This is often the case with *S. aureus* resistant to penicillin, ampicillin, methicillin, chloramphenicol, and erythromycin, and also with some types of resistance among Gram-negative organisms which are mediated by enzymic inactivation of antibiotics. Thus, until more information is available, it seems both rational and conservative to continue to use the overnight MIC result as the yardstick against which other procedures are judged.

CONCLUSION

There has been considerable technical progress in both the methodology and understanding of in vitro susceptibility tests in the past 10–15 years. There remain, however, several areas for further development. Probably the major technical needs are for agreed reference dilution procedures, and performance standards for their use, so that "minimum inhibitory concentration" has some absolute meaning, and results from one laboratory may be compared with more confidence with those from others. The acceptance of such standards would be of considerable help and encouragement to manufacturers in the development of new methods since an established basis for comparison would exist.

There is also need for better clinical laboratory correlative studies for which technical standardization is an essential prerequisite. The predictive value of susceptibility testing has been well demonstrated in a number of situations, for example, in the chemotherapy of tuberculosis, the relationship of gonococcal susceptibility to treatment failures, and with sulfonamide treatment of urinary tract infections. In many cases, however, the evidence is indirect or based on therapeutic experience, although treatment failures with species that have acquired resistant determinants are sufficiently common that their clinical significance seems clear. Certainly, trials of the clinical efficacy of established agents in infections due to organisms showing in vitro resistance could not now be done in man without extensive animal experimental evidence of a dissociation between in vitro and in vivo results.

It seems probable that future developments will involve relatively simple automated or mechanized systems using liquid media which will provide readings in 3 or 4 hr. These hold the promise of greater objectivity and of making reliable information available to the clinician at a time that may be of special benefit to the seriously ill patient.

LITERATURE CITED

1. Approved Standard ASM 2. Performance standards for antimicrobial disc susceptibility tests. Committee for Clinical Laboratory Standards, Villanova, 1975.
2. Balows, A. (Ed.) *Current Techniques for Antibiotic Susceptiility Testing*. Springfield, Ill.: Charles C Thomas, 1974.
3. Barry, A. L., Joyce, L. J., Adams, A. P., and Benner, E. J. Rapid determination of antimicrobial susceptibility for urgent clinical situations. *Am J Clin Pathol 59:*693–699, 1973.
4. Bauer, A. W., Kirby, W. M. M., Sherris, J. C., and Turck, M. Antibiotic susceptibility testing by a standardized single disk method. *Am J Clin Pathol 45:*493–496, 1966.
5. Bauer, A. W., and Sherris, J. C. The determination of sulfonamide susceptibility of bacteria. *Chemotherapia 9:*1–19, 1964.
6. Bell, S. M. The CDS disc method of antibiotic sensitivity testing (calibrated dichotomous sensitivity test). *Pathology* 7(Suppl)*:*1–48, 1975.
7. Boyle, V. J., Fancher, M. E., and Ross, R. W. Rapid modified Kirby-Bauer susceptibility test with single high concentration antimicrobial discs. *Antimicrob Agents Chemother 3:*418–424, 1973.
8. Branch, A., Starkey, D. H., Power, E. E., and Greenberg, L. Problems of standardization of manufactured dry penicillin discs, pp. 898–905, in *Antibiotics Annual 1956–57*. New York: Antibiotica, Inc., 1957.
9. Branch, A., Starkey, D. H., and Power, E. E. Diversification in the tube dilution test for antibiotic sensitivity of microorganisms. *Appl Microbiol 13:*469–472, 1965.
10. Brenner, V. C., and Sherris, J. C. Influence of different media and bloods on the results of diffusion antibiotic susceptibility tests. *Antimicrob Agents Chemother 1:*116–122, 1972.
11. Cooper, K. E., and Gillespie, W. A. The influence of temperature on streptomycin inhibition zones in agar cultures. *J Gen Microbiol 7:*1–7, 1952.
12. Ericsson, H. M., and Sherris, J. C. Antibiotic sensitivity testing. Report of an International Collaborative Study. *Acta Pathol Microbiol Scand Sect B, Suppl 217,* 1971.
13. Federal Register. Rules and regulations. Antibiotic susceptibility discs. *Fed Regist 37:*20525–20529, 1972.
14. Federal Register. Rules and regulations. Antibiotic susceptibility discs: correction. *Fed Regist 38:*2576, 1973.
15. Ferone, R., Bushby, S. R. M., Burchall, J. J., Moore, W. D., and Smith, D. Identification of Harper-Cawstor factor as thymidine phosphorylase and removal from media of substances interfering with susceptibility testing to sulfonamides and diaminopyrimidines. *Antimicrob Agents Chemother 7:*91–98, 1975.
16. Greenberg, L., Fitzpatrick, K. M., and Branch, A. The status of the antibiotic disc in Canada. *Can Med Assoc J 76:*194–198, 1957.
17. Greenwood, D., and O'Grady, F. Scanning electron microscopy of *Staphylococcus aureus* exposed to some common antistaphylococcal agents. *J Gen Microbiol 70:*263–270, 1972.
18. Hoffman, R. V., Jr., Jackson, G. G., and Turner, M. P. Reliability of antibiotic sensitivity tests as determined by a survey study. *J Lab Clin Med 51:*873–882, 1958.
19. Kluge, R. M. Accuracy of Kirby-Bauer susceptibility tests read at 4, 8, and 12 hours of incubation: comparisons with readings at 18 to 20 hours. *Antimicrob Agents Chemother 8:*139–145, 1975.
20. Lampe, M. F., Aitken, C. L., Dennis, P. G., Forsythe, P. S., Patrick, K. E., Schoenknecht, F. D., and Sherris, J. C. Relationship of early readings of minimal inhibitory

concentrations to the results of overnight tests. *Antimicrob Agents Chemother 8:*429–433, 1975.

21. Lennette, E. H., Spaulding, E. H., and Truant, J. P. (Eds.) *Manual of Clinical Microbiology,* ed. 2. Washington, D.C.: American Society of Microbiology, 1974.
22. Liberman, D. F., and Robertson, R. G. Evaluation of a rapid Bauer-Kirby antibiotic susceptibility determination. *Antimicrob Agents Chemother 7:*250–255, 1975.
23. Lorian, V. Some effects of subinhibitory concentrations of penicillin on the structure and division of staphylococci. *Antimicrob Agents Chemother 7:*864–870, 1975.
24. Lorian, V., Sabath, L. D., and Simionescu, M. Decrease in ribosomal density of *Proteus mirabilis* exposed to sub-inhibitory concentrations of ampicillin or cephalothin. *Proc Soc Exp Biol Med 149:*731–735, 1975.
25. MacLowry, J. D., Jaqua, M. J., and Selepak, S. T. Detailed methodology and implementation of a semi-automated serial dilution microtechnique for antimicrobial susceptibility testing. *Appl Microbiol 20:*46–53, 1970.
26. McCracken, L. M., and Palmer, P. H. Antibiotic sensitivity testing amongst the South Island laboratories; a survey. *NZ Med J 70:*390–393, 1969.
27. Metzler, C. M., and DeHaan, R. M. Susceptibility tests of anaerobic bacteria; statistical and clinical considerations. *J Infect Dis 130:*588–594, 1974.
28. Reller, L. B., Schoenknecht, F. D., Kenny, M. A., and Sherris, J. C. Antibiotic susceptibility testing of *Pseudomonas aeroginosa*. Selection of a control strain and criteria for magnesium and calcium content in media. *J Infect Dis 130:*454–463, 1974.
29. Report on antibiotic sensitivity test trial organized by the Bacteriology Committee of the Association of Clinical Pathologists. *J Clin Pathol 18:*1–5, 1965.
30. Rosenblatt, J. E., and Schoenknecht, F. D. Effect of several components of anaerobic incubation on antibiotic susceptibility test results. *Antimicrob Agents Chemother 1:*433–440, 1972.
31. Schoenknecht, F. D., and Sherris, J. C. New perspectives in antibiotic susceptibility testing, pp. 275–292, in S. C. Dyke (Ed.), *Recent Advances in Clinical Pathology*. Edinburgh: Churchill Livingstone, 1973.
32. Sherris, J. C. General considerations in *in vitro* susceptibility testing—a summation, pp. 128–137, in A. Balows (Ed.), *Current Techniques for Antibiotic Susceptibility Testing*. Springfield, Ill.: Charles C Thomas, 1974.
33. Sherris, J. C., Rashad, A. L., and Lighthart, G. A. Laboratory determination of antibiotic susceptibility to ampicillin and cephalothin. *Ann N Y Acad Sci 145:*248–265, 1967.
34. Steers, E., Foltz, E. L., Graves, B. S., and Roden, J. An inocula replicating apparatus for routine testing of bacterial susceptibility to antibiotics. *Antibiot Chemother 9:*307–311, 1959.
35. Survey of antibiotic sensitivity testing. Scientific Projects Committee, The College of Pathologists of Australia. *Med J Aust 2:*171–172, 1968.
36. Takatsy, G. The use of spiral loops in serological and virological micro methods. *Acta Microbiol Hung 3:*191–202, 1956.
37. Thornsbery, C., Gavan, T. L., Sherris, J. C., Balows, A., Matsen, J. M., Sabath, L. D., Schoenknecht, F., Thrupp. L. D., and Washington, J. A., II. Laboratory evaluation of a rapid, automated susceptibility testing system; report of a collaborative study. *Antimicrob Agents Chemother 7:*466–480, 1975.
38. Tilton, R. C., Lieberman, L., and Gerlach, E. M. Microdilution antibiotic susceptibility test; examination of certain variables. *Appl Microbiol 26:*658–665, 1973.

15

Significance of in Vitro Antimicrobial Susceptibility Tests in Care of the Infected Patient

W. EUGENE SANDERS, JR.

CHRISTINE C. SANDERS

Despite general availability of a variety of in vitro antimicrobial susceptibility tests, a surprising amount of controversy remains regarding their proper performance and interpretation. A disconcerting polarity of opinion exists concerning even the value of susceptibility tests in care of patients suffering from infectious diseases. At one extreme, some physicians encourage the laboratory to totally disregard identification of bacteria and report only results of antimicrobial susceptibility tests, which will then be used as precise guidelines for selection of therapy. At the other end of the spectrum of opinion, some physicians rarely order or use results of susceptibility tests because they believe that little or no correlation exists between sensitivity or resistance of bacteria in the laboratory and success or failure of antimicrobial therapy in their patients. Very likely, the truth may be found somewhere between these extreme viewpoints.

The merits or demerits of antimicrobial susceptibility tests are often debated in abstraction. Factors other than in vitro sensitivity or resistance of the infecting microorganism often exert profound influences upon the results of antimicrobial therapy. In certain instances, the host, microorganism and drug may interact to doom therapy, regardless of extreme sensitivity of the infectious agent to the drug, or drugs, administered. Paradoxically, drug therapy occasionally may appear to have been highly successful despite reports of in vitro resistance of the infecting microorganisms. Clearly, laboratory tests of antimicrobial susceptibility reveal nothing of pharmacokinetics of the various drugs or host factors that may interact to facilitate or impair the therapeutic response. In addition, those (including the authors) who advocate the regular use of susceptibility tests may share in responsibility for the continuing controversy. Perhaps, in faith, we have expounded too often the logic of the statement "In vitro sensitivity of the infectious agent correlates reasonably well with therapeutic success and in vitro resistance portends therapeutic failure" without reviewing the evidence or carefully enunciating exceptions to these very general rules.

The following discussion will attempt to (a) summarize studies designed to assay the predictive clinical value of in vitro susceptibility tests in general, (b) review the relative merits of available tests specifically and cite instances in which they may be of little or no value to the physician, (c) outline factors other than sensitivity or resistance of the infecting microorganism that may influence outcome of therapy, and (d) provide tentative indications for choice among the available in vitro tests.

VALUE OF SUSCEPTIBILITY TESTS GENERALLY IN PREDICTING OUTCOME OF THERAPY

Considering the increasing number of available antimicrobial agents and the vast array of microorganisms implicated in infectious diseases, there have been relatively few scholarly studies designed to assay the value of the various in vitro susceptibility tests in predicting outcome of therapy. Some antimicrobials and specific bacterial infections have escaped study altogether. Among the better published reports, there have been wide variations in method of in vitro testing, definitions of sensitivity or resistance, microorganisms studied, and criteria for success or failure of treatment. Nevertheless, the weight of evidence accumulated to date suggests that results of in vitro tests may be of value when applied in the context of one's knowledge of other factors of host, drug and microorganism that may influence clinical outcome. A few of the more representative studies are summarized below.

Bacteremia. In a retrospective study of 100 consecutive cases of staphylococcal bacteremia seen at Milwaukee County Hospital between 1952 and 1957, a favorable clinical response was observed in approximately 60% of patients treated with an antibiotic to which the organisms were judged sensitive (minimum inhibitory concentration, or MIC, of 1.5 μg/ml or less by tube dilution) (1). In contrast, no patients responded to treatment with a drug to which the organisms had been found resistant by the same assay method. The authors concluded that "the clinician should not give patients antibiotics to which his infecting staphylococci are resistant in vitro by the tube dilution method"; furthermore, "in vitro sensitivity by this method is no guarantee of clinical results and that host factors and complications must always be considered." In subsequent studies of 173 patients with Gram-negative bacillary bacteremia, two major determinants of death or survival were identified; the nature of the underlying noninfectious disease and administration of an antimicrobial agent to which the infecting microorganisms were sensitive (31, 32). In general, all therapeutic modalities exerted little or no beneficial effect among patients with immediately life-threatening underlying diseases. In patients with less severe or no underlying disease, survival rates were greater and correlated closely with administration of an antibiotic judged to be "effective" or "appropriate" by in vitro susceptibility tests. For example, among the group of patients with nonlethal underlying diseases, death ensued in (a) none of 49 patients who received appropriate antibiotic therapy, (b) 1 of 3 who received no therapy, (c) 2 of 16 treated with an antibiotic of inderminant activity in vitro, and (d) 3 of 13 treated with an antibiotic to which the infecting microorganism was resistant in vitro. The authors emphasized that use of an antibiotic judged to be ineffective in vitro did not preclude recovery, nor did use of an effective agent ensure recovery of all patients. Even more recently, survival of patients with pseudomonas bacteremia has been shown to correlate with sensitivity of the isolates to the bactericidal action of gentamicin if dosage of the drug was optimal and the patient's underlying disease was not immediately life-threatening (21).

Urinary Tract Infections. In the mid 1960s, at least three extensive studies with reviews of the literature emphasized the importance of both sensitivity of the infecting microorganism to the drug administered and host factors in determining the outcome of the initial infection (29, 33, 46). Relapses were often associated with resistance of the infecting bacteria, while recurrences (infection with different microorganisms) tended to occur in patients predisposed to development of repeated urinary tract infections. Several of these authors also emphasized the importance of interpretation of in vitro susceptibility tests based upon achievable urinary, rather than serum, levels of drug. A variety of subsequent investigators have confirmed these earlier observations and stressed the need for correction of underlying urinary tract abnormalities, if possible, to enhance likelihood of a satisfactory response to antimicrobial therapy and prevention of recurrences.

Bacterial Endocarditis. In few bacterial diseases has the need for use of antimicrobial agents to which the infecting microorganisms are sensitive been so well documented and widely accepted (5, 10, 14, 20, 23, 37, 39, 44). In addition, many have emphasized the importance of (a) selection of an antibiotic that is bactericidal in vitro for the infecting microorganism and (b) documentation of evidence for bactericidal activity in the patient's serum during administration of that drug. Poor therapeutic responses have been observed repeatedly in patients whose microorganisms were resistant to the drug administered or in whom high bactericidal concentrations of drug were not maintained in serum. More recent studies in patients with endocarditis due to Gram-negative bacilli have indicated an extraordinarily high mortality rate in patients treated with antimicrobial agents alone, despite in vitro sensitivity (5, 39). To date, the most effective therapy in intractable disease—especially with pseudomonas infection of the tricuspid valve—appears to be surgical excision *combined* with therapy with bactericidal agents to which the microorganisms are sensitive in vitro (39).

Gonorrhea. Shortly after the advent of penicillin, it was recognized that the drug was remarkably effective in treatment of gonorrhea. In ensuing decades, a gradual increase in in vitro resistance to penicillin was noted among isolates from infected patients (30, 45). Concurrently, an unexpected increase in treatment failures with penicillin was observed. The close correlation between these two events was noted as early as 1958 (6). More recent studies have confirmed the association of increasing in vitro resistance to penicillin with treatment failures (24, 34) and have observed a similar phenomenon with ampicillin (24) and the tetracyclines (50, 52). For example, an almost linear relationship has been observed between rate of therapeutic failures of methacycline and minimum inhibitory concentrations of the drug for isolates of gonococci obtained prior to initiation of treatment (52). The percentage of therapeutic failures according to the various minimum inhibitory concentrations of methacycline were as follows: 0, 31, 82, and 100% with MICs of equal to or less than 0.25, 0.5, 1.0, and greater than 1.0 μg/ml, respectively. Paradoxically, controversy remains regarding the value of in vitro susceptibility tests of spectinomycin for prediction of therapeutic success or failure (24, 34). This may result from the relatively low frequency of resistance to spectinomycin, less extensive use of the drug, variations in the method of susceptibility testing in vitro, or yet undetermined factors.

Sinusitis. Recently, an excellent correlation between in vitro antibiotic sensitivity and response to therapy was noted in patients with sinusitis of the maxillary antrum (8). Each of 14 patients infected with bacteria sensitive to the antibiotic given responded promptly and completely to therapy. On the other hand, 5 of 6 patients infected with organisms resistant to the antibiotic given failed to respond.

Anaerobic Infections. Correlation between in vitro sensitivity of isolates and success of therapy has been somewhat more difficult to substantiate in management of anaerobic infections (9, 15, 16, 48). This may result in part from (a) the important role of the underlying noninfectious disease in determining outcome, (b) the frequency of infections due to more than one organism, each with differing sensitivity to a variety of antimicrobial agents, (c) the relative lack of experience in management of disease due to anaerobes rather than aerobic or facultative bacteria, (d) the apparent dependence of anaerobes upon coexistent facultatively anaerobic bacteria for their pathogenicity or even survival in some infected tissues, and (e) the primary role of surgical intervention, and not antimicrobial therapy, in ensuring recovery from many soft tissue infections. Experience has shown that most anaerobic pulmonary infections will respond to penicillin G even in the presence of *Bacteroides fragilis,* which is highly resistant to penicillin in vitro. In contrast, optimal responses appear to occur in intraabdominal, genital, and nervous system infections when an agent known to be active against the involved anaerobe(s) is included in the therapeutic regimen.

Miscellaneous Infections. A number of early investigators attempted to correlate clinical response of a variety of infections with results of in vitro susceptibility tests (26, 36, 40, 49). Although analysis of data was difficult due to the wide range of infections and antimicrobial agents studied, the conclusions reached were similar; therapeutic failures were far more common among infections treated with antimicrobials to which the infecting microorganism was determined to be resistant in vitro.

Combination Therapy. A voluminous literature has accumulated on the simultaneous effects of two or more antimicrobial agents upon microorganisms in vitro and in vivo. Assay methods have been designed to determine whether drugs are indifferent, antagonistic, additive, or synergistic in their activity in vitro (13, 25). Results of these in vitro assays must be interpreted with extreme caution, as suggested by Garrett (11, 12). In many instances, data that would appear to indicate synergy may be artifactual and merely reflect additive or indifferent effects. In many instances when extrapolation of in vitro additive or synergistic effects to clinical situations was attempted, especially with fixed ratio combinations, no increase in response to therapy was observed, regardless of the criteria employed to measure success or failure (32, 33). In a *few* instances, combination therapy has been successful and results have correlated well with those obtained in vitro; examples are use of penicillin or ampicillin plus an aminoglycoside in enterococcal infections and a sulfonamide plus trimethoprim in urinary tract infections due to sensitive bacteria. As cited below, some, but not all, of the antagonisms between drugs observed in vitro may be directly applicable to clinical practice. Since the correlation between in vitro assays of the combined effects of drugs and the response of patients is so poor, physicians would be well advised to avoid clinical use of more than one drug until proof of efficacy and superiority of the combination is well established.

SPECIFIC IN VITRO ANTIMICROBIAL SUSCEPTIBILITY TESTS

Despite variations in methodology and populations studied, the overwhelming majority of investigators have concluded that there is a useful correlation between results of most in vitro antimicrobial susceptibility tests and the clinical response of patients. The degree of correlation or predictability was highest when extrapolating in vitro *resistance* to therapeutic *failure*. Although to a lesser degree, the correlation of in vitro *sensitivity* with a *favorable outcome* was sufficiently good to commend its use as an adjunct in management of infections that are complicated or due to an organism known *not* to be predictably sensitive to one or more safe and effective antimicrobial agents.

Accepting the value of in vitro susceptibility tests in general, what test(s) should be performed? Ideally, one would wish to measure easily and inexpensively the activity of the antimicrobial agent against the microorganism directly in the infected tissues; however this is clearly not possible, practical or necessary in most infections at present. Four categories of related tests may be performed. They are (a) measurement of antibiotic *concentration* in body fluids by bioassay or chemical methods; (b) assay of antibiotic *activity* against the infecting strain in body fluids by a dilution procedure, often referred to as the "Schlichter test" (44) or the "serum bactericidal test"; (c) agar or broth dilution tests to determine the minimum concentration of drug necessary to inhibit or kill the infecting microorganism in vitro; and (d) disc diffusion assays of antimicrobial activity in vitro. In general, in progression down this list, the tests become *less* precise, tedious and expensive and *more* practical to perform on a large scale. The details of procedure and applicability of many of these tests will be discussed elsewhere in this volume. Thus only major indications for use and limitations will be discussed here.

Disc Diffusion Tests. The disc diffusion assay is widely used. It is easy to perform, relatively inexpensive, and adaptable to use on a large scale. Much of the initial disenchantment with the value and usefulness of this test has been dispelled by increasing use of standardized, controlled procedures within and among microbiology laboratories (4). Nevertheless, there are important limitations to this procedure at the bench and in its interpretation. In use of the standardized test, preparation of the bacterial inoculum is the most cumbersome and time-consuming step. This is a critical step in the laboratory procedure because variations of 10-fold or greater in the inoculum size may profoundly alter results in many instances (3, 4, 26, 38). The endpoints (zone sizes) for interpretation of sensitivity or resistance have been standardized by extrapolation from results of dilution assays; thus the disc diffusion is the less precise of the two types of tests. For example, in assays of the activity of aminoglycosides against Gram-negative bacilli a relatively poor correlation between results of disc diffusion and dilution assays has been observed (41). In general, the two tests agree more often in identification of resistance rather than sensitivity. The error of the disc diffusion test in interpretation of sensitivity has been estimated to be as high as 25% when a dilution assay has been used as the standard of comparison (19, 22).

Other possible limitations of the disc diffusion assay are (a) the inability to obtain a bactericidal endpoint, (b) lack of adaptability for use with many slowly growing microorganisms, (c) failure to identify those "resistant" microorganisms that might respond therapeutically to larger than usual doses of drug, (d) inability to assay the combined effects of drugs, and (e) the so-called "intermediate" zone. Proper interpretation of the "gray area" of zone sizes, between those more precisely reflecting sensitivity or resistance of some bacterial isolates, continues to perplex many physicians and microbiologists. There is insufficient data to permit a determination of correlation of these intermediate results with clinical responses. Many have suggested that an intermediate susceptibility indicates that the infecting microorganism may respond to the drug in vivo if (a) large doses are administered, or (b) the infection is located in the urinary tract and the drug achieves high concentrations in urine. Although plausible, these suggestions remain to be proven by careful study. In practice, the physician confronted with a report of intermediate susceptibility would be well advised to select an alternative drug to which the infecting microorganism was clearly sensitive or request a more precise assay of susceptibility, such as a dilution test.

Agar and Broth Dilution Tests. Dilution assays have been less widely used than disc diffusion tests because they are more laborious and expensive to perform. They have been employed generally in evaluation of new antimicrobial agents, or in practice when greater precision or a bactericidal endpoint is desired. Although inherently more accurate, wide variations in results and interpretation have been encountered within and between various laboratories (7, 35, 41). Results may be profoundly altered by differences in size of bacterial inoculum, constitution of the medium, and conditions and time of incubation (7). In addition, there has not been general agreement upon interpretation of the results obtained. Recommended levels for definition of "sensitivity" have varied from two to sixteen fold below the concentration of drug that is attainable in the patient. Despite pleas for standardization and exhaustive evaluation of various procedures, no uniform, controlled method for dilution assays has been widely adopted. Without doubt, the value of these assays would be greatly enhanced were the procedure standardized, preferably in accordance with the guidelines established by the International Collaborative Study Group (7).

Situations in Which in Vitro Susceptibility Tests May Be of Little or No Value. Susceptibility of many isolates need not be determined in vitro if all strains of a given species are known to be uniformly sensitive or resistant to certain antimicrobial agents. For example, group A streptococci, pneumococci, and meningococci are so predictably sensitive to penicillin that routine testing is unnecessary. On the other hand, some microorganisms are predictably sensitive in vitro to certain drugs that seldom, if ever, are efficacious in therapy. Examples are aminoglycosides and *Salmonella typhi*, and cephalosporins and enterococci. Strains of *Nocardia* often will appear sensitive to many antibacterials in vitro; however therapeutic efficacy of drugs other than the sulfonamides has been difficult to demonstrate. Although sulfonamides inhibit most strains of group A streptococci in vitro and are efficacious in prophylaxis of infection; they *will not* erradicate the organisms in established infection *nor* prevent the late sequellae.

Recommendations for Use. Except for the most innocuous infections, specimens should be obtained for culture *prior to* initiation of antimicrobial therapy. With the exceptions noted above, in vitro tests of antimicrobial susceptibility should be performed. The standardized disc diffusion test is most appropriate for initial or routine use. If the microorganism is reported to be resistant, the result is likely to be highly reliable and the drug(s) should not be used if equally safe alternatives are available. If the isolate is reported to be sensitive, the result is somewhat less accurate but may be relied upon in most uncomplicated infections. In the presence of (a) doubt, (b) complicated or life-threatening infections, or (c) a limited choice from among potentially toxic agents, a carefully controlled dilution assay should be performed. The dilution test should also be used when determination of a bactericidal endpoint is desirable (bacterial endocarditis) or in conjunction with assays of antimicrobial concentrations in body fluids. The serum bactericidal assay is of value in monitoring antibacterial activity in bacterial endocarditis. Determinations of antimicrobial concentration should be reserved for severe or complicated infections in which precise knowledge of therapeutic or potentially toxic levels is necessary.

REASONS FOR FAILURE OF APPROPRIATE ANTIMICROBIAL THERAPY

As noted above, a favorable clinical response all too frequently does not occur following administration of antimicrobial agents to which the infecting microorganism is sensitive in vitro. These failures of seemingly appropriate therapy present some of the most vexing problems in management of patients with infectious diseases. Responsibility is often assigned to laboratory error or emergence of drug resistance among the infectious agents. The experience of the authors and others indicates that these are relatively infrequent explanations for drug failure in acute infections (42, 43, 51). More often the host, microorganism and drug have interacted to diminish the anticipated antimicrobial activity.

When confronted with such a problem, the diagnosis and laboratory findings should be reconsidered. If sensitivity of the infecting microorganism has been demonstrated only by disc diffusion test, confirmation by a dilution assay should be obtained. An attempt should then be made to measure antimicrobial activity or concentration in appropriate body fluids. If these procedures demonstrate sensitivity of the etiologic agent and less than anticipated drug (or activity) in the tissues, reasons for impaired absorption or accelerated elimination of drug should be sought. For example, many divalent or trivalent cations when ingested may impair absorption of the tetracyclines from the gastrointestinal tract. Patients with hypotension or diseases involving small vessels, such as diabetes mellitus, may poorly absorb drugs from sites of intramuscular injection. Elimination of some antimicrobials may be accelerated by induction of inactivating enzymes by other unrelated drugs. The list of adverse interactions of drugs in vivo or in solutions prepared for administration has grown too unwieldy to commit to memory. Perhaps all combinations should remain suspect, until compatibility has been confirmed by consultation of an appropriate reference source (2, 18). If sensitivity of the infecting microorganism and presence of adequate levels of drug are confirmed, other potentially remedial causes of drug failure must be sought.

Deep-seated abscess, obstruction to natural drainage, or presence of a foreign body at the nidus of infection often predispose to therapeutic unresponsiveness. Once these are detected, drainage of the abscess, relief of obstruction or removal of the foreign body will markedly enhance the likelihood of a favorable clinical response.

Emergence of genotypic drug resistance during therapy of acute infections is relatively uncommon. Resistance is most likely to emerge when most or all of the following factors are present: (a) treatment with aminoglycosides, carbenicillin, or bacteriostatic antibiotics; (b) infections due to Gram-negative bacilli or staphylococci; (c) persistence of abscess, obstruction or foreign body; and (d) chronic infections or those that tend to become chronic. In the presence of these predisposing factors, repeated cultures and in vitro susceptibility testing are essential for appropriate modification of therapy.

Noninvasive staphylococci or other beta lactamase-producing organisms may inactivate penicillin, thereby protecting a pathogen that is infecting adjacent or proximate tissues. Although infrequent, this phenomenon has been implicated in treatment failures in gonorrhea and in persistence of surface infections due to group A streptococci. When recognized, these persistent infections have responded promptly to therapy with an agent that is insusceptible to inactivation by beta lactamases.

Inhibition of the activity of bactericidal drugs, especially penicillin, by bacteriostatic agents has been observed repeatedly in vitro. Similar antagonism has been demonstrated in treatment of experimental infections and in assays of antimicrobial activity in serum of human volunteers. Evidence that antagonisms between antibiotics have adversely influenced the course of infectious diseases in man is meagre. To date, this phenomenon has been implicated in delayed therapeutic responses of chronic pyelonephritis, streptococcal pharyngitis, and pneumococcal meningitis (28, 33, 47). Antagonism of lincomycin by erythromycin, both bacteriostatic drugs, has been observed in vitro (17) and in vivo (27).

Clinically important superinfections may mimic persistence of the preceding infection (42). Diagnosis and proper management of superinfection depends upon (a) awareness of predisposing factors; (b) recognition of worsening signs and symptoms, often following early clinical improvement in patients receiving therapy appropriate for their initial infections; (c) demonstration of absence of other factors known to predispose to drug failure; and (d) laboratory identification and antimicrobial susceptibility testing of the superinfecting microorganism. Factors that may predispose patients to development of superinfection include (a) prematurity, infancy, advanced age or extreme debility; (b) infections of the tracheobronchial tree in the presence of underlying lung disease; (c) measles; (d) administration of broad spectrum antimicrobials; and (e) use of large doses of antibiotics.

Impairment of normal host defense mechanisms may significantly impede the response of infection to antimicrobial agents. Disseminated malignancies, uncontrolled diabetes, advanced liver disease and immunosuppressive therapy may each preclude an optimal response. Bactericidal antibiotics are clearly preferable in these settings, but success is not necessarily assurred by this choice. For example, the aminoglycosides which are usually bactericidal have been shown to be less active in the presence of leukopenia. Eradication of the infecting microorganisms may be possible only when the underlying disease process is controlled or remits. Occasionally, an underlying disease may mimic infection, thus obscuring a favorable response of the infectious process to apppropriate antimicrobial therapy. Drug fever is frequently mistaken for persistence of infection. It most often may be recognized by (a) improvement in other signs and symptoms of infection; (b) sustained nature of the fever, if unaltered by antipyretics; and (c) prompt defervescense after withdrawal of the suspected drug.

CONCLUSIONS

It has become increasingly apparent that in vitro antimicrobial susceptibility tests may be of great value in care of patients with infectious diseases. *However, there are important pitfalls in their performance and limitations to their interpretation*. The microbiologist should expend every effort to ensure proper performance of these tests and encourage standardization of assays among laboratories. The physician should request these tests as indicated and interpret them in light, and not in lieu, of a knowledge of other factors in his patient that may profoundly influence the response of infection to antimicrobial therapy.

LITERATURE CITED

1. Abboud, F. M., and Waisbren, B. A. Correlation between results of the tube dilution method for determining bacterial sensitivity to antibiotics and the results of the administration of these antibiotics to patients with staphylococcic bacteremia. *Antibiot Ann 1958–1959:*748–756, 1959.
2. American Pharmaceutical Association. Evaluations of Drug Interactions, listed. Washington, D.C.: American Pharmaceutical Association, 1973.
3. Barry, A. L., Garcia, F., and Thrupp, L. D. An improved single-disc method for testing the antibiotic susceptibility of rapidly growing pathogens. *Am J Clin Pathol 53:*149–158, 1970.
4. Bauer, A. W., Kirby, W. M. M., Sherris, J. C., and Turck, M. Antibiotic susceptibility testing by a standardized single disc method. *Am J Clin Pathol 45:*493–496, 1966.
5. Bryan, C. S., Marney, S. R., Alford, R. H., and Bryant, R. E. Gram negative bacillary endocarditis. Interpretation of serum bactericidal test. *Am J Med 58:*209–215, 1975.
6. Curtis, F. R., and Wilkinson, A. E. A comparison of the in vitro sensitivity of gonococci to penicillin with the results of treatment. *Br J Vener Dis 34:*70–82, 1958.
7. Ericsson, H. M., and Sherris, J. C. Antibiotic sensitivity testing, report of an international collaborative study. *Acta Pathol Microbiol Scand (B) Suppl 217:*1–90, 1971.
8. Evans F. O. Jr., Sydnor, J. B., Moore, W. E. C., Moore, G. R., Manwaring, J. L., Brill, A. H., Jackson, R. T., Hanna, S., Skarr, J. S., Holdeman, L. V., Fitz-Hugh, G. S., Sande, M. A., and Gwaltney, J. M., Jr. Sinusitis of the maxillary antrum. *N Engl J Med 293:*735–739, 1975.
9. Finegold, S. M. Management of anaerobic infections. *Ann Intern Med 83:*375–389, 1975.
10. Finland, M. Treatment of bacterial endocarditis. *N Engl J Med 250:*419–428, 1954.
11. Garrett, E. R. Drug action and assay by microbial kinetics. *Progr Drug Res 15:*271–352, 1971.
12. Garrett, E. R. The use of microbial kinetics in the quantification of antibiotic action. *Arzneim Forsch 16:*1364–1369, 1966.
13. Garrod, L. P., and Waterworth, P. M. Methods of testing combined antibiotic bactericidal action and the significance of the results. *J Clin Pathol 15:*328–338, 1962.
14. Geraci, J. E., and Martin, W. J. Antibiotic therapy of bacterial endocarditis. VI. Subacute enterococcal endocarditis; clinical, pathologic and therapeutic considerations of 33 cases. *Circulation 10:*173–194, 1973.
15. Gorbach, S. L., and Thadepalli, H. Clindamycin in pure and mixed anaerobic infections. *Arch Intern Med 134:*87–92, 1974.
16. Gorbach, S. L., and Bartlett, J. G. Anaerobic infections. *N Engl J Med 290:*1177–1184, 1974.
17. Griffith, L. J., Ostrander, W. E., Mullins, C. G., and Beswick, D. E. Drug antagonism between lincomycin and erythromycin. *Science 147:*746–747, 1975.
18. Hansten, P. D. *Drug Interactions*. Philadelphia: Lea & Febiger, 1975.
19. Hoffman, R. V., Jr., Jackson, G. G., and Turner, M. P. Reliability of antibiotic sensitivity tests as determined by a survey study. *J Lab Clin Med 51:*873–888, 1958.
20. Hunter, T. H. Bacterial endocarditis. *Mod Concepts Cardiovasc Dis 12:*172–173, 1953.
21. Jackson, G. G., and Riff, L. J. Pseudomonas bacteremia; pharmacologic and other bases for failure of treatment with gentamicin. *J Infect Dis Suppl 124:*S185–S191, 1971.
22. Jackson, G. G., and Finland, M. Comparison of methods for determining sensitivity of bacteria to antibiotics in vitro. *Arch Intern Med 88:*446–460, 1951.
23. Jackson, G. G., Rubenis, M., and Kennedy, R. P. Comparison of three penicillins against microorganisms causing bacterial endocarditis with some clinical observations on phenethicillin. *Antimicrob Agents Chemother 1961:*697–705, 1962.

24. Jaffe, H. W., Biddle, J. W., Thornsberry, C., Johnson, R. E., Kaufman, R. E., Reynolds, G. H., and Wiesner, P. J. National gonorrhea therapy monitoring study. In vitro antibiotic susceptibility and its correlation with treatment results. *N Engl J Med 294:*5–9, 1976.
25. Jawetz, E., Gunnison, J. B., Bruff, J. B., and Coleman, V. R. Studies of antibiotic synergism and antagonism. Synergism among seven antibiotics against various bacteria in vitro. *J Bacteriol 64:*29–39, 1952.
26. Kanazawa, Y. Clinical use of the disc sensitivity test. *Antimicrob Agents Chemother 1961:*926–942, 1962.
27. Kislak, J. W. Brief recording, type 6 pneumococcus resistant to erythromycin and lincomycin. *N Engl J Med 276:*852, 1967.
28. Lepper, M. H., and Dowling, H. F. Treatment of pneumococcic meningitis with penicillin compared with penicillin plus aureomycin. *Arch Intern Med 88:*489–494, 1951.
29. Lindemeyer, R. I., Turck, M., and Petersdorf, R. G. Factors determining the outcome of chemotherapy in infections of the urinary tract. *Ann Intern Med 58:*201–216, 1963.
30. Martin, J. E., Lester, A., Price, E. V., and Schmale, J. D. Comparative study of gonococcal susceptibility to penicillin in the United States, 1955–1969. *J Infect Dis 122:*459–461, 1970.
31. McCabe, W. R., and Jackson, G. G. Gram negative bacteremia. *Arch Intern Med 110:*856–864, 1962.
32. McCabe, W. R., and Jackson, G. G. Treatment factors of bacteremia caused by gram-negative bacteria. *Antimicrob Agents Chemother 1961:*133–141, 1962.
33. McCabe, W. R., and Jackson, G. G. Treatment of pyelonephritis. Bacterial, drug and host factors in success or failure among 252 patients. *N Engl J Med 272:* 1037–1044, 1965.
34. Pedersen, A. H. B., Wiesner, P. J., Holmes, K. K., Johnson, C. J., and Turck, M. Spectinomycin and penicillin G in the treatment of gonorrhea. A comparative evaluation. *JAMA 220:*205–208, 1972.
35. Petersdorf, R. G., and Sherris, J. C. Methods and significance of in vitro testing of bacterial sensitivity to drugs. *Am J Med 39:*766–779, 1965.
36. Petersdorf, R. G., and Plorde, J. J. The usefulness of in vitro sensitivity tests in antibiotic therapy. *Ann Rev Med 14:*41–56, 1963.
37. Quinn, E. L., and Colville, J. M. Subacute bacterial endocarditis. Clinical and laboratory observations in 27 consecutive cases treated with penicillin V by mouth. *N Engl J Med 264:*835–842, 1961.
38. Report on antibiotic sensitivity test trial organized by the Bacteriology Committee of the Association of Clinical Pathologists. *J Clin Pathol 18:*1–5, 1965.
39. Reyes, M. P., Patulke, W. A., Wylin, R. F., and Lerner, A. M. Pseudomonas endocarditis in the Detroit Medical Center, 1969–1972. *Medicine 52:*173–194, 1973.
40. Rodger, K. C., Branch, A., Power, E. E., Starkey, D. H., Gregory, E., Murray, R. D., and Harrop, J. Antibiotic therapy: correlation of clinical results with laboratory sensitivity tests. *Can J Med 74:*605–612, 1956.
41. Sanders, C. C., and Sanders, W. E., Jr. Effects of procedural variations on the in vitro activity of aminoglycosides. *Am J Clin Pathol 63:*438–445, 1975.
42. Sanders, W. E. Jr. Therapeutic applications of antimicrobial agents, pp. 3–17, in G. R. G. Monif(Ed.), *Infectious Diseases in Obstetrics and Gynecology*. Hagerstown, Md.: Harper & Row, 1974.
43. Sanders, W. E. Jr., and Jurgensen, P. F. Remedial causes of failure of "appropriate" antimicrobial therapy. *Postgrad Med 50:*161–165, 1971.
44. Schlichter, J. G., and MacLean, H. A method of determining the effective therapeutic level in the treatment of subacute bacterial endocarditis with penicillin. A preliminary report. *Am Heart J 34:*209–211, 1947.
45. Sparling, P. F. Antibiotic resistance in *Neisseria gonorrhoeae*. *Med Clin North Am 56:*1133–1144, 1972.
46. Stamey, T. A., Govan, D. E., and Palmer, J. M. The localization and treatment of urinary tract infections; the role of bactericidal urine levels as opposed to serum levels. *Medicine 44:*1–36, 1965.
47. Strom, J. The question of antagonism between penicillin and chlortetracycline illustrated by therapeutical experiments in scarlatina. *Antibiot Med 1:*6–12, 1955.
48. Thadepalli, H., Gorbach, S. L., Broido, P. W., Norsen, J., and Nyhus, L. Abdominal trauma, anaerobes, and antibiotics. *Surg Gynecol Obstet 137:*270–276, 1973.
49. Tunevall, G., and Ericsson, H. Sensitivity tests by disc method as a guide for chemo-

therapy. *Antibiot Chemother 4:*886–893, 1954.
50. Tyson, W. T., Jr., Roberts, F. L., and Fine, M. H. Evaluation of treatment of gonorrhea in males with single doses of minocycline. *J Tenn Med Assoc 64:*743–777, 1971.
51. Weinstein, L., and Dalton, A. C. Host determinants of response to antimicrobial agents. *N Engl J Med 279:*467–473, 524–531, 580–588, 1968.
52. Wiesner, P. J., Holmes, K. K., Sparling, P. F., Maness, M. J., Bear, D. M., Gutman, L. T., and Karney, W. W. Single doses of methacycline and docycycline for gonorrhea; a cooperative study of the frequency and cause of treatment failure. *J Infect Dis 127:*461–466, 1973.

16

The Significance of Antibiotic Blood Levels

L. D. SABATH

The level, or amount of an antibiotic present in the blood or serum of a patient receiving one or more antibiotics may be measured in one of two ways: either as "activity" or as the absolute concentration (in micrograms/ml) of the antimicrobial substance. The former approach has been in use for at least two decades in the form of determining serum bacteriostatic or serum bactericidal activity against the infecting organism. These measurements have been made especially in patients with bacterial endocarditis. On the other hand the practice of measuring the absolute amount of antibiotic in blood or serum has only recently been introduced into medicine on a rather wide scale although the ability to perform such tests have been available since the beginning of the antibiotic era and have been employed primarily in formal studies of the pharmacology of various antibiotics.

ANTIBIOTIC ACTIVITY IN BLOOD OR SERUM

The practice of determining the antibacterial activity of serum has been in existance for over 20 years, primarily in the regulation of therapy of bacterial endocarditis. It is often suggested that for effective therapy of bacterial endocarditis the serum bacteriocidal level or titer should be 1 to 8 or greater. Data substantiating that figure are relatively meager and the various ways that the test can be performed are so variable that it is essentially impossible on the basis of present data, to know how high the serum bactericidal level should be, what the definition of the bactericidal should be (99.9% killing is the frequently used figure), how long the incubation should be for determining the killing, whether the dilution should be in broth or in serum (to correct for problems of extent of serum binding of the antibiotic, or antibiotics, in use), and whether the serum for the tests should be drawn at the peak or the trough level. Thus, although the concept of knowing how much antibacterial activity and whether it is bacteriostatic or bactericidal is widespread, the variations in the way the tests could be done and the lack of a strong correlation with clinical results are such that at this time it is difficult to correlate activity, as so defined, with clinical results.

The only disease in which it has been clearly demonstrated that having bactericidal activity is desirable (in comparison with only bacteriostatic activity) is the disease bacterial endocarditis. The major proof that bactericidal activity is required is based on the fact that essentially no cures of bacterial endocarditis were noted until the introduction of bactericidal agents (first benzylpenicillin). The suggestion that bactericidal activity is desirable in immunosuppressed patients, neutropenic patients and possibly in meningitis has been made but convincing evidence to substantiate these impressions is lacking.

TESTS TO MEASURE ABSOLUTE ACTIVITY OF ANTIBIOTICS IN SERUM

A value or utility of the measurement of antibiotic levels in serum from patients obviously requires a system that would render an answer rapidly—the same day, or better yet the same half day—and that would yield the results with sufficient accuracy (with an error of 11 or 12% or less) that the information would be useful in regulating antibiotic therapy. A number of methods have been suggested and they are listed below:

1. Chemical assays
2. Microbiological assays
3. Potentiometric assays
4. Enzymatic assays
5. Radioimmunoassays.

The three methods that are most widely used now are the agar diffusion microbiological assay (7), enzymatic assay (for aminoglycoside antibiotics: acetylating (2) or adenylating (8) enzyme assay), and the radioimmunoassay (3). All three of these methods are presently in common use and fulfill the two major requirements for providing useful data: a rapid assay that is accurate. The latter two assays are more selective than the microbiological assay but require radioactive isotopes and also require material, either enzymes or antiserum, that have only recently become readily available. On the other hand, the microbiological assay requires no equipment not readily available in the average microbiology laboratory and requires no reagents that are not readily available. It does have the shortcoming of being less selective and this may present a problem in two settings. The first setting is when the patient is on multiple antibiotics and the second is when it is not known whether the patient is on multiple antibiotics or whether antibiotics recently administered may contribute to the activity measured in the assay.

The techniques used in the microbiological assay to circumvent these problems are to either use an organism resistant to all antibiotics except the one to be measured or to use an inactivating enzyme to destroy the activity of the other antibiotics present (the activity which is not desired to be measured). This enzymatic inactivation approach is readily achieved when the antibiotics to be inactivated are penicillins or cephalosporins, by use of β-lactamase II produced by *Bacillus cereus*. β-Lactamase II requires zinc for activation, to permit it to be a broad spectrum enzyme to inactivate all penicillins and cephalosporins currently in use. When the patients are on antibiotics other than β-lactam antibiotics, in addition to be an aminoglycoside or vancomycin (the most common antibiotics for which quantitation of the serum levels is required) the necessity to use multiply resistant organisms comes into play. The two most frequently used organisms for these situations are either a multiply resistant Klebsiella (4) or a multiply resistant *Staphylococcus epidermidis* (1).

SIGNIFICANCE OF SERUM BLOOD LEVELS

The significance and utility of having information on serum blood levels or antibiotics is primarily in relation to the use of antibiotics with narrow toxic to therapeutic ratios. This primarily concerns gentamicin and tobramycin of commonly used antibiotics but to a lesser extent kanamycin, streptomycin, neomycin, and vancomycin.

The toxic to therapeutic ratio with penicillin G may range between 1000 and 1 million, whereas with gentamicin and tobramycin the toxic therapeutic ratio may be as low as 2 and sometimes as high as 40 (Table 16.1). Because of the wide variation in serum blood levels achieved with a given dose, even in patients with normal renal function, one cannot be sure, a priori, that the patient is receiving adequate therapy. Rapid accurate assays permit the prescribing physician to know whether he is giving enough potentially toxic antibiotic or whether the dose should be raised or lowered. Riff and Jackson (5) have published data on the wide range in blood levels achieved with fairly similar levels of similar doses of gentamicin and other investigators have also noted this (Fig. 16.1). In addition, patients with abnormal renal function have a very special problem in that aminoglycoside antibiotics, and also many others, are not as readily excreted and would accumulate, very likely to toxic levels, if usual doses are continued. Formulas and nomograms for planning dosages in such patients with renal failure have been mainly useful only as approximations of what the dose should be but these must be verified with assays in order to assure that the levels are adequately above the minimum inhibitory concentration desired, and not in what many consider to be a toxic level. Removal of one aminoglycoside antibiotic (gentamicin) during hemodialysis was shown to be very irregular (6) (Fig. 16.2).

The possibility that serum levels would also permit toxic levels to be avoided has been considered by most workers in the fine control of antibiotic therapy but preliminary studies in our laboratory have failed to show a significant correlation between blood levels and appearance of

TABLE 16.1
Toxic to therapeutic ratios

	Penicillin G	Gentamicin
	μg/ml	*μg/ml*
Common blood level	0.2–100	1.5–8.0
Common MIC	0.0002–0.2	0.3–6.0
"Toxic" blood level	>200	>12
Toxic to therapeutic ratio	1,000 to 1,000,000	2 to 40

Figure 16.1

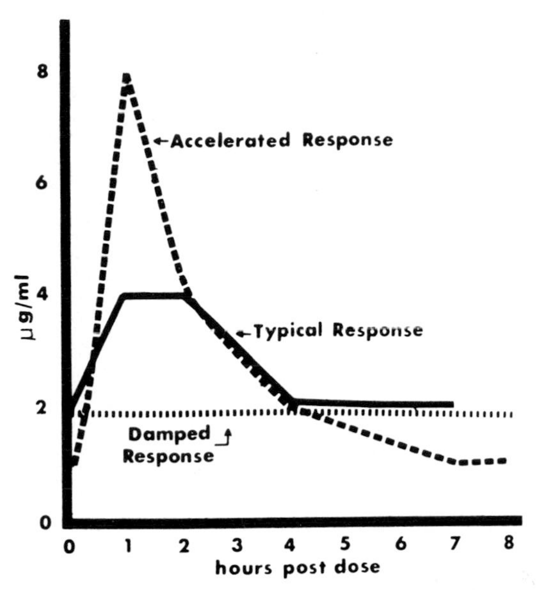

Responses to dosage in patients with normal renal function. Maintenance therapy was 3 mg of gentamicin/kg per day. (From L. J. Riff and G. G. Jackson: *Journal of Infectious Diseases, 124:* S98–S105, 1971; reproduced with permission.)

Figure 16.2. Effect of 8-hr hemodialysis on concentration of gentamicin in serum.

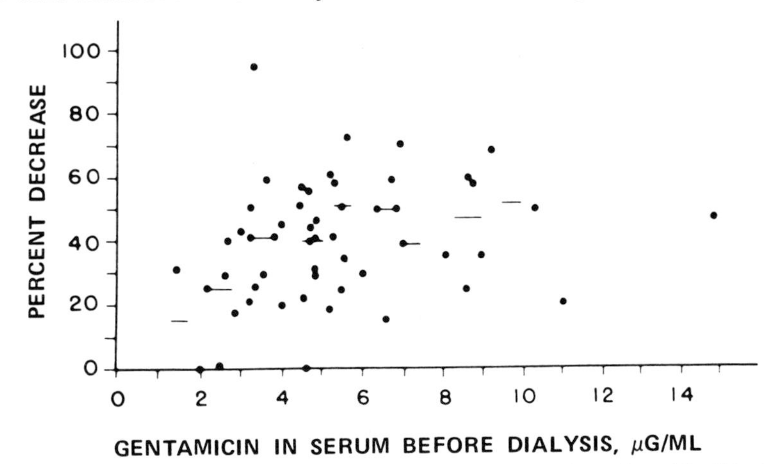

Each dot represents percentage fall in serum gentamicin level produced by 8-hr hemodialysis. The percent decrease in serum level is related to gentamicin level at outset of dialysis (abscissa). The horizontal lines represent mean decrease for the particular integer of predialysis levels. Note wide variation in percent gentamicin removed. (From L. D. Sabath: *Infectious Disease Reviews, 2:* 17–44, 1973; reproduced with permission.)

nephrotoxicity or VIII nerve toxicity. Others have also failed to show a significant relationship although they have stressed the fact that somewhat higher levels are seen in patients who develop toxicity.

Thus, the significance of measurement of antibiotic blood levels in regard to potentially toxic antibiotics such as aminoglycosides and vancomycin is an insurance that enough antibiotic is given to exceed what is necessary to inhibit the pathogen causing disease. The hope that measurement of blood levels would provide information that would be help in avoiding toxicity has not as yet been realized.

One other setting where it may be desirable to have assays for antibiotic levels in the blood is in patients who are taking oral antibiotics for relatively serious diseases such as osteomyelitis and (much less commonly) bacterial endocarditis. Because the absorption of antibiotic after oral administration tends to be much more variable it is obviously desirable to know if a given patient on a given regimen of oral medication is achieving the desired blood levels. If the assay indicates he is not it is thus possible to change the medication, add an adjuvant (such as probenecid in the case of penicillins or cephalosporins) or make arrangements for parenteral administration of antibiotics.

CONCLUSION

Thus, two general types of measurements of antibiotics in serum are in use: either measurements for activity (primarily in determining the serum bacteriocidal activity) or for absolute blood levels. The latter is useful for regulating therapy of potentially toxic antibiotics with narrow toxic therapeutic ratios, or for guidance with use of oral antibiotics for treatment of severe diseases.

LITERATURE CITED

1. Alcid, C. V., and Seligman, S. J. Simplified assay for gentamicin in the presence of other antibiotics. *Antimicrob Agents Chemother 3:*559–561, 1973.
2. Haas, M. J. and Davies, J. Enzymatic acetylation as a means of determining serum aminoglycoside concentrations. *Antimicrob Agents Chemother 4:*497–499, 1973.
3. Lewis, J. E., Nelson, J. C., and Elder, H. A. Radioimmunoassay of an antibiotic; gentamicin. *Nature 239:*214–216, 1972.
4. Lund, M. E., Blazevic, D. J., and Matsen, J. M. Rapid gentamicin bioassay using a multiple-antibiotic-resistant strain of *Klebsiella pneumoniae. Antimicrob Agents Chemother 4:*569–573, 1973.
5. Riff, L. J., and Jackson, G. G. Pharmacology of gentamicin in man. *J Infect Dis 124:*S98–S105, 1971.
6. Sabath, L. D. 1973. Precision in the clinical use of antibiotics; the indications and laboratory techniques for achieving it. *Infect Dis Rev 2:*17–44, 1973.
7. Sabath, L. D., Casey, J. I., Ruch, P. A., Stumpf, L. L., and Finland, M. Rapid microassay of gentamicin, kanamycin, neomycin, streptomycin and vancomycin in serum or plasma. *J Lab Clin Med 78:*457–463, 1971.
8. Smith, D. H., Van Otto, B., and Smith, A. L. A rapid chemical assay for gentamicin. *N Engl J Med 286:*583, 1972.

17

A Five-Hour Disc Antibiotic Susceptibility Test

VICTOR LORIAN

With the technical assistance of A. Waluschka, C. Carruth, R. Gavrilovich, S. Johann, S. Kurien, and F. Marcano

Most of the chapters in this book discuss the appropriateness, the efficiency, or the clinical meaningfulness of various bacteriological procedures, that is, the optimization of standard procedures. In contrast, this chapter presents a detailed method for rapid, accurate antibiotic susceptibility testing. There is general agreement that the task most important for patient care that is carried out by a contemporary clinical laboratory is selecting the antibiotic most active in vitro. Therefore, we felt justified in presenting this rapid method.

The Bauer-Kirby (BK) method has become widely accepted for determination of bacterial susceptibility to antibacterial agents (2). It requires an overnight incubation. Recent modifications of this disc method reduced the incubation period to 5–8 hr (7, 8). Such modified tests were recommended for use in urgent clinical situations (1). Results could be obtained more rapidly only by using costly instrumentation (5, 9).

This chapter introduces a simple method which furnishes antibiotic susceptibility data within 5 hr of isolation of bacteria in pure culture. The method requires only the ordinary diagnostic bacteriology equipment used by the BK method.

MATERIALS AND METHODS

Sensitivity tests were performed using plates 150 mm in diameter which contained either Mueller-Hinton agar (MH) or MH with 5% sheep blood (Scott Co., Fiskeville, R. I.). Plates were prewarmed at 37 C for 1 hr prior to inoculation. Brain-heart broth (Difco) was used for preparing the inoculum. High strength sensitivity discs (BBL) containing the antibacterial agents listed in Table 17.1 were used.

A total of 1039 strains (see Tables 17.2–17.4) isolated from specimens submitted to the Microbiology Laboratory were tested. A few colonies with similar characteristics were removed with a swab and suspended in 3 ml of

TABLE 17.1
Antibacterial agents tested*

Agent	Used for: Gram-positive cocci	Used for: Gram-negative rods	Used for: Interpretation: Sensitive at and above zone diameters
			mm
Penicillin	+*	−	15†
Ampicillin	+	+	13†
Carbenicillin‡	−	−	14
Oxacillin	+	−	12
Cephalothin	+	+	16
Erythromycin	+	−	16
Chloramphenicol	+	+	16
Tetracycline	+	+	17
Streptomycin	+	+	13
Kanamycin	+	+	16
Clindamycin	+	−	19
Colistin	−	+	10
Nitrofurantoin§	+	+	16
Nalidixic acid§	−	+	16
Gentamicin	+	+	13

* + = tested, − = not tested.
† Except *Staphylococcus aureus* for which zone is 24 mm.
‡ Only for *Pseudomonas aeruginosa*.
§ Only for strains isolated from urine samples.

TABLE 17.2
Sensitivity Results: Enterobacteriaceae

Species (Strains) and No. of Tests		Zone Diameter at 24 Hours Compared to 5 Hours: Larger zones 2–4 mm	Larger zones 4–6 mm	Smaller zones 2–4 mm	Smaller zones 4–6 mm	Sensitive at 5 Hours, Resistant at 24 Hours	Resistant at 5 Hours, Sensitive 24 Hours	Growth Inadequate for Zone Measurement*
Enterobacter cloacae	(22)	68	27	17	1	1	3	0
	214	(31.7)†	(12.6)	(7.9)	(0.4)	(0.4)	(1.4)	
Enterobacter aerogenes	(31)	82	21	23	2	8	1	1
	297	(27.6)	(7.0)	(7.7)	(0.6)	(2.6)	(0.3)	(0.3)
Klebsiella	(87)	243	76	71	2	5	12	6
	875	(27.7)	(8.6)	(8.1)	(0.2)	(0.5)	(1.3)	(0.6)
Proteus mirabilis	(123)	282	89	152	33	9	10	1
	1353	(20.8)	(6.5)	(11.2)	(2.4)	(0.6)	(0.7)	(0.07)
Escherichia coli	(357)	1427	549	118	2	6	19	31
	3717	(38.3)	(14.7)	(3.1)	(0.05)	(0.1)	(0.5)	(0.8)
Others‡	(38)	84	23	31	11	6	2	10
	384	(21.8)	(5.9)	(8.0)	(2.8)	(1.5)	(5.2)	(2.5)
Totals	(658)	2186	785	412	51	35	47	49
	6822	(32.0)	(11.5)	(6.0)	(0.7)	(0.5)	(0.6)	(0.7)

* Four (0.61) strains showed insufficient growth after 5 hr.
† Numbers in parenthesis are percentages.
‡ *Enterobacter agglomerans, Providencia stuartii,* Citrobacter, Serratia, *Proteus morganii, Proteus vulgaris,* and *Proteus rettgerii.*

TABLE 17.3
Sensitivity results: Gram-positive cocci

Species (Strains) and No. of Tests		Zone Diameter at 24 Hours Compared to 5 Hours				Sensitive at 5 Hours, Resistant at 24 Hours	Resistant at 5 Hours, Sensitive at 24 Hours	Growth Inadequate for Zone Measurement*
		Larger zones		Smaller zones				
		2-4 mm	4-6 mm	2-4 mm	4-6 mm			
Staphylococcus aureus	(128) 1430	489 (34.1)†	89 (6.2)	59 (4.1)	0	1 (0.06)	2 (0.1)	19 (1.3)
Staphylococcus epidermidis	(80) 864	231 (26.7)	165 (19.0)	70 (8.1)	14 (1.6)	11 (1.2)	1 (0.1)	23 (2.6)
Streptococci group D	(78) 895	114 (12.7)	43 (4.8)	110 (12.2)	2 (0.2)	16 (1.7)	10 (1.1)	58 (6.4)
Streptococci: not group A not group D	(19) 204	36 (17.6)	15 (7.3)	28 (13.7)	0	0	0	11 (5.3)
Totals	(305) 3303	870 (26.3)	312 (9.4)	267 (8.1)	16 (0.5)	28 (0.8)	13 (0.4)	111 (3.36)

* Eight (2.6) strains showed insufficient growth.
† Numbers in parenthesis are percentages.

TABLE 17.4
Sensitivity Results: Pseudomonas*

Species (Strains) and No. of Tests		Zone Diameter at 24 Hours Compared to 5 Hours				Sensitive at 5 Hours, Resistant at 24 Hours	Resistant at 5 Hours, Sensitive at 24 Hours	Growth Inadequate for Zone Measurement†
		Larger zones		Smaller zones				
		2-4 mm	4-6 mm	2-4 mm	4-6 mm			
Pseudomonas aeruginosa	(80) 240	49 (20.4)‡	9 (3.7)	8 (3.3)	0	1 (0.41)	4 (1.66)	68 (28.33)

* Tested for colistin, gentamicin, and carbenicillin only.
† Seventeen (21.25) strains showed insufficient growth after 5 hr.
‡ Numbers in parenthesis are percentages.

brain-heart broth. The suspension was diluted to a final turbidity of #1 McFarland standard by adding small amounts of broth. This suspension was then inoculated by streaking on MH plates. Gram-positive cocci were tested on MH with blood. Sensitivity discs were dispensed on the plates, and the plates were incubated at 37 C for 24 hr. The zones were examined at 5 and 24 hr, and diameters were measured with a caliper dial (Helios #17713, VWR Scientific, Pequannock, N. J. 07440).

Departures from the BK technique were: (a) the inoculum was not preincubated (1), (b) the inoculum was twice the turbidity recommended by BK (that is, #1 McFarland standard or 3.0×10^8 organisms/ml instead of 0.5 McFarland standard or 1.5×10^8 organisms/ml), and (c) blood was added when Gram-positive cocci were tested. These modifications do not significantly change diameters of the zones of inhibition or their interpretation (1, 3, 4, 8, 10). To confirm this finding, 100 strains of the most common isolates in this laboratory (see Table 17.5) were tested according to

TABLE 17.5
Comparison of diameters of zones of inhibition and their interpretation as sensitive or resistant—obtained with different inoculae. Measurements were done after 18 hr at 37 C.

Species	No. of Strains	No. of Zones	Smaller Diameters† (mm) with #1 Compared to 0.5 McFarland Standard					Smaller Diameters (mm) with #2 Compared to 0.5 McFarland Standard				
			1 mm	2	3	4	R‡	1 mm	2	3	4	R‡
Escherichia coli	20	200	26.2*	10.4	3.3	0	0	35.8	10.8	7.5	2.5	0.7
Staphylococcus aureus	20	220	24.0	5.9	0.4	0	0	24.5	22.7	14.0	4.0	0
Klebsiella	20	200	21.5	7.0	1.5	0	1.0	32.5	16.0	3.5	1.0	1.5
Proteus mirabilis	20	200	21.0	5.0	2.0	0	0	29.0	9.5	2.0	0.5	0
Streptococci group D	10	110	26.3	4.5	0.9	0	0	25.4	9.0	1.8	0.9	0
Pseudomonae aeruginosa	10	30	33.0	0	0	0	0	33.0	13.3	0	0	0
Total	100	960	25.33	5.46	1.35	0	0.16	30.03	13.55	5.76	1.48	0.36

* Numbers are percentages.
† Some zones were 1–2 mm larger.
‡ Resistant while sensitive with 0.5 McFarland standard.

the BK technique using the recommended inoculum (0.5 McFarland standard), as well as inocula equal to #1 and #2 McFarland standard.

The BK tables (2) were used as a guide to interpret inhibition zone diameters. Since the ability of "intermediary" values to predict clinical antibiotic effectiveness has not yet been substantiated, there was no reason to segregate the intermediate values into a separate group. In fact, only a small fraction of all tests give results in the "intermediary" range (6); illustrative data are presented in Table 17.6. Accordingly, values which should have received an "intermediary" BK designation were assigned to either the "sensitive" or to the "resistant" classification. Those close to the borderline of sensitive were designated "sensitive" (see Table 17.1), all others were classified as "resistant." A similar classification has been used by others (7).

RESULTS AND DISCUSSION

The data presented in Table 17.5 confirm previous observations that the number of organisms used to inoculate plates used for antibiotic susceptibility have only a slight effect on the diameters of the zones of inhibition (3). When an inoculum of twice the BK standard concentration was used, 1.35% of all zone diameters were reduced by 3 mm; if a 4-fold increase in inoculum concentration was used, 7.24% of all zones had diameters 3–4 mm smaller than the BK standard. Interpretation was changed from sensitive to resistant for only 0.36% of all zones (see Table 17.5). That is, for 99.64% of the zones, interpretation remained the same. A few zones were slightly larger; however, this did not influence interpretation. These findings support the direct comparability of the following result, obtained with 1039 clinical isolates, with those of the standard BK method.

The number of zones showing a change in diameter, a change in sensitivity classification, and the number of zones in which growth was inadequate for measurement are presented in Tables 2–4 and 7–9. After incubating 5 hr at 37 C growth was adequate and the zones of inhibition were clearly measurable in 99.3% of the tests for Enterobacteriaceae, 96.6% of the tests for Gram-positive cocci, and 71.6% of the tests for *Pseudomonas aeruginosa*. In fact, when Enterobacteriaceae were tested many zones were readable after only 4 hr of incubation.

Most often, inability to measure zones was due to insufficient growth of a given strain on the whole plate. However, in several instances, while the amount of growth was sufficient, the edges of some zones were too vaguely defined to make accurate measurement possible. Those tests in which either growth was insufficient or zones were ill defined are presented in the tables as "growth inadequate for zone measurement."

Classification as "sensitive" or "resistant" after 5 hr was the same after 24 hr in 98.9% of the tests for Enterobacteriaceae, 98.7% of the tests for Gram-positive cocci, and 97.9% of the tests for *P. aeruginosa*.

Comparing the diameters of the zones of inhibition observed at 5 hr with those observed at 24 hr, it was noted that at the later observation they were larger in 43.5% and smaller in 6.7% of the tests for Enterobacteriaceae,

TABLE 17.6
Percentage of 17,025 isolates giving intermediary zone diameters in antibiotic disc sensitivity tests (BK technique)*

Species (Percent from All Isolates)	Intermediary Zones (Percentage)													
	Ampicillin	Cephalothin	Chloramphenicol	Tetracycline	Gentamicin	Kanamycin	Streptomycin	Colistin	Nitrofurantoin	Erythromycin	Penicillin	Methicillin	Lincomycin	Total
Escherichia coli (24.96%)	1.48	9.24	0.89	2.72	0.04	2.30	4.56	0.16	0.97	—	—	—	—	2.48
Proteus mirabilis (7.15%)	0.08	0.14	5.10	0.24	0.24	1.31	2.05	0	14.3	—	—	—	—	2.73
Staphylococcus aureus (13.04%)	—	0.05	0.14	0.23	—	—	—	—	0	1.58	4.05	1.71	0.14	0.98
Enterococci (11.9%)	—	33.5†	15.9	1.97	—	—	—	—	1.62	8.73	83.47†	0.54	0.5	24.09
Total	1.6	43.2	21.2	5.2	0.3	3.6	6.6	0.2	16.9	10.3	87.5	2.2	0.6	

* Adapted from computer data supplied by Doctors L. Kunz and R. Moelleriny, Massachusetts General Hospital, Boston.

† Penicillin and cephalothin alone are well known for their therapeutical ineffectiveness against enterococci, therefore all these intermediary results should be moved to the resistant category.

TABLE 17.7
Enterobacteriaceae

Agent	Zone Diameter at 24 Hours Compared to 5 Hours				Zone the Same	Not Readable	Total Drug Tests
	Zone larger		Zone smaller				
	2-4 mm	4-6 mm	2-4 mm	4-6 mm			
Ampicillin	219 (33.2)	44 (6.6)	41 (6.2)	1 (0.1)	348 (52.8)	5 (0.7)	658
Chloramphenicol	209 (31.7)	234 (35.5)	41 (6.2)	3 (0.4)	166 (25.2)	5 (0.7)	658
Tetracycline	165 (25.0)	27 (4.1)	74 (11.2)	3 (0.4)	384 (58.3)	5 (0.7)	658
Streptomycin	213 (32.3)	110 (16.7)	14 (2.1)	1 (0.1)	314 (47.7)	6 (0.9)	658
Kanamycin	338 (51.3)	50 (7.5)	9 (1.3)	0	256 (38.9)	5 (0.7)	658
Cephalothin	167 (25.3)	36 (5.4)	67 (10.1)	11 (1.6)	372 (56.3)	5 (0.7)	658
Carbenicillin	223 (33.8)	141 (21.4)	20 (3.0)	2 (0.3)	267 (40.5)	5 (0.7)	658
Colistin	13 (1.9)	5 (0.7)	8 (1.2)	3 (0.4)	626 (94.9)	3 (0.4)	658
Gentamicin	348 (52.8)	79 (12.0)	11 (1.6)	1 (0.1)	215 (32.6)	4 (0.6)	658
Nitrofurantoin*	97 (21.5)	14 (3.1)	90 (20.0)	19 (4.2)	227 (50.4)	3 (0.6)	450
Nalidixic acid*	194 (43.1)	45 (10.0)	37 (8.2)	7 (1.5)	164 (36.4)	3 (0.6)	450
Total Tests	2186	785	412	51	3339	49	6822

* Only for strains isolated from urine samples.

larger in 35.7% and smaller in 8.6% of the tests for Gram-positive cocci, and larger in 24.1% and smaller in 3.3% of the tests for *P. aeruginosa*. Only 52% of all tests had the same diameter at both observations. Similar differences have been reported when observations were made at 7 and 20 hr of incubation (8).

There were differences in zone diameters at the two readings for all antibiotics. There were no large differences among the various drugs. As Table 17.7 shows, the extreme values were: 67.2% of the chloramphenicol tests increased in zone diameter, 24.2% of the nitrofurantoin tests decreased in diameter, and colistin gave the same size for 94.9% of the tests.

Zones in which the diameters became smaller are easily explained by bacterial overgrowth and are probably due to antibiotic-inactivating enzymes. However, in the zones which became larger, there was an apparent disappearance of growth. Changes in zone diameter were not related to specific drugs, although chloramphenicol generally produced larger diameters, nitrofurantoin smaller diameters, and colistin the smallest variation in diameters. Obviously, there could be some lysis produced by β-lactam antibiotics. But with the other antibiotics what actually occurred was that they continued to diffuse, eccentrically from the disc, throughout the incubation period. Therefore, the zone containing the minimum inhibitory concentration of antibiotic in the agar continued to expand. Since subse-

TABLE 17.8
Gram-positive cocci

Agent	Zone Diameter at 24 Hours Compared to 5 Hours				Zone the Same	Not Readable	Total Drug Tests
	Zone larger		Zone smaller				
	2–4 mm	4–6 mm	2–4 mm	4–6 mm			
Penicillin	34 (11.14)	17 (5.57)	44 (14.42)	2 (0.65)	199 (65.24)	9 (2.95)	305
Ampicillin	42 (13.77)	24 (7.86)	37 (12.13)	1 (0.3)	193 (63.27)	8 (2.62)	305
Oxacillin	86 (28.19)	16 (5.24)	15 (4.9)	1 (0.3)	176 (57.7)	11 (3.6)	305
Erythromycin	89 (29.1)	39 (12.7)	38 (12.4)	2 (0.6)	128 (41.9)	9 (2.9)	305
Chloramphenicol	99 (32.4)	17 (5.5)	41 (13.4)	0	137 (44.9)	11 (3.6)	305
Tetracycline	72 (23.6)	45 (14.7)	32 (10.4)	2 (0.6)	144 (47.2)	10 (3.2)	305
Streptomycin	58 (19.0)	13 (4.2)	4 (1.3)	2 (0.6)	215 (70.4)	13 (4.2)	305
Kanamycin	115 (37.7)	32 (10.4)	12 (3.9)	1 (0.3)	137 (44.9)	8 (2.6)	305
Cephalothin	94 (30.8)	44 (14.4)	24 (7.8)	2 (0.6)	132 (43.2)	9 (2.9)	305
Gentamicin	102 (33.4)	37 (12.1)	10 (3.2)	3 (0.9)	141 (46.2)	12 (3.9)	305
Clindamycin*	79 (31.2)	28 (11.0)	10 (3.9)	0	125 (49.4)	11 (4.3)	253
Total Tests	870	312	267	16	1727	111	3303

* Not tested for enterococci isolated from urine.

TABLE 17.9
Pseudomonas aeruginosa

Agent	Zone Diameter at 24 hours Compared to 5 Hours				Zone the Same	Not Readable	Total Drug Tests
	Zone larger		Zone smaller				
	2–4 mm	4–6 mm	2–4 mm	4–6 mm			
Carbenicillin	24 (30.0)	9 (11.2)	4 (5.0)	0	15 (18.7)	28 (35.0)	80
Colistin	7 (8.7)	0	1 (1.2)	0	55 (68.7)	17 (21.2)	80
Gentamicin	18 (22.5)	0	3 (3.7)	0	36 (45.0)	23 (28.7)	80
Total Tests	49	9	8	0	106	68	240

quent growth can occur only in areas of antibiotic concentration less than the minimum inhibitory concentration, the heavy growth seen at 24 hr was further from the disc than the thin 5-hr growth. Although the growth at 5 hr was sparse, the zones of inhibition were clearly measurable. By 24 hr, this sparse growth was overshadowed by the heavy growth. Furthermore, slight evaporative loss from the plate made the layer of 5-hr growth even thinner and less visible. If the plates are carefully examined in oblique light, the 5-hr growth zone can be seen (Fig. 17.1).

Figure 17.1

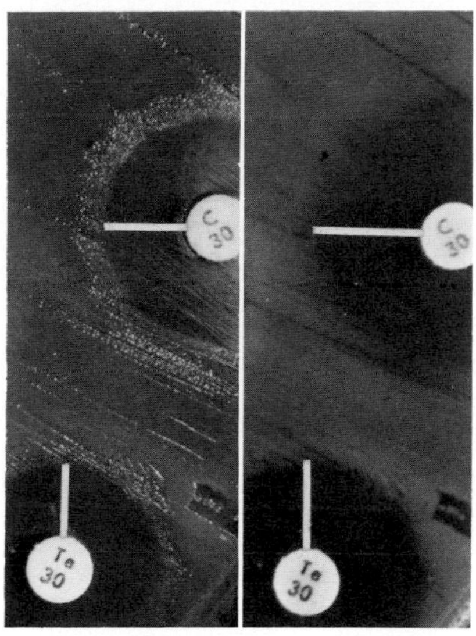

Zones of inhibition produced by discs of chloramphenicol and tetracycline in the growth of a strain of *Escherichia coli* after 24 hr of incubation. (*Left*) Illuminated laterally so that the 5-hr thin growth becomes visible. (*Right*) The same zones illuminated from the top showing only the 24-hr thick growth. (*White bars* emphasize differences in radii of zones.)

These changes in zone diameter measurements caused classification as "sensitive" or "resistant" to be changed in less than 2% of the tests. Similar results have been obtained by others (1, 8). In fact, it was shown long ago that the large majority of strains are either very resistant or highly sensitive (6).

The BK technique is widely accepted, although its value as a predictor for effective antibiotic therapy has not yet been firmly substantiated with clinical therapeutic evidence. The modified inoculum affects the zones of inhibition only slightly; therefore, the predictive value of the modified technique is exactly comparable to the original BK technique. This slight change in inoculum concentration provides the considerable advantage of speed without requiring expensive equipment. Results can be obtained within 5 hr (the same day) which, in most cases, is within 24 hr from the time a specimen is received by the microbiology laboratory.

There is unanimous agreement that the results of antibiotic susceptibility tests provide a valuable therapeutic guide. Patients with infections await antibiotic therapy; drug choice depends on the results of in vitro sensitivity testing. A method which provides such information within 5 hr—and with 98% accuracy—should be used routinely rather than only in the emergency situation. Indeed, when can treatment of septic patients be considered routine?

LITERATURE CITED

1. Barry, A. L., Joyce, l. J., Adams, A. P., et al. Rapid determination of antimicrobial susceptibility for urgent clinical situations. *Am J Clin Pathol 59:*693–699, 1973.
2. Bauer, A. W., Kirby, W. M., Sherris, J. C., et al. Antibiotic susceptibility testing by a standardized single disk method. *Am J Clin Pathol 36:*493–496, 1966.

3. Ericsson, H., Högman, C., and Wickman, K. A paper disc method for determination of bacteral sensitivity to chemotherapeutic and antibiotic agents. *Scan J Clin Lab Invest Suppl 11:*21, 1954.
4. Ericsson, H., and Sherris, J. C. Antibiotic sensitivity testing. Report of an international collaborative study. *Acta Pathol Microbiol Scand Suppl 217(B):*29, 1971.
5. Isenberg, H. P., Reicher, A., and Wiseman, D. Prototype of a fully automated device for the determination of bacterial antibiotic susceptibility in the clinical laboratory. *Appl Microbiol 22:*980–986, 1971.
6. Kirby, M. W., Yoshihara, G. M., Sunsted, K. S., et al. Clinical usefulness of a single disc method for antibiotic sensitivity testing. *Antibiot Ann 1956–1957:*892–897, 1957.
7. Kluge, R. M. Accuracy of Kirby-Bauer susceptibility tests read at 4, 8, and 12 hours of incubation; comparison with readings at 18 to 20 hours. *Antimicrob Agents Chemother 8:*139–145, 1975.
8. Liberman, D. F., and Robertson, R. G. Evaluation of a rapid Bauer-Kirby antibiotic susceptibility determination. *Antimicrob Agents Chemother 7:*250–255, 1975.
9. Thornsberry, C., Gavan, T. L., Sherris, J. C., et al. Laboratory evaluation of a rapid, automated susceptibility testing system; report of a collaborative study. *Antimicrob Agents Chemother 7:*466–480, 1975.
10. Washington, J. A., II. *Laboratory Procedures in Clinical Microbiology,* p. 289. Boston: Little, Brown & Co., 1974.

18

Medical Microbiology: Pediatric Aspects

ERWIN NETER

Fundamentally, the approaches to the microbiologic diagnosis of infectious diseases apply to all subjects, irrespective of age, sex, nutrition, and the presence or absence of underlying diseases. Nonetheless, significant differences exist in the details regarding the various patient groups. Thus, the selection of a given procedure, the particular specimen, and the interpretation of the findings may differ significantly between the newborn, young infant or child, and adults. It is the pediatric aspects of medical microbiology that are the subject of this brief presentation. Only a few illustrative examples are given to underline certain unique aspects of pediatric microbiology.

INFECTIOUS DISEASE AND INDICATIONS FOR MICROBIOLOGIC EXAMINATION

Certain infections occur with undue frequency in the newborn or child, and in other infections the classic features, as seen in the adult, may be absent in the pediatric age group. Thus, indications for certain microbiologic tests may differ substantially depending upon the age of the patient.

Septicemia and meningitis caused by group B hemolytic streptococci occur almost exclusively during the first 4 months of life (1, 7, 11, 12). The modes of transmission and the prophylactic aspects of this infection are quite unique. Although increasing number of cases are being reported in adults, *Haemophilus influenzae* type b meningitis is far more common in children under 4 years of age, and particularly between 3 months and 2 years of age, than in older children and adults (15). The striking effect of age on the occurrence of the various forms of bacterial meningitis is shown in Tables 18.1 and 18.2. Table 18.1 provides data on children and adults, and Table 18.2 gives information of recent experience at the Children's Hospital of Buffalo. It is of particular note that *Escherichia coli* meningitis, for unexplained reasons, has become a rarity even in the newborn period as compared to group B streptococcal meningitis that is seen far more frequently. Another pediatric infection is the so-called staphylococcal scalded skin syndrome due to the exfoliative toxin of phage group 2 strains of *Staphylococcus aureus*. Thus, when dealing with such patients, phage-typing may be of interest even in the absence of staphylococcal epidemics

TABLE 18.1
Age and bacterial meningitis*

Microorganisms	Approximate Percentage of Cases at Ages:			
	Newborn	2 Months to 5 years	5 to 40 Years	>40 Years
Haemophilus influenzae	0	50	5	2
Neisseria meningitidis	0	25	45	10
Streptococcus pneumoniae	10	12	20	50
Other *Streptococcus* sp.	8	3	5	5
Escherichia coli and other Enterobacteriaceae	55	0	0	10
Other microorganisms	27	10	25	23

* Adapted from Benner and Hoeprich (2).

TABLE 18.2
Bacterial meningitis and age (Children's Hospital, 1973–1975)

Microorganisms	No. of Patients at Ages:					Totals
	0–30 Days	1–4 Months	4–24 Months	2–6 Years	>6 Years	
Haemophilus influenzae	1	6	45	23	2	77
Streptococcus pneumoniae	1	4	11	3	2	21
Neisseria meningitidis	2	1	2	5	5	15
Streptococcus group B	10	5	1	0	1	17
Other *Streptococcus* sp.	0	0	1	0	0	1
Escherichia coli and other Enterobacteriaceae	1	0	0	0	0	1
Totals:	15	16	60	31	10	132

or nosocomial infection (5, 13). Patients with severe cystic fibrosis, a genetic disease with characteristic bacterial complications, have a limited life span and, therefore, are seen far more frequently by pediatricians than by internists. Mucoid strains of *Pseudomonas aeruginosa* are encountered almost exclusively in this condition (3, 4, 14). It follows from these few examples that the laboratory dealing with specimens from children may pay special attention to serogrouping of hemolytic streptococci of group B, serotyping of *Haemophilus influenzae* and determination of its antibiogram, phagetyping of staphylococci, and to mucoid strains of *P. aeruginosa*, which do not form the characteristic colonies on all culture media.

A few decades ago, severe outbreaks of diarrheal disease occurred among newborns and young infants. Certain serotypes of *E. coli* were found to be associated with some of these epidemics and were referred to as enteropathogenic. Thus, in pediatric institutions antigenic analysis of *E. coli* recovered from fecal specimens of infants with diarrhea became a routine procedure. During more recent times, however, such epidemics have become extraordinarily rare and the clinical disease very much milder. Thus, serotyping IS NO LONGER RECOMMENDED AS A ROUTINE DIAGNOSTIC AND EPIDEMIOLOGIC PROCEDURE. Enteritis of children and adults may be due to enterotoxin-producing or enteroinvasive strains of *E. coli*. Unfortunately, the determination of these characteristics has not yet become a routine procedure. It is known that the above mentioned antigenically characterized serotypes do not usually exhibit these characteristics, but it is not known which factors account for the apparent lesser virulence of current isolates (6, 8).

Indications for bacteriologic examination often differ between pediatrics and internal medicine because the characteristic classic features of a given infection seen in adults may be absent in children. For example, septicemia may exist in infants even in the absence of fever or leukocytosis, and blood cultures may be indicated merely because the child is ill and a definitive diagnosis, such as pneumonia or enteritis, is not justified. Similarly, the classic symptoms and signs of meningitis may be absent in young infants. Thus, examination of cerebrospinal fluid is often indicated when an unexplained acute illness is present in such an infant, since it is imperative to diagnose bacterial meningitis without delay, even of hours. One more example: Bacterial infection of the urinary tract in the infant and young child may progress and result in serious renal damage without the characteristic urinary symptoms and thus escape detection; failure of the child to grow and thrive may be the incisive clue to the diagnosis, provided by appropriate microbiologic examination of properly obtained urine to establish the etiology of the disease and as a guide to chemotherapy.

SPECIMEN AND BACTERIOLOGIC METHODS

Certain adaptations have to be made regarding specimens from infants and young children as compared to adults. For example, it is not feasible to ask for four blood specimens of 10 ml each for cultural purposes, and the examination for bacteremia has to be carried out with minimal amounts of blood and inherent shortcomings. In addition, contamination of such blood cultures, particularly when obtained from the inguinal region or the umbilical cord, is encountered somewhat more frequently and differentiation between bacteremia and contamination can be difficult. Another example of the particular problems encountered in pediatrics pertains to bacteriologic diagnosis of pneumonia in young infants. Clearly, sputum cannot be obtained in the usual manner; nasopharyngeal swabs or gastric aspirates are at best an inadequate substitute; blood cultures are often negative; and specimens directly obtained from the lung lesion are rarely available, thus compounding the difficulties in providing an etiologic diagnosis. Two additional examples regarding the special problems in specimen selection may be mentioned. *Bordetella pertussis* may be demonstrated by the fluorescent antibody (FA) technique in nasopharyngeal smears rather than by the cultural examination of sputum (cough plate). The procurement of aseptically obtained urine specimens in the young infant also presents problems. When a definitive laboratory diagnosis of urinary tract infection is needed and doubt exists regarding the reliability of the voided urine specimen, suprapubic aspiration of the bladder has proved to be highly reliable.

NORMAL MICROBIAL FLORA

For the interpretation of microbiologic data information on the normal flora is particularly relevant when dealing with specimens such as sputum or feces. Thus, striking differences exist in the normal flora between the newborn and other age groups. Unless an infectious disease process, such as syphilis, is taking place during pregnancy, the fetus lives in an entirely sterile area. During delivery, exposure to organisms of the genital tract of the mother represents the first host-microorganism confrontation. More often than not, the acquisition of various microorganisms does not produce any ill effects. A particularly incisive example of an unusual initial bacterial flora has come to light by the study of newborn infants harboring group B hemolytic streptococci in the upper respiratory tract, the ear canal, the feces, and/or on the skin. Some of these infants are clinically entirely well, although in other instances sepsis and meningitis may develop within the first few weeks of life (7). Similarly, in the feces of newborn babies, staphylococci may be present, and Enterobacteriaceae, notably *E. coli*, may be absent. These findings must not be misinterpreted as evidence of the existence of staphylococcal enterocolitis nor as an abnormal flora. Fecal contamination in the genital region is regularly encountered in untrained infants, and cultures of the pelvic skin, vulva, etc., in this age group usually provide a different picture than that seen in older children and adults. Thus, correct interpretation of bacteriologic data often requires correlation with the age of the patients.

IMMUNOLOGIC DIAGNOSIS

Two important considerations arise in the study of the antibody response of infants and young children as compared to that of adults: (a) obviously, only small amounts of blood may be available for such investigations, particularly when dealing with seriously ill premature infants, and (b) more importantly, antibodies to microbial pathogens are found often in healthy subjects, and these titers must not be automatically applied to those of infants and children. The "normal" antibody titers often are substantially lower in the young. For example, a titer of antibodies at the higher limit of normal for the adult may be quite diagnostic in young children. Thus, a cold agglutinin titer of 1:64 has an entirely different connotation when found in a 2-year-old child than a 25-year-old adult and may be quite diagnostic. Similarly, an increase in the titer of antibodies from 1:2 to 1:16, a rise which is highly significant, will be overlooked if the lowest serum dilution examined is 1:20. In addition, it is important to realize that IgG antibodies are essentially quantitatively transferred from mother to child and, therefore, present in the newborn at levels identical with those found in adults. In contrast, IgM antibodies in the newborn are either absent or present in very low titers. IgM antibodies in significant amounts in cord blood, therefore, are usually diagnostic of fetal infection (9).

PRELIMINARY REPORT

Adequate communication between physicians and microbiology laboratories is mandatory under all circumstances. Such communication is particularly important when dealing with serious or potentially serious diseases and those population groups at great risk of nosocomial infection. The newborn, and particularly the premature infant, belongs to this group. In our laboratory, we send preliminary reports providing information on initial microscopic and cultural findings. These preliminary reports are then replaced with the final reports. In addition, findings that may be particularly important are communicated by telephone. Daily rounds with Residents and Fellows have established a unique communication system.

MINIMAL, INTERMEDIATE, AND MAXIMAL EXAMINATION

So far as patient care is concerned, it is unrealistic to expect that all patients with definitive or suspected infections be studied microbiologically. Such an approach would neither be productive nor financially and technically feasible. The following important questions face all diagnostic microbiology laboratories: In what detail should certain specimens be examined and the isolates characterized, and, further, which examinations should be expedited? Clearly, a standard for all laboratories cannot possibly be devised.

The question of minimal, intermediate, or maximal examination will be answered differently depending upon the particular purpose: (a) patient care, including guidance to therapy; (b) epidemiology, with particular reference to health benefits for family, community, or the nation, including infection control (c) research, such as investigations of laboratory procdures or studies dealing with prophylactic and therapeutic measures; and (d) teaching, particularly of technologists or medical students.

Minimal examination solely for the benefit of the patient may require only a microscopic examination of urine, for example, or a film for the demonstration of gonococci in an obviously infected individual. Similarly, identification by a few critical tests of a strain of *E. coli* from the urine is adequate without the total battery of characteristics. Of course, one could visualize, but not endorse, the determination of an antibiogram without identification of the organism as a guide to therapy. If the physician were to recognize on clinical grounds alone the problem cases and those whose infection does not terminate spontaneously or can be treated successfully without laboratory examination, then, the laboratory would carry out more complete rather than minimal examinations of significantly fewer specimens than are processed today.

So far as patient care is concerned, antibiograms are not needed for certain microorganisms, such as hemolytic streptococci of group A, provided that the drug to be used by the physician is effective almost without exception (penicillin vs. tetracycline). Species or serotype identification of salmonellae, other than *Salmonella typhi*, when isolated from the feces of patients with mild gastroenteritis, will not provide meaningful information for the care of the particular patient, but may provide indispensable data for epidemiologic investigation. These considerations also apply to many other pathogens, such as staphylococci, meningococci, and *P. aeruginosa*. To give only one example from the pediatric age group: a suspected staphylococcal infection, even minor in nature, of a newborn infant should be studied bacteriologically and the isolate preserved; when the occasion arises, this strain can be compared with other isolates from infants born in the same institution or admitted to the same nursery. At times it is only by inquiry about discharged infants that the existence of a staphylococcal epidemic in a nursery comes to light, since the incubation period may be longer than routine hospitalization. Thus, serotyping or serogrouping, phagetyping, colicintyping, etc., may provide the needed information to determine whether infection of a few patients represents an epidemic or unrelated cases. In case of doubt, it is reasonable to preserve isolates in the laboratory until the epidemiologic situation has been clarified. This procedure will avoid unnecessary examinations without making such information unavailable when needed.

Certain additional studies on isolates, which are not among the routine procedures, may clarify the nature of the disease. For example, serotyping of *E. coli* may provide the definitive evidence that, what appears to be clinically a relapse of a urinary tract infection, in reality is another infection by an entirely different serotype of *E. coli*. Needless to say, the implications regarding therapy of the existing infection and prevention of a second infection by an unrelated microorganism are substantial (10).

The methods for the identification of Enterobacteriaceae have improved to an extraordinary extent in the last few years and thus provide the basis for definitive etiologic diagnosis of pathogens formerly identified as members of the ill-defined group of paracolon bacilli. To give one more example: recently, we observed a second attack of pneumococcal meningitis a few months after an initial episode of pneumococcal meningitis and, by means of typing of the pneumococci, could establish the fact that, indeed, both attacks were due to the identical type 23. Clearly, such a finding suggests that, more likely than not, a focus of infection existed in the patient requiring special attention.

Maximal examinations often require the contributions of other laboratories, such as State Laboratories, the Center for Disease Control, or laboratories of certain researchers. Acknowledgment of the significant contributions of these laboratories is in order.

It is also clear that the research of yesterday may become routine today. Since it was shown that *Haemophilus influenzae* type b, causing meningitis and epiglotitis, frequently associated with bacteremia, may be resistant to ampicillin, the determination of the antibiogram, formerly unnecessary, has become a routine procedure.

It is evident that the diagnostic laboratory has to take into account new information emerging from on-going research to provide the information needed by the physician for optimal patient care. The microbiologists may wish to remember the advice given to physicians: "The physician should treat what the patient has got, and he should not decide the patient has got what he likes to treat."

More complete, rather than minimal examination, is often required in teaching institutions and in support of certain clinical research programs.

One problem facing clinical laboratories requiring further consideration relates to the financial support of these examinations that are not required for the care of individual patients. For example, funding of studies on colonization of hospitalized subjects in connection with control of nosocomial infections as well as more complete examinations in support of education and/or on-going research all too often presents problems. It is progress in medicine and microbiology and unresolved questions that are a challenge to the diagnostic microbiologist in revising the scope and nature of microbiologic testing, be it minimal, intermediate, or maximal.

LITERATURE CITED

1. Baker, C. J., and Barrett, F. F. Transmission of group B streptococci among parturient women and their neonates. *J Pediatr 83:*919–925, 1973.
2. Benner, E. J., and Hoeprich, P. D. Acute bacterial meningitis, pp. 931–944, in P. D. Hoeprich (Ed.), *Infectious Diseases*. Hagerstown, Md.: Harper & Row, 1972.
3. Diaz, F., Mosovich, L. L., and Neter, E. Serogroups of *Pseudomonas aeruginosa* and the immune response of patients with cystic fibrosis. *J Infect Dis 121:*269–274, 1970.
4. Doggett, R. G., Harrison, G. M., and Wallis, E. S. Comparison of some properties of *Pseudomonas aeruginosa* isolated from infections in persons with and without cystic fibrosis. *J. Bacteriol 87:*427–431, 1964.
5. Melish, M. E., and Glasgow, L. A. The staphylococcal scalded-skin syndrome. Development of an experimental model. *N Engl J Med 282:*1114–1119, 1970.
6. Neter, E. Enteropathogenicity of *Escherichia coli*. *Am J Dis Child 129:*666–667, 1975.
7. Neter, E. 1975. Meningitis and bacteremia due to group B hemolytic streptococci, pp. 125–137, in A. von Graevenitz and T. Sall (Eds.), *Pathogenic Microorganisms from Atypical Clinical Sources*, Vol. 1. New York: Marcel Dekker, Inc., 1975.
8. Neter, E. *Escherichia coli* as a pathogen. *J Pediatr 88:*000–000, 1976.
9. Neter, E., and Milgrom, F. (Eds.) Maturation of the immune system, pp. 14–63, in *The Immune System and Infectious Diseases*. Basel: S. Karger, 1975.
10. Neter, E., Oberkircher, O. R., Rubin, M. I., Steinhart, J. M., and Krzeska, I. Patterns of antibody response of children with infections of the urinary tract. *Pediatr. Res. 4:*500–509, 1970.
11. Quirante, J., Ceballos, R., and Cassady, G. Group B β-hemolytic streptococcal infection in the newborn. I. Early onset infection. *Am J Dis Child 128:*659–665, 1974.
12. Reid, T. M. S. Emergence of group B streptococci in obstetric and perinatal infections. *Br Med J 2:*533–536, 1975.
13. Rogolsky, M., Wiley, B. B., and Glasgow, L. A. Phase group II staphylococcal strains with chromosomal and extrachromosomal genes for exfoliative toxin production. *Infect Immun 13:*44–52, 1976.
14. Seidmon, E. J., Mosovich, L. L., and Neter, E. Colonization by Enterobacteriaceae of the respiratory tract of children with cystic fibrosis of the pancreas and their antibody response. *J Pediatr 87:*528–533, 1975.
15. Sell, S. H. W., and Karzon, D. T. (Eds.) *Hemophilus Influenzae*. Nashville, Tenn.: Vanderbilt University Press, 1973.

19

Significance of Environmental Microbiology in Nosocomial Infections and the Care of the Hospitalized Patient

HENRY D. ISENBERG

Hospital-acquired infectious disease is as old as the first hospital. Attempts to understand and control this dread complication date back to almost the same period. As soon as microorganisms were recognized as purveyors of infectious disease they became implicated in the spread of contagion in institutions (88).

There are several considerations basic to the discussion of the role of environmental microorganisms in the care of the hospitalized patient. It is accepted that protista are involved in infectious diseases primarily and secondarily. Similarly, the frequency of transmission is proportional to the density of susceptible hosts (29). No one should contest the assumption that a hospital is intended for the care of patients. It is also assumed—without fear of contradiction—that sick persons are more susceptible to infectious complications following microbial colonization (5, 12, 23, 29, 30, 32, 33, 38, 56, 83, 90–92).

Close scrutiny of these "self-evident clues" leads to the realization that mechanisms of infection and infectious disease are not understood as well as assumed a priori (80). Certain infections fall very easily within the scope of Koch's postulates. Many do not. It would appear that the more modern view of these circumstances, namely that disturbances of the host/parasite equilibrium lead to clinical disease, that the host's health and experience are the determinants of the overt clinical manifestations, that many if not all so-called pathogenic microorganisms do not cause disease at all times and that members of the "normal" microbiota may participate serendipically in disease production when the individual is ill and compromised, must figure prominently in consideration of infections especially those termed nosocomially or hospital acquired (35).

Acknowledgment of the greater susceptibility of the patient must be coupled with the appreciation that the hospital environment is the repository of microbial particles representing lesions as well as the normal microbiota of patients, hospital staff, and visitors creating an ecosystem which differs qualitatively and quantitatively from the intimate biosphere of each person and certainly that of the newly admitted patient (11, 12, 14, 24, 39, 40, 43, 45, 70, 71, 73, 79, 81–85, 87, 90, 92, 93). In addition, the various treatments accorded the ill individual, the procedures required to restore her or his health, and the pharmacological and manipulative exposures he or she must experience are regarded as opportunities to increase the contact between the patient and the hospital's microbiota (1, 2, 9, 10, 15, 16, 22, 32–34, 42, 47–53, 55, 59, 63–65, 67–69, 75, 89). The great advances in medical science can correct or palliate conditions which would have led to the person's demise a few years ago. Very frequently, these outstanding accomplishments are accompanied by requirements of great compromise of the patient's specific and nonspecific defense systems increasing the opportunity for complication by the microbial residents of the hospital environment as well as by the patient's endogenous microorganisms. Invariably, the situation increases the opportunity for disease-passaged organisms to join the environmental microbiota. Thus, the institutional environment is not only a repository but also a dynamic reservoir permitting selective and general multiplication of many of the microorganisms (29).

The mere presence of many diverse protista is but a first plateau in the sequence of patient colonization by microorganisms not representing his own microflora. All strategies directed at patient protection against institutional contamination must take into account the endless opportunities for transmission provided by air (10, 13, 27, 28, 37, 55, 59, 77, 78, 81, 84, 94), fomites (12, 14, 16, 23, 25, 40, 43, 58, 66, 70–72, 76, 79, 82, 87, 88, 90, 91, 93), personnel (7, 8, 12, 13, 15, 20, 24, 25, 29, 30, 32, 34, 40, 42, 45, 47, 54, 56, 62, 63, 69, 70, 71, 73, 76, 83, 85, 87, 89, 92, 93), instruments (1, 2, 5, 7, 8, 15, 24, 25, 32, 38, 45, 49, 50, 53, 54, 59, 64, 65, 68, 69, 71, 75, 89, 93), medications (8, 9, 20, 25, 44, 48, 49, 51, 52, 55, 59, 63, 69, 93), procedures, contact with other patients (74), food (39, 57, 74), to name but a few. Also, it cannot be overemphasized that each patient will respond very individually to the intrusion of "foreign" biological particles into his intimate environment. While the total number and the specific type of colonizers may influence the transition from mere "new resident" to accomplice in disease production, it is this susceptibility of the individual patient at any given moment during hospitalization which will play the decisive role in the transformation of one or more microbial hospital or autochthonous representatives to active participants in infection.

The controversy concerning the proper method of recognizing the environmental microbiota with the potential to complicate the patient's recovery has not been resolved (3, 4, 19, 25, 26, 31, 32, 38, 83, 87, 91–93). Basic to any realistic appreciation of the significance of environmental microbiota in the care of patients must be the ability of the microbiologist to recognize these organisms. Unfortunately, this is a formidable task made the more difficult by the lack of similarity between institutions. It would appear that the individuality of each medical facility can be compared to the individual responses of one patient as compared to another. Thus, the microbiologist is confronted with a task for which he has some but not complete guidance from the experience of his colleagues in other institutions (5, 7, 8, 33, 46, 66, 85–87). The state of each individual medical facility is very dynamic vis-à-vis the microorganisms to be considered residents of an institution. All people who work or come in contact with the institution, the goods and supplies which enter and are used in the institution, the nature of the patients serviced by the institution, and the larger community with all of its multifarious practices influence in the final analysis, the type of microorganisms which predominate in the hospital environment. This consideration must be viewed in conjunction with the known ability of endogenous microorganisms carried by each patient to become involved in infectious complications. It has been suggested (34) that bacteria, fungi, and other protista may exist as minority groups among the amphibionts resident on the individual. These minority groups become dominant whenever selective pressures are exerted be they in the form of community practices or therapy within the institution. The complexity of the situation is increased further by a variety of personnel actions within institutions which tend to select those representatives of the microbiota resistant to the modalities of treatment employed within the institution. The hospital finds itself, therefore, at the very apex of the selective pressure pyramid which is based in the community but which provides the best opportunity within the confines of the medical facility. The role of plasmid-associated multiple resistance in the equation of hospital acquired infections (21) cannot be ignored.

The acknowledged role of endogenous and hospital multiple-resistant microorganisms or drug selected protista in nosocomial disease complicates the recognition of those particles thought to originate from the hospital environment. While many of these organisms regardless of their origin do not often lead to infectious disease in the community, there is no guarantee that isolants can be classified as to their origin with impunity. The only means at the disposal of the microbiologist vis-à-vis this very complicated problem is the change that may occur in the microbiota of the patient tested on admission to the hospital and again at some time during his hospitalization. One must hasten to admit that such data are very sparse if available at all. This lack of hard facts concerning the designation of environmental microbiota is magnified by a paucity of information concerning the actual mode of transmission from the environment to the patient despite the many studies performed (9, 12–14, 16, 18, 22–25, 27–30, 37, 38, 40, 42, 43, 45–47, 51–73, 75–79, 81–94). Space technology's concern with contamination of extraterrestrial devices has allowed detailed understanding of the control of microbial contamination under conditions of extreme isolation (46, 61, 66, 85, 86). Nevertheless, it seems unlikely that such extreme conditions can ever be achieved or, for that matter, that they would be desirable for the care of all patients under terrestrial conditions.

It is understandable, however, that this basic appreciation of the conditions which lead to the dissemination of microorganisms in a specific environment would lead to the suggestion that the entire institution should be monitored continuously and constantly. Very obviously this is an impossible task to perform. It is just as obvious that such information may not clarify the problem of hospital microorganisms and their rate of colonization and disease production in a patient. While it would seem logical to substitute an occasional environmental spot check in order to sequester information concerning the nature and quantity of environmental protista, these procedures would be even less productive in providing pertinent facts as has been stated recently by the American Public Health Association (4) and the American Hospital Association (3). However, it is just as unacceptable to any group of scientists and other interested individuals to state that no knowledge of the hospital microbiota can be gained at any time or that the hospital and its microorganisms may not play a role in the contagious diseases of its patient population. At the risk of redundancy it must be stated again that while a number of very excellent tomes exist which direct attention and procedures to various practices geared at the control of nosocomial disease, these guides assign minor considerations to the microbial hospital residents. It is, therefore, impossible at this time to provide anything other than suggestions to approaches which may permit the study and understanding of the role of the hospital microbiota in infectious diseases of patients, their recognition and measures which could be used to control them.

The recognition of potentially dangerous hospital microorganisms rests squarely on the capabilities of the microbiology laboratory to identify microorganisms at the species level and under certain circumstances to biotype or variant subdivision. Justifiable concern about the need to identify all microorganisms isolated from clinical specimens to this extent has been expressed recently in very strong terms (6). However, those who wish to appreciate the role of environmental microorganisms in colonization and infection of hospitalized patients must modify the analyses of their specimens at least to the level where repeated appearances of certain organisms will lead to the extension of identification procedures. One might suggest (33, 34) that these categories of microorganisms do not encompass all types recoverable from the specimen but are limited to certain members of the family Enterobacteriaceae, staphylococci, representatives of the nonfermenting Gram-negative rods, certain yeasts and fungi, and, on occasion, representatives of the genus *Bacillus*. Greater selection can even be exercised by applying this type of identification to select clinical specimens. Certainly such clinical material as blood submitted for culture, spinal fluids, and postoperative wound cultures will fall within this category without argument from anyone. Urine specimens from patients who have been exposed to indwelling catheters as well as aspirates from individuals with protracted exposure to respiratory assist machinery will be considered within this grouping. Actually, the age old dictum that a specimen cannot be interpreted properly unless a proper history concerning the exposure of the patient, his recent experiences vis-à-vis travel, treatment, and clinical suspicion are provided, cannot be ignored in the attempt to recognize the origins of the microbial particles isolated. It is also at this juncture that the microbiologist must be informed if it is intended to compromise the patient by a variety of treatment modalities. Then, other clinical specimens may be analyzed from the viewpoint of establishing a baseline by screening for certain microbial representatives reflecting potentially dangerous microbial particles in the intimate biosphere of that individual. The constant involvement of the hospital epidemiologist or his equivalent in gathering this information will allow proper interpretation of microbiological findings and enhance the recognition of organisms representing the hospital environment. This type of "now" epidemiology appears to be the most useful fashion of obtaining proper clues to the role the environment plays in colonizing patients and recognizing those individuals in whom this colonization might proceed or has proceeded to disease production.

The laboratory should be especially alert for specimens originating from what might be termed critical areas and sites. As already mentioned, specimens emanating from postoperative wounds, from various compromised patients including newborn and premature infants, from intravenous and urinary catheters and/or from patients who had recent experiences with such instrumentation, as well as specimens from the various implants fall into this category. Any organism isolated from such materials must be carefully noted and the likelihood of its environmental origin considered. While the decision process remains difficult, when the organisms isolated fall into the categories mentioned above, the possibility of environmental interaction must be entertained. We have found that careful analysis of the microbiota recovered at postmortem examinations is helpful in alerting the microbiologists as well as the epidemiologists to the presence of organisms involved in nosocomial disease production and which reflect in large measure resident microbiota of the institution (31, 36). While this study involved bacteria which displayed multiple resistance or protista not usually found in the intimate human biosphere, they were organisms which corresponded to microbiota recoverable from the environment. It cannot be denied that some of these protista might have been members of the individual's endogenous microbiota. However, by all measures available, they could be compared to organisms sequestered from water reservoirs, certain machinery, sinks, etc. What is of great interest is that while routine clinical specimens submitted for microbiological analysis yielded these organisms to the level of at most 5%, they were recovered at a significantly higher rate from the organs at autopsy. Careful analysis of the histories of such individuals helped to pinpoint potential vectors of transmission. In order to understand the mechanisms of colonization of patients, specific investigations to delineate the various microbial groups seem justified. Frequently, attempts to recognize microbiota from the environment with the potential to complicate the recovery of patients is stimulated by small epidemics which lead to investigations of practices, fomites, medications, resulting in the recognition of common source reservoirs responsible for these complications (9). Clustering of such organisms to specific areas of the institution also serves to alert personnel that an investigation should be conducted.

There are a large number of studies reported in which the environment seems to have played a direct role in complicating a patient's recovery and prolonging his stay within the institution. The ubiquity of microbial particles everywhere in the institution involving any surface or material be they water reservoirs, floors, walls, oxygen tanks, food, bathroom facilities, etc., makes it impossible and unnecessary to monitor such objects for the presence of microorganisms. There is no need to confirm by microbiological analysis the presence of grossly detectable soil and dirt. At the same time, it must be recognized that a sterile area cannot exist for long in a filthy environment. Instead, standard procedures must be established in each institution which permit for no deviation and are directed at a variety of practices intended to maintain the hospital environment in a sanitized condition. To sanitize is meant to convey that the environment or object is not sterile but carries the smallest achievable number of microbial particles, hopefully at a level which would not constitute an infective dose when patients are exposed to it. The rules and regulations carrying the imprimatur of the highest administrative official should address the proper modes of cleaning the institution, specify the uses of the various disinfectants employed, indicate the need for sterilization of contaminated fomites and materials, assure the least amount of contamination of the delivery of water and the disposal of waste so that the caricatures of plumbing frequently encountered within medical facilities are corrected and puddling on various joints avoided, insist on proper treatment of air with special concern directed at air conditioning, exhaust, circulation as well as the vortices created by laundry chutes and elevators. In addition, procedures concerning the handling of laundry material both clean and contaminated and the disposal of other wastes must be established. All housekeeping practices should be defined to the smallest detail with frequent administrative inspection on an ongoing basis establishing compliance with such directives. The proper and sanitary handling of foods on receipt, during storage, during preparation and delivery to patients and personnel as well as the return of all such materials and their proper care cannot be allowed to exist haphazardly. The general health and sanitary habits of all personnel involved with the dietary services must be under scrutiny on an ongoing basis. Even policies governing visitors to patients should receive attention in order to minimize the dangers inherent in permissive actions during visits.

Air and other physical factors play an undeniable part in transporting environmental microbiota to the susceptible patient. However, the role played by institutional personnel in the nosocomial disease equation cannot be denied and may equal or exceed any other factor in the delivery of environmental microorganisms to the patient. Obviously, close and constant scrutiny of all hospital employers with respect to their potential as vectors is not possible. But rules of conduct which each and every individual employed in the institution is expected to observe regardless of his station can effect a reduction of the opportunities of staff and personnel to serve as the final step of seeding the patient with hospital microorganisms. These rules can be effective only against a background of ongoing education (38). The example should be set by the professional members and by an attitude which emphasizes the significant role of all hospital personnel in protecting patients. The greatest emphasis might be placed on the most neglected practice in hospitals—proper handwashing after close contact with persons or patient-contaminated materials.

There are within institutions a number of areas considered critical in terms of the opportunities for patients to become colonized or contaminated by organisms from the environment or instruments. These areas include the operating suites, the delivery room, the various nurseries, intensive care units for various purposes, dialysis units and the central supply area in addition to isolation rooms, physical therapy and special areas in which intravenous or hyperalimentation medications may be prepared. The best means of assuring the safety of these highly critical sections of the medical facility is once more to insist on strict adherence to procedures which should be outlined for each area and directed at minimizing the opportunity to introduce microorganisms into patients while they reside within these specialized sections. The statements of Dineen (15) concerning the need for a disciplined approach to the operating room in order to curtail potential colonization and subsequent infection of patients during surgical procedures can be expanded to cover all of the other areas with similar purposes and with patient clientele similarly subject to various invasive or compromising techniques. Certainly, the interaction between such patients and personnel, the effect of clothing in these specialized areas, even the use of stethoscopes as well as medication and specialized equipment have been implicated in introducing environmental microorganisms into patients during procedures or prolonged stays within these areas. However, even in these selected sections of the hospital continuous and constant monitoring would be the only way to know precisely at what time and by what means actual delivery of such particles into the patient occurred. Since such surveillance could not be practiced on an ongoing basis, it is necessary to resort to inculcating all personnel with an abiding respect for the potential dangers deviation from outlined protocol presents to the patient. It goes without saying that strict adherence to the various types of isolation procedures must be insisted on.

There are a series of procedures and equipment which must be monitored at regular intervals. There is no argument that biological testing of sterilizers, be they steam autoclaves or ethylene oxide sterilizers, be performed at frequent intervals. Similarly, solutions prepared for hyperalimentation should undergo periodic testing. The care of inhalation equipment should be monitored especially by the in use test devised by Sanford and his group (64, 65). The practice of testing catheters on removal has probably no significance in the case of urinary cathethers. However, many individuals continue to submit and to accept such specimens for analysis. It would seem that a urine culture performed through a catheter just prior to its removal would yield more significant information than the device itself after it passed through contaminated areas such as the distal urethra. The care exercised in the removal of intravenous catheters determines the utility of the data obtained. Proper skin disinfection must be practiced prior to the removal of the intravenous catheter. It should be removed gingerly and carefully with the tip severed only with sterile instruments. The number of positive results diminished precipitously in our institution when the aforementioned concern was exercised during the removal of the device. Some institutions continue to monitor the biological acceptability of babies' formula despite the fact that in most institutions such preparations are supplied by a commercial source. While some of this may be as the result of the insistence on outmoded governmental regulations, it seems foolhardy to examine these products on a routine basis.

The assessment of the role environmental microorganisms play in complicating the recovery of patients in institutions is not possible under present conditions. This statement in no way diminishes the suspicion that the environment and its microorganisms may indeed participate actively in a process of complicating the recovery of patients. The increasing awareness of the dangers of acquiring infections attending each hospital admission and the recognition of a responsibility by institutions as a result of judicial opinions (17) emphasizes that hospitals are expected to exercise special cautions to protect patients from such infection. Standard IV of the Joint Commission on the Accreditation of Hospitals (1971) indicates the concern of the hospital and medical professions which deal with this problem. The mere isolation and identification of microorganisms are inadequate indicators of the potential role the environment plays in endangering a patient. The multifarious aspects which have been discussed above all contribute, by no means equally, to the total equation which may result in disease manifestations in a patient. With limited knowledge at our disposal it is the education of all personnel in an institution which may curtail the dangers inherent in hospitalization. An ongoing educational program for all individuals who work within medical facilities as well as insistence on strict adherence to all procedural outlines appears the only way of minimizing the effect of the environment and its microbiota on the patient. The history of nosocomial disease reflects the dynamic aspects of hospital microbiota and in a sense underlines the futility of directing efforts at specific microorganisms. While the value of knowing the major contributors to complicating infections arising from environmental microbiota is pertinent for the treatment of the patient, whose recovery is complicated thus, the futility of directing efforts at curbing one or two members of the institutional microflora becomes apparent as one considers the inexhaustible pool of microorganisms extant in nature. As one or the other major agent of infectious complications is controlled others take its place. It seems only proper that all efforts at excluding microorganisms be made and that effective barriers against microbial contamination be erected and enforced at all times. This practice should ensure protection for the patient regardless of the type and nature of the microbial populations waiting to occupy a special ecological niche made available by the successful battle against one of the major constituents of the hospital environment.

There is an understandable tendency on part of hospital administrators to have available concrete data to demonstrate institutional concern with the prevention of complications in patients from the hospital microbiota. Regardless of merits, ongoing microbiological spot checks of the physical facility provide the means to persuade legal and accrediting agencies' concerns. However, the objections of physicians and scientists are not directed against all environmental surveillance but against routine and random testing of the hospital plant and the far reaching conclusions drawn from results. The divergent views expressed by individuals (5, 7, 8, 19, 26, 31, 92, 93) and committees (3, 4, 13) reflect apparent rather than actual differences. No one among the experts will disagree with the concept that directed assessment of environmental microbiota during the investigation of a specific complication or of a series of such incidents should not or must not encompass the study of the environment. The attraction of routine microbiological sampling programs for administrators, legislators, judges, and accrediting agencies lies with its regularity. What the detractors among the professionals have failed to emphasize is that the same information, obtained haphazardly during routine testing, will be obtained in a meaningful fashion within the context of interpretable data with a frequency which matches if it does not exceed that of the routine testing during the investigation of hospital-acquired infectious complication. It is the preoccupation with retrospectively gathered, epidemiological data, presentable in statistical terms, which tends to obfuscate the significance of immediate and pertinent epidemiological analyses of nosocomial complications. But it is immediate exploration of all factors including environmental samplings which must accompany the detection of a suspect microorganism in a specimen. Such complete analysis should more than satisfy cautious administrators. It is the recognition of unexpected symptoms in a patient by alert clinicians which must stimulate epidemiological as well as microbiological attempts to focus on the potential role of "hospital" microbiota. It is the continued efforts of the infectious disease specialist, the epidemiologist, and the microbiologist which permit the recognition of the sequences which led to the infection, the measure required to treat the specific patient, and those steps which will prevent a recurrence.

Such investigations must time and again encompass the study of the various and sundry factors thought to be significant in the establishment and dissemination of the microbial hospital residents. The methods for testing and characterizing the hospital microbiota have been summarized (41) and can be followed in all clinical microbiology laboratories. The degree of interdigitation of efforts by the infectious disease specialist, epidemiologist, and microbiologist cannot be prescribed; it would seem that the overriding concern each faction professes for the welfare of patients would lead quickly to a productive relationship which would insure rapid investigation of individual and clustered problems and curtail threats from environmental microorganisms to patient safety efficiently and quickly. Constant and ongoing epidemiology coupled with proper reporting, good

TABLE 19.1
Suggested guide to environmental microbiological analysis

Item/Procedure	Frequency	Method	Comments
		I. Compulsory	
Autoclaves, gas sterilizers	Required at weekly intervals	Described by manufacturer	More frequent testing may be required; indoctrination of personnel in proper placing of test strips mandatory
		II. Environmental "Quality Control" Testing	
A. Respiratory Assist Service/ Anesthesiology			
1. Sanitizing devices such as "Pasteurmatic," "Cidamatic"	At least monthly	Random treated material flushed with sterile 0.9% NaCl and analyzed	This includes the various rubber or plastic tubings and attachments usually washed and prepared in these devices for patient use
2. "In Use" test of assembled equipment	At least monthly	See (64, 65, 69, 75)	Preferably performed by environmentalist or epidemiologist; may be executed by respiratory function personnel or anesthesiologists
B. "Critical Areas"			
1. Operating rooms, delivery rooms			
a. Sinks	Monthly	See (41, 86, 88)	
b. Sterilized instruments, packs, etc.	Every 2–3 Months	See (41, 86, 88)	Especially if prolonged storage of such items is common
c. Soap dispensers	Monthly	See (41, 86, 88)	Not applicable if individual packages are used
d. Suction bottles and related equipment	Monthly	See (41)	Not applicable, if disposable
2. Nurseries			
a. Sinks	Monthly		See B.1.a
b. Soap dispensers	Monthly		See B.1.c
c. Humidification	Monthly	See (41)	Especially significant in isolettes
d. Medications, creams, etc.	Monthly	See (41)	If not packaged as single-use items
3. Dialysis units: salt solution	Monthly	See (41)	Depending on the unit used, the separate units or general reservoirs should be tested
4. Hyperalimentation units	Monthly	See (41)	Initially, each batch should be tested until acceptability has been established. Change in ingredients should be accompanied by monitoring
		III. Epidemiologically Determined Environmental Testing	

A. Examples:
1. Nursery outbreaks:
   a. Skin diseases
   b. Diarrhea
2. Postsurgical wound infections involving common microbial agent
3. Infections during or following catheterization (urinary or intravenous) with identical microorganism
4. Compromised patients infected with presumptively the same microorganism(s)
5. Food associated diseases

B. Suggested Rules:
1. Test environmental role only after establishing:
   a. Common microbial denominator
   b. Mode of transmission not apparent
   c. Role of common human vector ruled out by epidemiological analysis
2. Establish priorities by epidemiological analysis
3. Test "common sources" sequentially in keeping with priorities
4. Use selective media to sequester specific agents from specimens
5. Individualize approach to each problem to obtain maximum information
6. Enlist help of reference laboratories to establish relationship between various isolates
7. Attempt establishing ultimate source of microorganisms and means of preventing recurrence

personal relationships between all interested parties, observations backed by irrefutable data, special evaluation of the introduction of new drugs, new instruments and new procedures as well as the time honored practice of frequent handwashing should do much to minimize the effects of microbial coinhabitants of medical facilities.

Appendix: Suggestions for Environmental Surveillance

These suggestions are written on the insistence of the Editor to illustrate one possible approach to the evaluation of environmental microorganisms and their role as potential purveyors of disease. I am aware that each institution represents a very individual set of circumstances and that differences concerning environmental monitoring would arise in any one institution advised by more than one person.

These suggestions are based on the limited experience in one medical center. I have tried to take into account the caveats of committees (3, 4) and of acknowledged authorities (92, 93). At the level of the individual hospital I perceive 3 major categories (see Table 19.1) for environmental microbiological analysis. The compulsory category requires no additional comment. The second group represents suggestions. It encompasses practices and hospital areas which were sources of difficulties in the past. They constitute potential reservoirs of opportunistic and selected microbiota for the colonization of patients. Their maintenance in an acceptable state is one measure of quality control of the hospital environment. This is, in a sense, an extension of concepts widely used in the delivery of health care to patients, as well as in the performance of laboratory examinations. While the "relevance" to patient care of these practices may not be immediate, unfavorable results serve to alert all personnel concerned to improper practices or neglect which should lead to immediate remedial measures as well as review of all procedures and personnel practices used to maintain the institutional environment in an acceptable condition.

The major effort of environmental microbiological analysis should be reserved for the third category listed. In my opinion, one must guard against haphazard environmental monitoring even during these investigations to prevent the extension of small problem clusters. Epidemiological evidence must play the decisive part in the selection of devices, fixtures, medications, etc., to be included in the testing scheme.

The actual testing methods are described in the references cited. It is unfortunate that standardized procedures have not been developed and advocated. Such agreement by diverse groups would be a real service to the understanding and interpretation of the role of the environmental microbiota in the care of the hospitalized patient.

LITERATURE CITED

1. Altemeier, W. A. Some epidemiological considerations of surgical infections, pp. 118–125, in *National Conference on Institutionally Acquired Infections*. Washington, D. C.: U. S. Government Printing Office, 1964.

2. Altemeier, W. A. Current infection problems in surgery, pp. 82–87, in P. S. Brachman and T. C. Eickhoff (Eds.), *Proceedings of the International Conference on Nosocomial Infections*. Chicago: American Hospital Association, 1971.
3. American Hospital Association. *Microbiological Sampling in the Hospital*. Chicago: American Hospital Association, 1975.
4. American Public Health Association. Environmental microbiological sampling in the hospital. *Nation's Health*, October 1975.
5. Bartlett, R. C. Control of hospital-associated infection. *Progr Clin Pathol 4:*259–282, 1972.
6. Bartlett, R. C. *Medical Microbiology: Quality Cost and Clinical Revelance*, New York: John Wiley & Sons, 1974.
7. Bartlett, R. C. Infection surveillance and control, pp. 841–845, in E. H. Lennette, E. H. Spaulding, and J. P. Truant (Eds.), *Manual of Clinical Microbiology*, Ed. 2. Washington, D. C.: American Society for Microbiology, 1974.
8. Bartlett, R. C., Groschel, D. H. M., Mackel, D. C., Mallison, G. F., and Spaulding, E. H. Microbiological surveillance, pp. 845–851, in E. H. Lennette, E. H. Spaulding, and J. P. Truant (Eds.), *Manual of Clinical Microbiology*, Ed. 2. Washington, D. C.: American Society for Microbiology, 1974.
9. Bassett, D. J. C. Common-source outbreaks. *Proc Roy Soc Med 64:*980–986, 1971.
10. Blakemore, W. S., McGarrity, G. J., Thurer, R. J., Wallace, H. W., MacVaugh, H., III, and Coriell, L. L. Infection by air-borne bacteria with cardiopulmonary bypass. *Surgery 70:*830–838, 1971.
11. Callia, F. M., Wolinsky, E., Mortimer, E. A., Abrams, J. S., and Rammelkamp, C. H., Jr. Importance of the carrier state as a source of *Staphylococcus aureus* in wound sepsis. *J Hyg 67:*49–57, 1969.
12. Central Health Services Council. *Staphylococcal Infections in Hospitals,* pp. 35. London: Her Majesty's Stationery Office, 1959.
13. Committee on Operating Room Environment. Definition of surgical microbiologic clean air. American College of Surgeons January 1976 Bulletin, pp. 19–23. 1976.
14. Cooke, W. B., and Foter, M. J. Fungi in used bedding material. *Appl Microbiol 6:*169–173, 1958.
15. Dineen, P. Influence of operating room conduct on wound infections. *Surg Clin North Am 55:*1283–1287, 1975.
16. Doig, C. M. The effect of clothing on the dissemination of bacteria in operating theatres. *Br J Surg 59:*878–881, 1972.
17. Dornette, W. H. L. Legal aspects of hospital-acquired infections. *J Legal Med 1:*37–41, 1973.
18. Drewett, S. E., Payne, D. J. H., Tuke, W., and Verdon, P. E. Eradication of *Pseudomonas aeruginosa* infection from a special-care nursery. *Lancet I:*946–948, 1972.
19. Eickhoff, T. C. Role of environmental sampling, p. 265, in *Proceedings of the International Conference on Nosocomial Infections*. P. S. Brachman and T. C. Eickhoff, (Eds.) Chicago: American Hospital Association, 1971.
20. Edmonds, P., Suskind, R. S., MacMillan, B. G., and Holder, I. A. Epidemiology of *Pseudomonas aeruginosa* in a burns hospital: surveillance by a combined typing system. *Appl Microbiol 24:*219–225, 1972.
21. Falkow, S. *Infectious Multiple Drug Resistance*. London: Pion Ltd., 1975.
22. Gage, A. A., Dean, D. C., Schimert, G., and Minsky, N. *Aspergillus* infection after cardiac surgery. *Arch Surg 101:*384–387, 1970.
23. Garrod, L. P. Some observations on hospital dust. *Br Med J 1:*245–247, 1944.
24. Gerken, A., Cavanagh, S., and Winer, H. I. Infection hazards from stethoscopes in hospital. Lancet *1:*1214–1215, 1972.
25. Gibson, G. L. *Infection in Hospital*, Ed. 2. Edinburgh: Churchill Livingstone, 1974.
26. Greene, V. W. Role of environmental sampling, pp. 266–267, in P. S. Brachman and T. C. Eickhoff (eds.), *Proceedings of the International Conference on Nosocomial Infections*. Chicago: American Hospital Association, 1971.
27. Greene, V. W., Vesley, D., Bond, R. G., and Michaelson, G. S. Microbiological contamination of hospital air. I. Quantitative studies. *Appl Microbiol 10:*561–566, 1962.
28. Greene, V. W., Vesley, D., Bond, R. G., and Michaelson, G. S. Microbiological contamination of hospital air. II. Qualitative studies. *Appl Microbiol 10:*567–571, 1962.
29. Haldeman, J. C. The effect of the hospital environment on the spread of infection, pp. 21–26, *National Conference on Institutionally Acquired Infections. P.H.S. Publication 1188*. Washington, D.C.: U.S. Government Printing Office, 1964.

30. Herman, L. G. Environmental microbiology and infection control programs in health care facilities: introductory remarks. *Health Lab Sci 11:*69–70, 1974.
31. Herman, L. G., and Hart, L. J. The role of environmental sampling, pp. 268–269, in P. S. Brachman and T. C. Eickhoff (Eds.) *Proceedings of the International Conference on Nosocomial Infections*. Chicago: American Hospital Association, 1971.
32. Howe, C. W., and Mozden, P. J. Postoperative infections: current concepts. *Surg Clin North Am 43:*859–882, 1963.
33. Isenberg, H. D. Laboratory diagnosis of nosocomial disease. *Infect Dis Rev 3:*1–20, 1974.
34. Isenberg, H. D., and Berkman, J. I. The role of drug-resistant and drug-selected bacteria in nosocomial disease. *Ann NY Acad Sci 182:*52–58, 1971.
35. Isenberg, H. D. and Painter, B. G. Indigenous and pathogenic microorganisms of man, pp. 45–58, in E. H. Lennette, E. H. Spaulding, and J. P. Truant (Eds.), *Manual of Clinical Microbiology*, Ed. 2. Washington, D.C.: American Society for Microbiology, 1974.
36. Isenberg, H. D., Painter, B. G., Berkman, J. I., Philipson, L., and Tucci, V. The postmortem microbiological analysis as an indicator of nosocomially-significant microorganisms in the hospital environment. *Health Lab Sci 11:*85–89, 1974.
37. Kethey, T. W. Air: its importance and control, pp. 35–46, in *National Conference on Institutionally Acquired Infections. P. H. S. Publication 1188*. Washington, D.C.: U.S. Government Printing Office, 1964.
38. Knight, V. Instruments and infection. *Hospital Practice 2:*82–95, 1967.
39. Kominos, S. D., Copeland, C. E., Grosiak, B., and Postic, B. Introduction of *Pseudomonas aeruginosa* into a hospital via vegetables. *Appl Microbiol 24:*567–570, 1972.
40. Koren, H. Environmental hazards in hospitals. *J Environm Health 37:*122–126, 1974.
41. Lennette, E. H., Spaulding, E. H., and Truant, J. P. *Manual of Clinical Microbiology*, ed. 2. Washington, D.C.: American Society for Microbiology, 1974.
42. Lindbom, G., and Laurell, G. Studies on the epidemiology of staphylococcal infections. 5. Importance of environmental factors during the early postoperative phase. *Acta Pathol Microbiol Scand 69:*246–263, 1967.
43. Litzky, B. Y., and Litzky, W. Bacterial shedding during bed-stripping of reusable and disposable linens as detected by the high volume air samples. *Health Lab Sci 8:*29–34, 1971.
44. Maki, D. G., Rhame, F. S., Goldman, D. A., and Mandell, G. L. The infection hazard posed by contaminated intravenous infusion fluids, pp. 76–91, in A. C. Sonnenwirth (Ed.), *Bacteremia. Laboratory and Clinical Aspects*. Springfield, Ill.: Charles C Thomas, 1973.
45. Mangi, R. J., and Andriole, V. T. Contaminated stethoscopes: a potential source of nosocomial infections. *Yale J Biol Med 45:*600–604, 1972.
46. McDade, J. J., Favero, M. S., Michaelson, G. S., and Vesley, D. Environmental microbiology and the control of microbial contamination, pp. 51–86, in *Spacecraft Sterilization Technology*. Washington, D.C.: National Aeronautics and Space Administration, 1966.
47. McNeill, I. F., Porter, I. A., and Green, C. A. Staphylococcal infection in a surgical ward. *Br Med J 17:*798–802, 1964.
48. Meers, P. D., Calder, M. W., Mazhar, M. M., and Laurie, G. M. Intravenous infusion of contaminated dextrose solution. *Lancet II:*1189–1192, 1973.
49. Mertz, J. J., Scharer, L., and McClement, J. H. A hospital outbreak of *Klebsiella pneumoniae* from inhalation therapy with contaminated aerosol solutions. *Am Rev Respir Dis 95:*454–460, 1967.
50. Moffet, H. L., and Williams, T. Bacteria recovered from distilled water and inhalation therapy equipment. *Am J Dis Child 114:*7–12, 1967.
51. Moffet, H. L., Allan, D., and Williams, T. Survival and dissemination of bacteria in nebulizers and incubators. *Am J Dis Child 114:*13–20, 1967.
52. Moffet, H. L., and Allan, D. Colonization of infants exposed to bacterially contaminated mists. *Am J Dis Child 114:*21–25, 1967.
53. Moore, B., and Forman, A. An outbreak of urinary *Pseudomonas aeruginosa* infection acquired during urological operations. *Lancet II:*929–931, 1966.
54. Morehead, C. D., and Houck, P. W. Epidemiology of *Pseudomonas* infection in a pediatric intensive care unit. *Am J Dis Child 124:*564–570, 1972.
55. Mortensen, J. D., Hurd, G., and Hill, G. Bacterial contamination of oxygen used clinically—importance and one method of control. *Dis Chest 42:*567–572, 1962.
56. Nahmias, A. J., and Eickhoff, T. C. Staphylococcal infections in hospitals. *N Engl J*

*Med 265:*74–81, 120–128, 177–182, 1961.
57. Newsom, S. W. B. Hospital infection from contaminated ice. *Lancet II:*620–622, 1968.
58. Newsom, S. W. B. Microbiology of hospital toilets. *Lancet II:*700–703, 1972.
59. Olds, J. W., Kisch, A. L., Eberle, B. J., and Wilson, J. N. *Pseudomonas aeruginosa* respiratory tract infection acquired from a contaminated anesthesia machine. *Am Rev Respir Dis 105:*628–632, 1972.
60. Pettit, F., and Lowbury, E. J. L. Survival of wound pathogens under different environmental conditions. *J. Hyg 66:*393–406, 1968.
61. Phillips, G. B., and Lewis, K. H. Microbiological barrier techniques, pp. 105–136, in *Spacecraft Sterilization Technology*. Washington, D.C.: National Aeronautics and Space Administration, 1966.
62. Phillips, I., Eykyn, S., Curtis, M. A., and Snell, J. J. S. *Pseudomonas cepacia (multivorans)* septicemia in an intensive care unit. *Lancet I:*375–377, 1971.
63. Phillips, I., Eykyn, S., and Laker, M. Outbreak of hospital infection caused by contaminated autoclaved fluids. *Lancet I:*1258–1260, 1972.
64. Pierce, A. K., and Sanford, J. P. Bacterial contamination of aerosols. *Arch Intern Med 131:*156–159, 1973.
65. Reinarz, J. A., Pierce, A. K., Mays, B. B., and Sanford, J. P. The potential role of inhalation therapy equipment in nosocomial pulmonary infection. *J Clin Invest 44:*831–839, 1965.
66. Riemensnider, D. K. Quantitative aspects of shedding of microorganisms by humans, pp. 97–104, in *Spacecraft Sterilization Technology*. Washington, D.C.: National Aeronautics and Space Administration, 1966.
67. Rifkind, D., Marchioro, T. L., Schneck, S. A., and Hill, R. B., Jr. Systemic fungal infections complicating renal transplantation and immunosuppressive therapy. *Am J Med 43:*28–38, 1967.
68. Roberts, F. J., Cockcroft, W. H., Johnson, H. E., and Fishwick, T. The infection hazard of contaminated nebulizers. *Can Med Assoc J 108:*53–56, 1973.
69. Roberts, R. B. The anesthetist, cross-infection and sterilization techniques. A review. *Anaesth Intensive Care 1:*400–406, 1973.
70. Rosendorf, L. L., Daicoff, G., and Baer, H. Sources of gram-negative infection after open heart surgery. *J Thorac Cardiovasc Surg 67:*195–201, 1974.
71. Rountree, P. M., and Beard, M. A. Sources of infection in an intensive care unit. *Med J Aust. 1:*577–582, 1968.
72. Rubbo, S. D. The role of textiles in hospital cross-infection, pp. 231–250, in R. E. O. Williams and R. A. Shooter (Eds.), *Infections in Hospitals: Epidemiology and Control*. Philadelphia: F. A. Davis Co., 1963.
73. Salzman, T. C., Clark, J. J., and Klemm, L. Hand contamination of personnel as a mechanism of cross-infections with antibiotic-resistant *Escherichia coli* and *Klebsiella-Aerobacter*. *Antimicrob Agents Chemother 1967:*97–100, 1968.
74. Schroeder, S. A., Aserkoff, B., and Brachman, P. S. Epidemic salmonellosis in hospitals and institutions. *N Engl J Med 279:*674–678, 1968.
75. Schulze, T., Edmonson, E. B., Pierce, A. K., and Sanford, J. P. Studies of a new humidifying device as a potential source of bacterial aerosols. *Am Rev Respir Dis 96:*517–519, 1967.
76. Seth, V., Ray, B. G., Walia, B. N. S., and Ghai, O. P. Role of environmental factors in the spread of staphylococcal infection in the maternity ward. *Indian J Med Res 61:*910–917, 1973.
77. Shaffer, J. G. Airborne infection in hospitals. *Am J Public Health 54:*1674–1682, 1964.
78. Shaw, D., Doig, C. M., and Douglas, D. Is airborne infection in operating theatres an important cause of wound infection in general surgery? *Bull Soc Int Chir 33:*35–41, 1974.
79. Sidwell, R. W., Dixon, G. J., and McNeil, E. Quantitative studies on fabrics as disseminators of viruses. *Appl Microbiol 14:*55–59, 1966.
80. Smith, H., and Pearce, J. H. *Microbial Pathogenicity in Man and Animals,* Cambridge: Cambridge University Press, 1972.
81. Speers, R., Jr., and Shooter, R. A. Shedding of bacteria to the air from contaminated towels in paper sacks. *Lancet II:*301–302, 1967.
82. Taplin, D., and Mertz, P. M. Flower vases in hospitals as reservoirs of pathogens. *Lancet II:*1279–1281, 1973
83. Thoburn, R., Fekety, F. R., Cluff, L. E., and Melvin, V. B. Infections acquired by hospitalized patients. *Arch Intern Med 121:*1–10, 1968.

84. Thom, B. T., and White, R. G. The dispersal of organisms from minor septic lesions. *J Clin Pathol 15:*559–562, 1962.
85. Ulrich, J. A. Skin carriage of bacteria in the human, pp. 87–96, in *Spacecraft Sterilization Technology*. Washington, D.C.: National Aeronautics and Space Administration, 1966.
86. Vesley, D. Survey of microbiological techniques for recovery from surfaces, pp. 147–154, in *Spacecraft Sterilization Technology*. Washington, D.C.: National Aeronautics and Space Administration, 1966.
87. Virtanen, S., and Casten, O. Contamination of new hospital premises. Hospital environmental and personnel as the source of nosocomial infections. *Public Health 86:*175–181, 1972.
88. Walter, C. W. Surfaces, their importance and control, pp. 27–34, in *National Conference on Institutionally Acquired Infections, P.H.S. Publication 1188*. Washington, D.C.: U.S. Government Printing Office, 1964.
89. Washington, J. A., II, Senjem, D. H., Haldorson, A., Schutt, A. H., and Martin, W. J. Nosocomially acquired bacteriuria due to *Proteus rettgeri* and *Providencia stuartii*. *Am J. Clin Pathol 60:*836–838, 1973.
90. Watson, A. G., and Koon, C. E. *Pseudomonas* on the chrysanthemums. *Lancet II:*91, 1973.
91. Whitby, J. L., and Rampling, A. *Pseudomonas aeruginosa* contamination in domestic and hospital environments. *Lancet I:*15–17, 1972.
92. Williams, R. E. O. Changing perspectives in hospital infection, pp. 1–10, in P. S. Brachman and T. C. Eickhoff (Eds.), *Proceedings of the International Conference on Nosocomial Infections*. Chicago: American Hospital Association, 1971.
93. Williams, R. E. O., and Shooter, R. A. *Infection in Hospitals. Epidemiology and Control*. Philadelphia: F. A. Davis Co., 1963.
94. Wolf, H. W., Harris, M. H., and Dyer, W. R. *Staphylococcus aureus* in air of an operating room. *JAMA 169:*1983–1987, 1959.

20

A Sample of Bacterial Isolates in the United States and Their Susceptibilities to Antimicrobial Agents

BARBARA ATKINSON

GEORGE MOORE

The incidence of bacteria in a hospital and the antibiotic susceptibility of these isolates are major concerns of the microbiologist, the epidemiologist, and the clinician. This chapter presents the 1971–1975 incidence trends for bacteria isolated from hospital patients throughout the United States. Antibiotic susceptibility patterns of the most frequently found organisms are also shown.

The shifting trends presented here represent the population dynamics of both patients and bacterial species. Furthermore, the susceptibility patterns reflect not only the proportion of resistant strains, but also the changing vogues in antibiotic choices. Overall, the trend data gives a sense of the ebb and flow of bacterial populations.

This data can be useful for microbiologists in assessing their own clinical laboratories. The incidence trends can serve as reference for comparing the isolation patterns found in individual laboratories. By knowing what the most common species are and where to find them, the clinical microbiologist can adjust budgetary, personnel, and technical priorities. Those preparing for proficiency tests will find the geographical incidence listing and the total organism listing helpful in determining which organisms are rare and which are common. Both the examiner and the microbiologist can use these lists to indicate what common organisms can be recovered and what rare organisms their clinical laboratories should be capable of identifying or isolating.

Data in the following tables were obtained from acute care hospitals of 100 beds or more. During this 5-year period the number of participants ranged between 150 and 300 institutions. The geographical distribution of these hospitals is shown in Figure 20.1. Bac-Data Medical Information Systems, Inc. (120 Brighton Road, Clifton, N. J. 07012), a national bacteriologic monitoring service for hospital laboratories, provided both the inci-

Figure 20.1. Survey hospitals. Includes 6.4% of United States acute care hospitals of 100 beds or more.

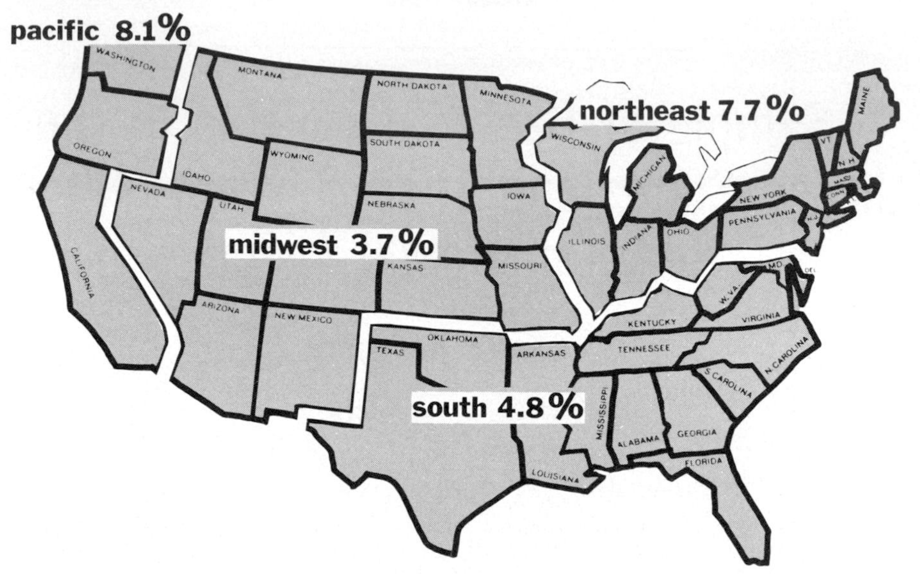

Data selected from the Bac-Data Medical Information Systems data bank for this survey include only hospitals using the Bauer-Kirby method of susceptibility testing. An average of 68,000 hospital beds are represented over the 5-year survey period.

dence and susceptibility data and the statistical analyses. For this survey, non-parametric statistics were used and statistical significance was assigned to differences varying at a $P < 0.01$.

Each month subscribing institutions submit laboratory results to a centralized computer system. All hospitals selected for this survey stated that they follow the Kirby-Bauer testing procedure for rapidly growing pathogens. Because this method is not recommended for anaerobic or fastidious aerobic bacteria, it is assumed that the identification of such isolates and susceptibility testing of fastidious bacteria were derived from conventional and appropriate methods.

Subscribing institutions receive monthly analytical reports from Bac-Data. It is assumed that because these institutions desire accurate information, they perform their own data collection and reporting carefully. The extremely large sampling size also tends to minimize any laboratory or clerical error.

Nearly 3 million isolates were used to establish the overall incidence trends presented in these tables. Because of space limitations, antibiotic susceptibility data are given only for the most frequently (incidence $\geq$ 0.5%) isolated organisms. Only truly sensitive isolates are called sensitive; isolates giving intermediate Kirby-Bauer results are called resistant.

The first series of tables present incidence information. General incidence trends of microbial flora for the years 1971–1975 is presented in total; then, the incidence information is subdivided. For the year 1975, geographical regions of the United States are compared; pediatric patients are compared to all patients; and the relative incidence of organisms found in blood, cerebrospinal fluid (CSF), urine, and wounds is compared.

The antibiotic susceptibilities of the most common (incidence ≥0.5%) bacterial species are presented in a second series of tables. The antibiotic susceptibilities of each organism isolated from the body sites listed above provide the microbiologist with a reference for evaluating his own hospital's experience. Perhaps, most importantly, a series of tables showing the susceptibility trends from 1971 to 1975 indicates the possible therapeutic effectiveness that various antibiotics could have. Increased resistance indicates that a drug is becoming less useful. This alerts the pharmaceutical industry to areas of need for developing new drugs.

INCIDENCE TRENDS

The incidence trends presented in the first series of tables were made from a total of 2,976,371 isolates of 77 species. Eighty to 85% of all isolates belonged to 11 species: 25 species accounted for 96% of all isolates (Fig. 20.2).

Table 20.1*B* does not include organisms such as *Yersinia enterocolitica* (67, 75, 81), *Vibrio parahaemolyticus* (7, 56, 64), *Aeromonas hydrophila* (15, 33), *Actinobacillus actinomyctemcomitans* (65, 66), and *Listeria monocytogenes* (8) because each laboratory which reports these to Bac-Data uses his own code number instead of reporting as genus and species.

Gram-negative organisms (57%) continued to be seen more often than Gram-positive organisms (43%). However, the proportion of Gram-negative organisms has decreased since 1971 and the most common Gram-positive organisms have been isolated more often. Although *Escherichia coli* continues to have the highest overall incidence, the general shift toward an increasing incidence of Gram-positive organisms is due to a relative decline in *E. coli*, *Proteus* sp. and *Klebsiella* sp., and a relative increase in *Staphylococcus aureus, Staphylococcus epidermidis, Candida albicans,* and *Corynebacterium* sp.

Gram-negative

The common organisms *E. coli, Proteus* sp., *Klebsiella* sp., and *Enterobacter* sp. have all decreased in incidence; this accounts for the overall decline in incidence for Gram-negative organisms. Later tables will show that *E. coli* is decreasing as the predominant organism in bacteremia of newborns and is being replaced by streptococcus group B (Chapter 18).

Figure 20.2 Incidence of microbial flora isolated from hospital patients in the United States: Species isolated from all body sites 1971–1975

TOTAL NUMBER OF ISOLATES ...	2,976,371
TOTAL NUMBER OF SPECIES	77
80–85% OF ALL ISOLATES*	11 SPECIES
96–97% OF ALL ISOLATES*	25 SPECIES
98–99% OF ALL ISOLATES*	48 SPECIES

* Range over 5 years.

TABLE 20.1
Trend of incidence of microbial flora isolated from hospital patients in the United States, 1971–1975

A. Frequency >0.5%

Species (Total Isolates Reported (77 Species))	Percentage of Total Isolates Reported					Trend
	1971 (464,018, 100%)	1972 (419,618, 100%)	1973 (532,914, 100%)	1974 (704,697, 100%)	1975 (855,124, 100%)	
Gram-negative						
Escherichia coli	24.3	23.3	22.5	21.0	20.1	↓ 13%
Proteus sp.	09.5	10.1	08.6	08.5	08.0	↓ 16%
Klebsiella sp.	09.1	09.2	08.7	08.1	07.7	↓ 15%
Pseudomonas sp.	07.7	07.5	07.8	07.4	07.5	NSC
Enterobacter sp.	05.7	05.7	05.2	04.8	04.7	↓ 18%
Haemophilus influenzae	00.9	01.2	01.6	01.6	01.4	↑ 56%
Haemophilus sp.	00.8	01.3	01.4	01.6	01.3	↑ 63%
Serratia sp.	00.8	00.8	00.8	00.9	01.1	↑ 38%
Citrobacter sp.	00.5	00.6	00.8	00.9	00.9	↑ 80%
Acinetobacter sp.	00.8	00.8	00.8	00.9	00.8	NSC
Neisseria sp. (not *N. gonorrhoeae*)	–	–	–	00.6	00.7	–
Bacteroides sp.	00.5	00.8	00.8	00.8	01.0	↑ 100%
Neisseria gonorrhoeae	00.4	00.3	00.3	00.4	00.5	NSC
Gram-positive						
Staphylococcus aureus	10.9	11.3	11.9	11.7	12.0	↑ 10%
Staphylococcus epidermidis	06.2	06.1	06.1	07.1	07.3	↑ 18%
Streptococcus faecalis	06.4	06.9	07.6	07.0	06.4	NSC
Streptococcus-α-hemolytic	04.5	04.1	03.6	03.7	03.4	↓ 24%
Candida albicans	01.2	01.6	02.0	02.2	02.5	↑ 108%
Streptococcus, β hemolytic						
Non-group A	02.3	02.3	02.5	02.5	02.3	NSC
Group A	01.8	01.6	02.0	02.1	01.9	NSC
Streptococcus pneumoniae	01.4	01.4	01.5	01.6	01.6	NSC
Corynebacterium sp.	00.0	00.0	00.0	00.5	01.0	↑ 100%
Streptococcus, non-hemolytic	–	–	–	00.6	00.9	–
Streptococcus, group D (not *S. faecalis*)	–	–	–	00.5	00.8	–
Streptococcus, group B	–	–	–	00.2	00.6	–
B. Frequency <0.5% (Gram-negative and -positive)						
Clostridium sp.	00.3	00.3	00.3	00.2	00.3	NSC
Providentia sp.	00.4	00.3	00.3	00.3	00.3	NSC
Salmonella sp.	00.2	00.4	00.3	00.2	00.2	NSC
Streptococcus, anaerobic	00.3	00.4	00.3	00.2	00.1	↓ 67%
Shigella sp.	00.2	00.2	00.2	00.2	00.1	NSC
Neisseria meningitidis	00.05	00.03	00.04	00.06	00.07	NSC
Moraxella sp.	00.07	00.05	00.04	00.05	00.07	NSC
Alcaligenes sp.	00.2	00.2	00.1	00.1	00.06	↓ 70%
Mycobacterium tuberculosis	00.01	00.01	00.05	00.04	00.04	↑ 300%
Edwardsiella sp.	00.02	00.02	00.02	00.01	00.03	NSC
Mycobacterium sp.	00.01	–	00.03	00.02	00.03	NSC
Pectobacterium	00.07	00.06	00.02	00.01	00.01	↓ 86%
Arizona group	00.02	00.01	00.01	00.01	00.01	NSC
Mycoplasma sp.	–	–	00.01	00.01	–	–
Other	02.6	01.1	01.8	01.5	02.6	NSC
Gram-negative	64.0	64.0	62.0	59.0	57.0	↓ 11%
Gram-positive	36.0	36.0	38.0	41.0	43.0	↑ 19%

Several of the Gram-negative organisms are increasing in frequency. *Haemophilus influenzae* has been increasing as a cause of meningitis in both children (57, 61) and adults (84). Other reports have indicated that *H. influenzae* is increasing as a cause of epiglottitis (70), arthritis (63), and cellulitis (68). The increased frequency of *Bacteroides* sp. probably reflects the increase in interest due to improved methods of isolation and identification (21, 36, 77). Increases in *Serratia* sp. isolates are especially related to hospital-acquired infections (29, 86) and contaminated catheterization equipment (16), although, in some hospitals, it now represents over 10% of the urinary isolates (50). *Citrobacter* sp. also has been increasing, particularly in urinary isolates, although this increase may reflect recent improvements in identification and characterization methods (22, 25).

Gram-positive

Increased incidence of *S. aureus* in blood cultures may be due to increases in patients using parenteral drugs and those diabetic patients who inject insulin (82, 83). A significant increase in *S. aureus* has been associated with cessation of hexachlorophene use in some nurseries (19, 42).

S. epidermidis, traditionally considered a contaminant (46), has recently received attention as a pathogen in urinary tract infections (3, 51). Often, the *S. epidermidis* and *Corynebacterium* sp. are considered blood culture contaminants (Chapter 4); however, subacute bacterial endocarditis has been caused by these organisms in immunologically compromised patients and those with prosthetic implants (6, 51, 76). Increases in *Corynebacteria* sp. may also be due to improved identification and thus increased reporting (56).

Candida sp. has been seen with increasing frequency in drug addicts with endocarditis (46). The rise in *Candida albicans* incidence can be attributed to the advent of broad spectrum antibiotics, corticosteriods, and antitumor agents (46).

Current interest in group B streptococci as a leading agent of sepsis in newborns (2) as well as the development of techniques for rapid identification (34) explain the appearance and apparent increase of these organisms during 1974 and 1975. Group B streptococci, although well established as opportunistic invaders in compromised hosts, have only recently been recognized as increasing in frequency as the cause of neonatal sepsis (23, 39), meningitis (4, 32), as the most common agent in puerperal sepsis, and in septic abortion (23).

Among the more infrequent organisms—those with an incidence of 0.5% or less—the significant declines in *Alcaligenes* sp. (46) and *Pectobacterium* (17, 22, 26, 27) are probably due to earlier misidentifications. Also, *Pectobacterium* has been renamed *Enterobacter agglomerans* (26–28).

Geographical Regions

E. coli and *S. aureus* show the highest incidence in all regions (see Table 20.2). Although there were no significant differences among the geographical regions, there were relatively more Gram-negative organisms isolated

TABLE 20.2
Incidence of microbial flora in four geographic regions of the United States, 1975*

Species (Total Isolates Reported (77 Species))	North East (243,090, 100%)	South (99,311, 100%)	Midwest (24,161, 100%)	Pacific (35,443, 100%)
Gram-negative				
Escherichia coli	20.5	18.9	21.0	20.8
Proteus sp.	07.5	08.1	05.9	06.9
Klebsiella sp.	08.0	07.8	07.3	07.3
Pseudomonas sp.	07.2	07.8	06.8	08.1
Enterobacter sp.	04.7	05.2	04.8	04.6
Haemophilus influenzae	01.3	01.4	01.9	02.2
Citrobacter sp.	00.9	00.9	01.5	01.1
Bacteroides sp.	00.8	00.8	01.2	01.2
Gram-positive				
Staphylococcus aureus	12.0	10.2	12.1	13.9
Staphylococcus epidermidis	07.2	08.1	09.2	05.3
Streptococcus faecalis	06.8	06.9	05.3	05.1
Streptococcus, α-hemolytic	03.8	03.2	05.4	01.8
Candida albicans	02.4	02.7	01.9	02.5
Streptococcus, β-hemolytic, non-group A	02.3	01.9	01.6	02.7
Streptococcus, β-hemolytic, group A	01.7	01.7	01.6	03.8
Streptococcus pneumoniae	01.5	01.6	01.7	01.6
Streptococcus, group D (not *S. faecalis*)	00.7	01.0	00.9	00.5
% Gram-negative	57.0	58.0	57.0	60.0
% Gram-positive	43.0	42.0	43.0	40.0

* The geographic regions are composed of U. S. census regions as follows: Northeast, census region 1, 2, and 4; South, census regions 3, 5, and 7; Midwest,- census regions 6 and 8; Pacific, census region 9.

on the Pacific Coast than in any other geographic region. On the Pacific Coast the incidence of α-hemolytic streptococcus is lower than in other areas of the country; however streptococcus group A are seen more often than in other regions. *H. influenzae* appears to show a slight increase in frequency as one travels from the east coast toward the west.

Pediatrics

Despite the availability of a vast array of antimicrobial agents, colonization and disease by the Gram-negative bacilli, *S. aureus*, and, more recently, streptococcus group B, remains a major problem in the management of newborns (48). Unlike the total patient population where *E. coli* is most frequent, *S. aureus* is the organism most often isolated from pediatric patients (see Table 20.3).

S. aureus infections increased after hexachlorophene bathing of babies was discontinued (19, 43). In addition, virulent strains of *S. aureus*, not 80/81, have emerged (19, 48).

The pediatric group also shows a higher incidence of *S. epidermidis* than the total patient population. The higher incidence of *S. epidermidis* could be related to an increased incidence in blood and CSF samples which are more difficult to obtain from pediatric patients than from most adults, therefore, they are more likely to be contaminated with *S. epidermidis*.

Immediately after sulfonamides and penicillin were introduced, streptococcal bacteremia declined, but since the 1960s the incidence has been increasing steadily (31). Perhaps due to parental fear of rheumatic fever, throat cultures are more frequent and group A streptococci are seen more often in pediatric patients (Table 20.3).

Concurrent with a decline of *E. coli* in bacteremia and diarrhea of neonates, there has been an increase in group B streptococcus. Table 20.3 shows the pediatric group with a higher incidence of β-hemolytic streptococcus, non-group A.

Within the last 10 years, pediatric meningitis has been caused less often by *E. coli* and more often by group B streptococci (Chapter 18). The pediatric incidence of *H. influenzae* is double the incidence seen in all patients. This is probably because, in children under 4 years of age, *H. influenzae* type B is the most frequent cause of meningitis (70).

Pseudomonas sp. is seen in children with cystic fibrosis (20) and in leukemic patients (Chapter 4). This may account for the slight increase of *Pseudomonas* sp. frequency seen in the pediatric group (Table 20.3).

Since urinary infections are less common in pediatric patients, and since

TABLE 20.3
Incidence of microbial flora: pediatric vs. all patients, 1975

Species (Total Isolates Reported (77 Species))	Percentage of Total Isolates Reported	
	Pediatric patients* (26,352, 100%)	All patients (855,124, 100%)
Gram-negative		
Escherichia coli	13.3	20.1
Pseudomonas sp.	08.6	07.5
Klebsiella sp.	05.2	07.7
Proteus sp.	03.3	08.0
Enterobacter sp.	03.1	04.7
Haemophilus influenzae	03.0	01.4
Bacteroides sp.	00.1	01.0
Gram-positive		
Staphylococcus aureus	16.5	12.0
Staphylococcus epidermidis	11.3	07.3
Streptococcus, β-hemolytic, group A	04.2	01.9
Streptococcus, α-hemolytic	04.0	03.4
Streptococcus, β-hemolytic, non-group A†	04.0	02.9
Streptococcus faecalis	03.0	06.4
Streptococcus pneumoniae	02.0	01.6
% Gram-negative	48.0	57.0
% Gram-positive	52.0	43.0

* Pediatric data based on 1975 experience of 5 children's hospitals geographically dispersed across the United States.

† Includes Streptococcus, group B.

most *Proteus* sp. are found in urinary infections it is not surprising that *Proteus* sp. occurs less often in the pediatric group (Table 20.3).

INCIDENCE TRENDS BY BODY SITE

Blood

A total of 74,461 blood culture isolates were analyzed. Considering that about one of every seven blood cultures is positive (13), this material is the result of about half a million blood cultures.

Of the 74,461 isolates, the Gram-positive cocci *S. epidermidis* and *S. aureus* predominate in blood cultures (see Table 20.4). Increased incidence of recovery of staphylococcus from blood cultures is probably due to increased drug abuse (60, 69, 82), contaminated intravenous products (49), and *S. epidermidis* colonization of atrioventricular shunts (38) or prosthetic heart valve endocarditis (44, 87).

E. coli is the second most frequent and the predominant Gram-negative organism seen in blood cultures. The incidence of *E. coli* bacteremia has been declining for over 35 years (30) (Chapter 18) and is now rare in the newborn compared to streptococci (Chapter 18).

The largest increase in frequency occurred with *H. influenzae* (Table 20.4). This confirms earlier reports that both *H. influenzae* and *H. parainfluenzae* are increasing in blood cultures (61).

During 1970–1971 there was a national epidemic with *Enterobacter agglomerans* or *Enterobacter* sp. caused by contamination of commercial intravenous products (49). The apparent decline in *Enterobacter* sp. (Table 20.4) reflects an actual increase during the epidemic; the *Enterobacter* sp.

TABLE 20.4
Trend of incidence of microbial flora from blood, 1971–1975

Species (Total Isolates Reported (77 Species))	Percentage of Total Isolates Reported					Trend
	1971 (8,603, 100%)	1972 (7,786, 100%)	1973 (12,433, 100%)	1974 (18,538, 100%)	1975 (27,101, 100%)	
Gram-negative						
Escherichia coli	19.1	17.1	16.5	16.0	16.3	↓ 15%
Klebsiella sp.	09.5	08.7	07.0	05.9	06.5	↓ 32%
Pseudomonas sp.	05.3	05.6	05.3	04.7	04.6	↓ 13%
Proteus sp.	05.0	05.0	04.3	04.1	04.0	↓ 20%
Bacteroides sp.	02.8	03.7	02.9	02.7	03.2	NSC
Enterobacter sp.	04.6	03.0	03.3	02.7	02.8	↓ 39%
Haemophilus influenzae	00.7	01.4	01.6	01.7	01.7	↑ 143%
Gram-positive						
Staphylococcus epidermidis	12.4	14.2	16.4	19.3	20.4	↑ 65%
Staphylococcus aureus	11.4	12.8	16.4	13.6	12.5	↑ 10%
Streptococcus, α-hemolytic	06.3	06.2	05.5	05.0	04.6	↓ 27%
Streptococcus faecalis	04.3	04.6	04.0	03.9	03.9	NSC
Streptococcus pneumoniae	04.4	04.0	04.1	04.1	03.3	↓ 25%
Streptococcus, β-hemolytic, non-group A	01.3	01.2	01.7	01.3	01.0	NSC
Streptococcus, β-hemolytic, group A	01.2	01.1	01.1	01.0	00.8	↓ 33%
% Gram-negative	56.0	54.0	49.0	47.0	46.0	↓ 18%
% Gram-positive	44.0	46.0	51.0	53.0	54.0	↑ 23%

incidence has not changed appreciably since 1972, the year following the epidemic.

Changing practices in the use of antibiotics can account for the slight decreases seen for *Proteus* sp. (1) and *Pseudomonas* sp. (10) incidences.

Urine

The incidence trends of 1,112,802 urinary isolates are presented in Table 20.5. Since only about 10% of urine samples contain flora, this information was obtained from over 10 million cultures!

The organisms found most frequently in the urinary tract are *E. coli, Proteus* sp., and *Streptococcus faecalis* (Table 20.5). *Pseudomonas* sp. appears to be increasing as a hospital-acquired infection; its increased frequency may reflect an increase in the number of immunodeficient patients (14) and an increase in urologic manipulations such as cystoscopy, transurethral surgery, and catheterization (10).

Laboratory practice accounts for some of the apparent incidence trends. The reports to Bac-Data state that urine specimens are not routinely checked for anaerobes (5); the apparent decrease in *Bacteriodes* sp. could be due to reduced recovery and reporting rather than an actual decline. Similarly, the apparent increase in *Streptococcus pneumoniae* is probably due to an increased identification rate rather than a real increase.

More stringent nursing care and controlled use of antibiotics (74) may explain the decreases seen in nosocomial urinary tract infections due to *Klebsiella* sp., *Proteus* sp., and *Enterobacter* sp. However *S. faecalis* is also a common hospital-acquired pathogen, and its incidence has increased.

TABLE 20.5
Trend of incidence of microbial flora from urine, 1971–1975

Species (Total Isolates Reported (77 Species))	Percentage of Total Isolates Reported					Trend
	1971 (189,187, 100%)	1972 (167,123, 100%)	1973 (194,421, 100%)	1974 (259,129, 100%)	1975 (302,942, 100%)	
Gram-negative						
Escherichia coli	36.6	36.0	36.7	34.9	34.1	↓ 7%
Proteus sp.	15.2	14.6	13.6	12.3	11.6	↓ 24%
Klebsiella sp.	09.5	09.6	09.2	08.5	08.1	↓ 15%
Pseudomonas sp.	06.8	06.8	07.1	07.1	07.5	↑ 10%
Enterobacter sp.	05.4	04.8	04.1	03.7	03.7	↓ 31%
Bacteroides sp.	00.10	00.08	00.06	00.05	00.03	↓ 70%
Haemophilus influenzae	00.01	00.01	00.01	00.02	00.02	NSC
Gram-positive						
Streptococcus faecalis	09.1	10.1	12.1	11.1	10.1	↑ 11%
Streptococcus epidermidis	05.6	05.8	05.4	06.1	05.8	NSC
Staphylococcus aureus	02.0	01.8	01.9	02.0	02.0	NSC
Streptococcus, α-hemolytic	02.6	02.6	02.0	01.9	01.7	↓ 35%
Streptococcus, β-hemolytic, non-group A	01.0	01.0	01.4	01.5	01.6	↑ 40%
Streptococcus, β-hemolytic, group A	00.2	00.2	00.2	00.2	00.1	↓ 50%
Streptococcus pneumoniae	00.05	00.04	00.05	00.06	00.07	↑ 40%
% Gram-negative	79.0	77.0	76.0	73.0	73.0	↓ 8%
% Gram-positive	21.0	23.0	24.0	27.0	27.0	↑ 29%

Cerebrospinal Fluid (CSF)

The most frequent organism found in CSF is *S. epidermidis* (Table 20.6). Its increased incidence probably reflects increased rate of reporting because previously it was always considered to be a contaminant. *S. aureus* and *Enterobacter* sp. are the most frequent causes of postoperative infections of the central nervous system. α-Hemolytic streptococcus may cause meningitis following ear, nose, and throat infections, trauma, and surgery (47). Group B streptococcus is a common cause of neonatal meningitis (4, 32, 47); its low position in Table 20.6, where it is included as streptococcus, β-hemolytic, non-group A, probably reflects the fact that Bac-Data has only recently begun to receive group B identifications. Although *S. pneumoniae* is often considered to be the most common cause of bacterial meningitis, it appears as the sixth most frequent organism in Table 20.6.

Although *Neisseria meningitidis* is not among the tabulated CFS isolates, it should not be overlooked as a cause of meningitis. Purulent meningitis is primarily a pediatric infection now that *N. meningitidis* epidemics have disappeared from the civilian population of the United States (47).

Wounds

S. aureus is by far the most common pathogen found in wound infections (Table 20.7). The general decrease in Gram-negative bacilli in wound cultures may be due to more stringent control in the prevention of nosocomial infections. The significant increase of *H. influenzae* seen in wounds

TABLE 20.6
Trend of incidence of microbial flora from cerebrospinal fluid, 1971–1975

Species (Total Isolates Reported (77 Species))	Percentage of Total Isolates Reported					Trend
	1971 (1,238, 100%)	1972 (1,229, 100%)	1973 (1,589, 100%)	1974 (1,980, 100%)	1975 (3,338, 100%)	
Gram-negative						
Haemophilus influenzae	13.6	11.6	12.8	14.2	9.9	↓ 27%
Escherichia coli	10.0	09.1	07.6	07.3	06.9	↓ 31%
Pseudomonas sp.	05.9	05.4	04.0	03.5	03.4	↓ 42%
Klebsiella sp.	04.6	06.0	03.8	02.9	02.7	↓ 41%
Enterobacter sp.	02.3	02.9	02.5	01.8	01.9	NSC
Proteus sp.	01.7	01.6	02.3	01.4	01.4	NSC
Bacteroides sp.	00.3	00.7	00.8	00.3	01.0	NSC
Gram-positive						
Staphylococcus epidermidis	15.3	18.9	22.5	25.3	25.2	↑ 65%
Staphylococcus aureus	10.1	12.4	10.7	09.0	09.5	NSC
Streptococcus, α-hemolytic	05.7	04.3	05.2	04.0	06.2	NSC
Streptococcus pneumoniae	05.9	05.4	07.1	06.9	04.1	NSC
Streptococcus faecalis	04.2	03.6	04.5	03.6	02.7	↓ 36%
Streptococcus, β-hemolytic, non-group A	01.3	01.1	02.2	02.4	02.6	↑ 100%
Streptococcus, β-hemolytic, group A	00.5	00.7	00.4	00.5	00.3	NSC
% Gram-negative	49.0	51.0	46.0	45.0	45.0	↓ 8%
% Gram-positive	51.0	49.0	54.0	55.0	55.0	↑ 8%

TABLE 20.7
Trend of incidence of microbial flora from wounds, 1971–1975

Species (Total Isolates Reported (77 Species))	Percentage of Total Isolates Reported					Trend
	1971 (60,189, 100%)	1972 (60,976, 100%)	1973 (90,712, 100%)	1974 (123,998, 100%)	1975 (153,576, 100%)	
Gram-negative						
Escherichia coli	16.2	15.5	14.5	13.3	12.4	↓ 23%
Proteus sp.	10.4	09.6	09.0	08.5	08.2	↓ 21%
Pseudomonas sp.	09.7	08.6	08.6	08.1	07.5	↓ 23%
Klebsiella sp.	06.9	06.9	06.1	05.7	05.3	↓ 23%
Enterobacter sp.	05.8	05.9	05.4	04.9	04.8	↓ 17%
Bacteroides sp.	01.3	01.8	02.1	02.1	02.4	↑ 85%
Haemophilus influenzae	00.09	00.1	00.1	00.2	00.2	↑ 100%
Gram-positive						
Staphylococcus aureus	21.8	22.9	22.2	22.9	23.6	↑ 8%
Staphylococcus epidermidis	07.7	07.5	08.5	10.1	10.6	↑ 38%
Streptococcus faecalis	06.5	06.9	07.7	07.3	07.1	↑ 9%
Streptococcus, α-hemolytic	02.3	02.3	02.4	02.4	02.4	NSC
Streptococcus, β-hemolytic, group A	02.6	01.9	02.5	02.3	02.1	↓ 19%
Streptococcus, β-hemolytic, non-group A	01.8	01.9	01.9	02.0	02.5	↑ 39%
Streptococcus pneumoniae	00.2	00.1	00.1	00.2	00.2	NSC
% Gram-negative	55.0	54.0	52.0	48.0	46.0	↓ 16%
% Gram-positive	45.0	46.0	48.0	52.0	54.0	↑ 20%

may be due to increased awareness and improved isolation techniques. *Bacteroides* sp. are significantly increasing in incidence in wound cultures also.

ANTIBIOTIC SUSCEPTIBILITY PRECENTAGES AND TRENDS

Gram-negative

Antibiotic susceptibilities of the two most common Gram-negative organisms, *E. coli* and *Proteus mirabilis*, are given according to the body site from which they are cultured (Table 20.8). There were no significant differences in antibiotic susceptibilities for organisms isolated from different body sites (Table 20.8). Kanamycin appeared to produce more resistant organisms in wound infections, for which it has been used extensively. The 100% susceptibility which occurred with CSF isolates of both organisms is of dubious value. *E. coli* isolates were susceptible to three antimicrobial agents used specifically for urinary tract infections. CSF isolates of *P. mirabilis* account for less than 0.1% of all *P. mirabilis* isolates; therefore, the 100% susceptibility of these isolates to gentamicin and nalidixic acid is questionable because of small sampling size. *Proteus* non-*mirabilis* isolated from urine appears to be generally more drug resistant (Table 20.8). *H. influenzae* does not show a great difference in resistance according to body site (Table 20.8).

Overall, *E. coli* does not appear to be becoming more resistant (Table 20.9). It is most resistant (26%) to tetracycline and sulfa.

Proteus mirabilis susceptibility appears to be stable (Table 20.9), whereas, the non-*mirabilis* species (Table 20.9) are becoming resistant.

TABLE 20.8
Antibiotic susceptibility of microbial flora (Gram-negative) isolated from hospital patients in the United States by source of specimen, 1975

Species	No. of Isolates*	Percentage Susceptible											
		Ampicillin	Carbenicillin	Cephalosporin	Chloramphenicol	Tetracycline	Gentamicin	Colistin	Kanamycin	Sulfa	TMP-SMZ†	Nalidixic acid	Nitrofurantoin
Escherichia coli													
All sites	171,794	76	81	81	95	74	99	97	91	74	96	97	95
Blood	4,406	77	81	76	95	75	98	97	90	72	96	96	94
CSF	229	80	80	82	95	78	99	98	90	77	100	100	100
Urine	103,310	77	81	81	95	73	99	97	91	73	96	97	95
Wound	19,104	75	80	79	95	72	99	97	88	75	96	97	94
Proteus mirabilis													
All sites	48,370	91	94	91	89	4	99	2	94	84	93	92	13
Blood	782	89	92	90	88	3	99	1	93	76	91	89	14
CSF	33	82	92	90	86	10	100	—	96	66	75	100	33
Urine	24,429	90	94	91	89	3	99	3	95	84	93	92	13
Wound	9,032	90	94	91	89	4	99	2	93	85	94	91	14
Proteus (not *mirabilis*)													
All sites	10,105	21	79	17	71	37	92	5	86	57	75	84	19
Blood	177	19	73	12	79	43	92	2	83	56	—	87	27
CSF	11	—	—	—	—	—	—	—	—	—	—	—	—
Urine	5,485	21	72	17	59	27	87	5	82	50	68	79	17
Wound	1,957	19	85	16	77	44	95	4	89	69	89	95	21
Klebsiella sp.													
All sites	65,989	6	11	87	92	84	98	96	92	75	88	90	70
Blood	1,752	5	8	84	87	80	94	97	87	69	85	89	65
CSF	90	4	6	92	87	83	95	100	89	77	100	94	80
Urine	24,595	5	10	84	89	80	97	96	91	71	87	89	69
Wound	8,155	6	12	86	92	84	97	96	90	77	90	92	70
Enterobacter sp.													
All sites	40,603	16	74	21	93	82	98	92	93	80	91	91	72
Blood	749	15	72	15	90	80	98	89	91	86	—	—	66
CSF	65	16	72	22	93	87	100	86	89	—	—	—	—
Urine	11,368	15	63	27	90	77	97	90	89	76	89	—	68
Wound	7,251	15	76	16	93	82	99	92	92	82	94	—	66
Pseudomonas aeruginosa													
All sites	47,186	1	76	1	9	8	94	96	5	27	10	4	1
Blood	969	1	77	1	7	8	94	97	5	30	14	5	4
CSF	75	2	82	4	13	9	91	91	11	50	33	10	—
Urine	16,750	1	75	1	9	7	93	96	5	21	8	3	1
Wound	8,576	1	78	1	8	7	95	96	5	33	14	5	2
Haemophilus influenzae													
All sites	12,306	90	—	—	99	—	—	—	—	—	—	—	—
Blood	451	95	—	—	100	—	—	—	—	—	—	—	—
CSF	332	91	—	—	99	—	—	—	—	—	—	—	—
Urine	59	78	—	—	95	—	—	—	—	—	—	—	—
Wound	231	87	—	—	99	—	—	—	—	—	—	—	—

* No. of isolates represent total incidence. Actual isolates tested with each antibiotic ranged between 100% and 75% of the total except for nitrofurantoin, nalidixic acid, and sulfonamide. The latter were tested on 30–70% of the isolates.

† TMP-SMZ = trimethoprim-sulfamethoxazole.

TABLE 20.9
Trend of antibiotic susceptibility of microbial flora (Gram-negative) isolated from hospital patients in the United States, 1971–1975

Species	Year	No. of Isolates*	Percentage Susceptible											
			Ampicillin	Carbencillin	Cephalosporin	Chloramphenicol	Tetracycline	Gentamicin	Colistin	Kanamycin	Sulfa	TMP-SMZ†	Nalidixic acid	Nitrofurantoin
Escherichia coli	1971	112,735	73	81	82	94	71	98	92	91	68	—	95	95
	1972	97,648	74	80	81	95	72	98	92	89	72	—	95	96
	1973	119,644	77	80	83	95	72	99	94	90	73	—	96	95
	1974	147,986	78	80	82	96	73	99	96	91	74	96	97	96
	1975	171,794	76	81	81	95	74	99	97	91	74	96	97	95
Trend‡			↑ 4%	NSC	NSC	NSC	↑ 4%	NSC	↑ 5%	NSC	↑ 9%	NSC	↑ 2%	NSC
Proteus mirabilis	1971	26,629	86	92	89	83	5	98	3	93	72	—	88	25
	1972	23,258	89	93	90	85	4	98	2	92	79	—	90	21
	1973	30,663	90	94	92	86	4	99	3	94	81	—	90	14
	1974	41,786	91	94	92	89	3	99	2	95	84	93	91	14
	1975	48,370	91	94	91	89	4	99	2	94	84	93	92	13
Trend			↑ 6%	↑ 2%	↑ 2%	↑ 7%	NSC	NSC	NSC	NSC	↑ 17%	NSC	↑ 4%	↑ 48%
Proteus (not *mirabilis*)	1971	5,569	42	85	44	76	25	96	6	90	62	—	84	35
	1972	4,972	35	86	36	76	30	96	5	89	61	—	87	29
	1973	6,253	29	84	26	70	33	95	5	89	59	—	86	19
	1974	8,675	25	81	22	70	35	95	5	89	60	81	86	21
	1975	10,105	21	79	17	71	37	92	5	86	57	75	84	19
Trend			↓ 50%	↓ 7%	↓ 61%	↓ 7%	↑ 48%	↓ 4%	NSC	↓ 4%	↓ 8%	NSC	NSC	↓ 46%
Klebsiella sp.	1971	42,169	10	23	79	87	76	98	88	86	68	—	90	77
	1972	38,549	10	16	82	89	79	98	89	87	75	—	90	76
	1973	46,140	8	12	86	91	82	99	93	91	78	—	90	73
	1974	56,847	8	11	87	92	84	99	95	92	77	91	90	74
	1975	65,989	6	11	87	92	84	98	96	92	75	88	90	70
Trend			↓ 40%	↓ 52%	↑ 10%	↑ 6%	↑ 11%	NSC	↑ 9%	↑ 7%	↑ 10%	NSC	NSC	↓ 9%
Enterobacter sp.	1971	27,486	19	64	36	89	74	98	85	90	70	—	90	74
	1972	24,447	19	66	31	91	78	98	84	90	74	—	90	72
	1973	28,365	18	70	25	92	80	98	87	92	80	—	92	69
	1974	34,419	18	71	22	93	81	98	89	93	80	95	92	71
	1975	40,403	17	72	22	93	81	98	92	92	80	92	91	68
Trend			NSC	↑ 13%	↓ 39%	↑ 4%	↑ 9%	NSC	↑ 8%	NSC	↑ 14%	NSC	NSC	↓ 8%

TABLE 20.9 continued

Species	Year	No. of Isolates*	Percentage Susceptible											
			Ampicillin	Carbencillin	Cephalosporin	Chloramphenicol	Tetracycline	Gentamicin	Colistin	Kanamycin	Sulfa	TMP-SMZ†	Nalidixic acid	Nitrofurantoin
Pseudomonas aeruginosa	1971	11,109	1	70	1	15	13	95	95	8	43	—	8	1
	1972	12,886	1	73	1	14	12	95	95	6	43	—	4	1
	1973	25,321	1	69	1	11	8	95	95	5	38	—	4	1
	1974	36,207	1	71	1	11	9	94	96	5	32	8	4	2
	1975	47,186	1	76	1	9	8	94	96	5	27	10	4	1
Trend			NSC	↑ 9%	NSC	↓ 40%	↓ 38%	NSC	NSC	NSC	↓ 37%	NSC	NSC	NSC
Haemophilus influenzae	1971	3,947	90	—	—	99	—	—	—	—	—	—	—	—
	1972	4,976	89			99	—	—	—	—	—	—	—	—
	1973	8,729	88	—	—	99	—	—	—	—	—	—	—	—
	1974	11,258	89	—	—	94	—	—	—	—	—	—	—	—
	1975	12,306	90	—	—	99	—	—	—	—	—	—	—	—
Trend			NSC	—	—	NSC	—	—	—	—	—	—	—	—

* No. of isolates represent total incidence. Actual isolates tested with each antibiotic ranged between 100% and 75% of the total except for nitrofurantoin, nalidixic acid, and sulfonamide. The latter were tested on 30% to 70% of the isolates.

† TMP-SMZ = trimethoprim-sulfamethoxazole.

‡ Trend indicators: ↑ 00% = percent increase in susceptibility 1971–1975, ↓ 00% = percent decrease in susceptibility 1971–1975, NSC = no significant change 1971–1975.

Resistance to gentamicin and kanamycin may be due to extensive use of these drugs (73). Awareness that treatment with oral tetracyclines results in fecal excretion of resistant bacteria may have led to decreased use of the drug (35).

Klebsiella sp. resistance to ampicillin and carbenicillin continues to increase (Table 20.9). The increasing incidence of indole-positive *Klebsiella*, sp. which is known to have a high antibiotic sensitivity, could account for the increased susceptibility of *Klebsiella* sp. to other antibiotics (45).

Enterobacter sp. is becoming increasingly resistant to cephalosporin (Table 20.9).

An increase in serious *Pseudomonas* sp. infections has been reported (12), although our survey does not corroborate such an increase (Table 20.9). During this 5-year study period, there has been no significant change in *Pseudomonas aeruginosa* susceptibility to colistin and gentamicin, the drugs of choice. This finding stands in contrast to reports that gentamicin resistance emerges in association with hospital stay (37), length of stay (40), and that cross resistance among kanamycin, gentamicin, and tobramycin has developed. The larger percent of resistant strains which is not susceptible to gentamicin are often susceptible to tobramycin (40).

Since 1963 ampicillin has been regarded as the drug of choice for treating *Haemophilus* meningitis in children. However, within the past 2 years, ampicillin-resistant strains have been isolated from critically ill patients (18, 42, 78). This has made rapid and accurate sensitivity tests for *H. influenzae* mandatory (80) and has led some to use chloramphenicol for initial therapy (58).

The problem of ampicillin resistance is widespread (41); however, the data show no change in *H. influenzae* sensitivity to ampicillin or to chloramphenicol (Table 20.9). Because of the fastidious growth requirements, the conditions for making antibiograms of *Haemophilus* are poorly standardized. The 1-min β-lactamase test which can rapidly detect ampicillin-resistant forms of *H. influenzae* (24, 79) may be more reliable.

Overall, resistance to nitrofurantoin is increasing in all organisms except *E. coli*. Cephalosporin resistance is appearing in *Proteus* non-*mirabilis* and *Enterobacter* sp. A marked increase in resistance to ampicillin is occurring in *Klebsiella* sp. and *Proteus* non-*mirabilis*.

Gentamicin susceptibility is unchanged (52, 85) and tetracycline shows increased susceptibility, perhaps due to controlled use. Susceptibility to sulfa seems to be increasing for all Gram-negative organisms except *Proteus* non-*mirabilis*, the only organism showing multiple increased resistance.

Gram-positive

A comparison of antibiotic susceptibilities shows that no significant differences exist among isolates from differing body sites (Table 20.10). Some susceptibilities are of questionable value. For example, nitrofurantoin is used only for urinary tract infections; its 100% effectiveness against CSF isolates is clinically meaningless. Since cephalosporin and clindamycin (71) do not cross the blood-brain barrier well (59), drug susceptibilities

TABLE 20.10
Antibiotic susceptibility of microbial flora (Gram-positive) isolated from hospital patients in the united states by source of specimen, 1975

Species	No. of Isolates*	Percentage Susceptible											
		Penicillin	Nafcillin/Oxacillin	Ampicillin	Cephalosporin	Chloramphenicol	Tetracycline	Clindamycin/lincomycin	Erythromycin	Gentamicin	Kanamycin	Streptomycin	Nitrofurantoin
Staphylococcus aureus													
All sites	102,487	17	95	19	99	98	85	96	93	99	95	89	95
Blood	3,376	19	95	21	99	98	88	95	94	98	94	89	99
CSF	318	25	93	25	99	96	82	93	94	100	96	94	100
Urine	6,012	27	88	31	96	95	70	90	87	97	91	84	92
Wound	36,299	16	95	18	99	98	86	96	93	99	96	88	96
Staphylococcus epidermidis													
All sites	62,610	32	81	37	97	93	54	84	76	98	80	77	94
Blood	5,517	29	82	33	98	92	58	82	74	98	77	71	94
CSF	841	43	82	48	98	93	65	88	83	98	83	87	100
Urine	17,449	37	80	47	96	94	49	86	77	97	83	80	92
Wound	16,262	30	79	34	96	91	56	80	74	98	77	74	96
Streptococcus faecalis													
All sites	54,329	40	6	94	34	90	23	9	63	59	19	4	92
Blood	1,051	48	10	94	40	90	31	16	65	61	16	3	90
CSF	90	48	3	91	41	91	43	13	66	71	18	6	100
Urine	30,659	39	5	95	33	90	20	7	64	57	18	4	92
Wound	10,897	38	4	93	33	90	26	9	58	64	18	3	90
Streptococcus β-hemolytic, non-group A													
All sites	19,462	93	84	93	97	98	47	93	96	38	15	9	91
Blood	279	92	84	94	96	95	43	92	94	45	9	9	90
CSF	44	94	91	96	100	100	29	91	94	39	7	6	100
Urine	3,486	91	86	92	96	98	32	91	94	31	14	7	93
Wound	3,042	93	87	93	98	98	54	93	95	46	17	9	85

* No. of isolates represent total incidence. Actual isolates tested for an antibiotic ranged between 100% and 75% except for nitrofurantoin and nafcillin/oxacillin. The latter were tested on 30% to 70% of the isolates.

of CSF isolates to these drugs have no practical significance. Furthermore, the number of CSF isolates is less than 1% of the total isolates for all organisms tested except *S. epidermidis* for which CSF isolates were 1.3% of the total.

S. faecalis is commonly found in urine and blood cultures. It is generally resistant to antibiotics (Table 20.10). All the β-hemolytic streptococcus, not group A, are susceptible to penicillin.

The high incidence of resistance of *S. aureus* to penicillin G—usually 85–95%—has been known for many years (71). *S. epidermidis* has become even more antibiotic resistant than *S. aureus* (11, 51, 87). Both organisms show increased resistance to penicillin G and ampicillin (55), and *S. epidermidis* also shows increasing resistance to erythromycin and kanamycin (Table 20.11*A*).

Oxacillin susceptibility appears to be increasing for *S. aureus*. This may be due to errors in speciation (9), since *S. epidermidis* is more resistant (54). Also, oxacillin discs used in the Bauer-Kirby test are more stable than

TABLE 20.11
Trend of antibiotic susceptibility of microbial flora (Gram-positive) isolated from hospital patients in the United States, 1971–1975

Species	Year	No. of Isolates*	Percentage Susceptible											
			Penicillin	Nafcillin/oxacillin	Ampicillin	Cephalosporin	Chloramphenicol	Tetracycline	Clindamycin/lincomycin	Erythromycin	Gentamicin	Kanamycin	Streptomycin	Nitrofurantoin
Staphylococcus aureus	1971	50,672	30	91	33	98	97	81	94	92	97	95	82	92
	1972	47,528	26	92	30	98	98	83	95	92	97	94	86	95
	1973	63,411	22	95	24	99	98	85	96	93	98	96	88	95
	1974	82,491	20	94	21	99	98	85	95	93	99	96	90	94
	1975	102,487	17	95	19	99	98	85	96	93	99	95	89	95
Trend†			↓ 43%	↑ 4%	↓ 42%	NSC	NSC	↑ 5%	NSC	NSC	NSC	NSC	↑ 9%	NSC
Staphylococcus epidermidis	1971	28,871	45	83	52	96	93	52	86	85	96	90	79	93
	1972	25,539	41	84	49	97	94	50	86	84	96	88	81	95
	1973	32,423	38	82	43	97	94	52	86	81	97	87	79	94
	1974	49,897	36	82	41	97	94	53	84	78	98	84	80	94
	1975	62,610	32	81	37	97	93	54	84	76	98	80	77	94
Trend			↓ 29%	NSC	↓ 29%	NSC	NSC	↑ 4%	NSC	↓ 11%	NSC	↓ 11%	NSC	NSC
Streptococcus faecalis	1971	29,487	59	15	90	63	91	29	13	77	60	32	10	91
	1972	28,830	52	10	89	55	92	26	11	73	54	22	7	91
	1973	40,300	49	9	93	47	91	24	11	70	54	19	6	91
	1974	49,139	39	6	93	37	90	24	9	66	60	19	5	91
	1975	54,329	40	6	94	34	90	23	9	63	59	19	4	92
Trend			↓ 32%	↓ 60%	↑ 4%	↓ 46%	NSC	↓ 21%	↓ 31%	↓ 18%	NSC	↓ 41%	↓ 60%	NSC
Streptococcus β-hemolytic, non-group A	1971	10,616	91	77	94	97	97	57	91	96	57	24	16	92
	1972	9,456	88	76	90	97	97	52	91	96	48	18	14	95
	1973	13,204	90	79	92	97	97	46	91	95	44	15	13	93
	1974	17,717	92	82	93	97	97	46	92	96	39	15	12	94
	1975	19,462	93	84	93	97	98	47	93	96	38	15	9	91
Trend			↑ 2%	↑ 9%	NSC	NSC	NSC	↓ 18%	↑ 2%	NSC	↓ 33%	↓ 37%	↓ 44%	NSC

* No. of isolates represent total incidence. Actual isolates tested for an antibiotic ranged between 100% and 75% except for nitrofurantoin and nafcillin/oxacillin. The latter were tested on 30% to 70% of the isolates.

† Trend indicators: ↑ 00% = percent increase in susceptibility 1971–1975, ↓ 00% = percent decrease in susceptibility 1971–1975, and NSC = no significant change 1971–1975.

the methicillin discs used previously (72). *S. aureus* susceptibility to tetracycline and streptomycin has increased (Table 20.11).

S. faecalis is becoming more resistant to most of the antibiotics (Table 20.11), especially to penicillin G, oxacillin, cephalosporin, clindamycin (62), kanamycin, and streptomycin (53, 62).

β-Hemolytic streptococcus, non group A, is not changing in drug susceptibility pattern. Its sensitivity to penicillin G, the agent most commonly used for streptococcus infections, has not deteriorated (Table 20.11). In fact, susceptibility has remained unchanged or increased slightly for ampicillin and cephalothin.

Resistance to tetracyclines was first noticed for group A in 1963 and the group B streptococci have shown similar increases in resistance (31). Up to 85% of group B streptococci are resistant to tetracycline and up to 100% are resistant to kanamycin (2). In this survey, large increases in resistance occurred only with those antibiotics which were already ineffective.

REMARKS

The microbiologist is faced with hard decisions in the areas of collection, processing, reporting, and interpretation of data. He must distinguish between pathogens, normal flora, and contaminants. The effective microbiologist knows what organisms to expect, what is unexpected, and what organisms are opportunistic in the immunosuppressed patient. The data included in this chapter will help both epidemiologist and clinician to evaluate data and establish guidelines for improved patient care.

From this survey we conclude:

1. There is an overall reversal in incidence. Gram-positive organisms constitute a larger percentage increase of all isolates, and the percentage of Gram-negative organisms is declining. Overall, the organisms most frequently isolated during 1975 were *E. coli* (20.1%), *S. aureus* (12.0%), *Proteus* sp. (8.0%), *Klebsiella* sp. (7.7%), and *S. epidermidis* (7.3%).

2. The most important incidence increases in Gram-positive organisms occurred with *S. aureus, S. epidermidis,* and *Candida albicans*.

3. The incidences of the most common Enterobacteriaceae, *E. coli, Proteus* sp., *Klebsiella* sp., and *Enterobacter* sp., are decreasing. However, some less frequent Gram-negative organisms have been appearing more often. These include *H. influenzae, Haemophilius* sp., *Citrobacter* sp., and *Serratia* sp.

4. Fifty-two percent of all pediatric isolates are Gram-positive. Gram-negative bacteria are the most frequent in the "total" patient group. Pediatric patients have a much higher incidence of *H. influenzae*, all staphylococci, and all streptococci except *S. faecalis*.

5. Incidence trends do not vary geographically.

6. Isolation frequency varied with the body site of the infection. *S. aureus, E. coli,* and *S. epidermidis* were most frequently isolated from blood and wounds. *H. influenzae* replaced *E. coli* as the second most frequent isolate for CSF. Urine isolates were: *E. coli, S. faecalis,* and *Klebsiella* sp.

7. Gram-positive organisms are most sensitive to penicillin G.

8. Gram-positive organisms are most resistant to tetracycline and streptomycin.

9. *Proteus* non-*mirabilis* is becoming more resistant; notably, cephalosporin, ampicillin and nitrofurantoin are less active.

10. There is a general increase of Gram-negative organisms resistance to nitrofurantoin.

11. *Klebsiella* shows a marked increase in resistance to ampicillin.

12. Because there is differential drug susceptibilities between the two groups, speciation of group D into enterococcal and non-enterococcal streptococci is important.

LITERATURE CITED

1. Adler, J. L., Burke, J. P., Marlin, D. F., et al. Proteus infections in a general hospital. Some clinical and epidemiological charactersitics with an analysis of 71 cases of Proteus bacteremia. *Ann Intern Med 75:*531–536, 1971.
2. Anthony, B. F., and Conception, N. F. Group B Streptococcus in a general hospital. *J Infect Dis 132:*561–567, 1975.
3. Bailey, R. P. Significance of coagulase-negative staphylococci in urine. *J Infect Dis 127:*179–182, 1973.
4. Baker, C. J., and Barrett, F. F. Group B streptococcal infections in infants. *JAMA 230:*1158–1160, 1974.
5. Barry, A. L., Smith, P. B., and Turck, M. *Laboratory Diagnosis of Urinary Tract Infections. Cumitech 2.* Washington, D.C.: American Society for Microbiology, 1975.
6. Bartlett, J. C., Ellner, P. D., and Washington, J. A., II. *Blood cultures. Cumitech 1.* Washington, D.C.: American Society for Microbiology, 1974.
7. Baross, T., and Liston, J. Occurrence of *Vibrio parahaemolyticus* and related Vibrios in marine environment of Washington State. *Appl Microbiol 20:*179–186, 1970.
8. Bassan, R. Bacterial endocarditis produced by *Listeria monocytogenes. Am J Clin Pathol 63:*522–527, 1975.
9. Bauer, A. W., Kirby, W. M., Sherris, J. C., and Turck, M. Antibiotic susceptibility testing by a standardized disk method. *Am J Clin Pathol 45:*493–496, 1966.
10. Bennett, J. V. Nosocomial infections due to Pseudomonas. *J Infect Dis 130*(Suppl):S1-S166, 1974.
11. Bentley, D. W., Hahn, J. J., and Lepper, M. H. Transmission of chloramphenical-resistant *Staphylococcus epidermidis*; epidemiologic laboratory studies. *J Infect Dis 122:*365–375, 1967.
12. Blair, D. C., Fekety, F. R., Jr., Bruce, B., Silva, J., and Archer, G. Therapy of *Pseudomonas aeruginosa* infections with tobramycin. *Antimicrob Agents Chemother 8:*22–29, 1975.
13. Blazevic, D. J., Stemper, J. E., and Matsen, J. M. Comparison of macroscopic examination, routine gram stain, and routine subcultures in the initial detection of positive blood cultures. *Appl Microbiol 27:*537–539, 1974.
14. Blazevic, D. J., Stemper J. E., and Matsen, J. M. Organisms encountered in urine cultures over a 10-year period. *Appl Microbiol 23:*421–422, 1972.
15. Bulger, R. J., and Sherris, J. C. The clinical significance of *Aeromonas hydrophila*; report of two significant cases. *Arch Intern Med 118:*562–654, 1948.
16. Center for Disease Control. *Morbid Mortal Weekly Rep 24:* No. 45, Nov. 8, 1975.
17. Center for Disease Control. Septicemias associated with contaminated intravenous fluids. *Morbid Mortal Weekly Rep 22:*99, 1973.
18. Center for Disease Control. *Morbid Mortal Weekly Rep 24:*205–206, 1975.
19. Dixon, R. E., Kaslow, R. A., Mallison, G. F., and Bennett, J. V. Staphylococcal disease outbreaks in hospital nurseries in the United States—December 1971 through March 1972. *Pediatrics 51:*413–416, 1973.
20. Doggett, R. G., Harrison, G. M., and Wallis, E. S. Comparison of Some Properties of *Pseudomonas aeruginosa* isolated from infections in persons with and without cystic fibrosis. *J Bacteriol 87:*427–431, 1964.
21. Dowell, V. R., Jr., and Hawkins, T. M. Laboratory methods in anaerobic bacteriology. *CDC Laboratory Manual.* Atlanta: Center for Disease Control, 1973.

22. Edwards, P. R., and Ewing, W. H. *Identification of Enterobacteriaceae,* Ed. 3. Minneapolis: Burgess Publishing Co., 1972.
23. Eickhoff, T. C., Klein, J. O., Daly, A. K., Ingall, D., and Finland, M. Neonatal sepsis and other infections due to group B β-hemolytic streptococci. *N Engl J Med 271:*1221-1228, 1964.
24. Escamilla, J. Susceptibility of *Haemophilus influenzae* to ampicillin as determined by use of a modified, one-minute β-lactamase test. *Antimicrob Agents Chemother 9:*196-198, 1976.
25. Ewing, W. H., and Davis, B. R. Biochemical, characterization of *Citrobacter diversus* (Burkey) Werkman and Gillen and designation of the neotype strain. *Int J Syst Bacteriol 22:*12-18, 1972.
26. Ewing, W. H., and Fife, M. A. *Enterobacter agglomerans* (Beijerinck) comb. Nov. (the Herbicola-Lathyri bacteria). *Int J Syst Bacteriol 22:*4-11, 1972.
27. Ewing, W. H., and Fife, M. A. *Enterobacter agglomerans,* the herbicola-lathyri bacteria. Atlanta: Center for Disease Control, 1971.
28. Ewing, W. H., and Fife, M. A. Biochemical characterization of *Enterobacter agglomerans*. Atlanta: Center for Disease Control, 1972.
29. Ewing, W. H., Johnson, J. G., and Davis, B. R. The occurrence of *Seratia marcescens* in noscomial infections. Atlanta: Center for Disease Control, 1962.
30. Finland, M. Excursions into epidemiology; selected studies during the past four decades at Boston City Hospital. *J Infect Dis 128:*76-120, 1973.
31. Finland, M., Garner, C., Wilcox, C., and Sabath, L. D. Susceptibility of β-hemolytic streptococci to 65 antibacterial agents. *Antimicrob Agents Chemother 9:*11-19, 1976.
32. Franciosi, R. A., Knostman, J. D., and Zimmerman, R. A. Group B streptococcal neonatal and infant infections. *J. Pediatr 82:*707-718, 1973.
33. Gilardi, G. L., Battone, E., and Birnbaum, M. Unusual fermentative, gram-negative bacilli isolated from clinical specimens. II. Characterization of *Aeromonas* sp. *Appl Microbiol 20:*156-159, 1970.
34. Hill, H. R., Riter, M. E., Menge, S. K., Johnson, D. K., and Matsen, J. M. Rapid identification of group B streptococci by counterimmunoelectrophoresis. *J Clin Microbiol 1:*188-191, 1975.
35. Hirsch, D. C., Burton, G. C., and Blendon, D. C. Effect of oral tetracycline on the occurrence of tetracycline-resistant strains of *Escherichia coli* in the intestinal tract of humans. *Antimicrob Agents Chemother 4:*69-71, 1973.
36. Holdeman, L. V., and Moore, W. E. C. *Anaerobic Laboratory Manual*. Blacksburg, Va.: Virginia Polytechnic Institute and State University, 1972.
37. Holmes, R. K., Minshaw, B. N., Gould, K., and Sanford, J. P. Resistance of *Pseudomonas aeruginosa* to gentamicin and related aminoglycoside antibiotics. *Antimicrob Agents Chemother 6:*253-262, 1974.
38. Holt, R. The classification of staphylococci from colonized ventriculoatrial shunts. *J Clin Pathol 22:*475-482, 1969.
39. Horn, K. A., Meyer, W. T., Wyrich, B. C., and Zimmerman, R. A. Group B streptococcal neonatal infection. *JAMA 230:*1165-1167, 1974.
40. Jackson, G. G., and Riff, L. J. Pseudomonas bacteremia; pharmacologic and other bases for failure of treatment with gentamicin. *J Infect Dis 124*(Suppl):S185-S191, 1971.
41. Jorgensen, J. H., and Jones, P. M. Simplified medium for ampicillin susceptibility testing of *Haemophilus influenzae*. *Antimicrob Agents Chemother 7:*186-190, 1975.
42. Kammer, R. B., Preston, D. A., Turner, J. B., and Hawley, L. C. Rapid detection of ampicillin-resistant *Haemophilus influenzae* and their susceptibility to sixteen antibiotics. *Antimicrob Agents Chemother 8:*91-94, 1975.
43. Kaslow, R. A., Dizon, R. E., Martin, S. M., Mallison, G. E., Goldmann, D. A., Linsey, J. D. II, Rhame, F. S., and Bennett, J. V. Staphylococcal disease related to hospital nursery bathing practices—a nationwide investigation. *Pediatrics 51:*418-427, 1973.
44. Kaye, D. Changes in the spectrum, diagnosis and management of bacterial and fungal endocarditis. *Med Clin North Am 57:*941-956, 1973.
45. Klein, D., Spindler, J. A., and Matsen, J. M. Relationship of indole production and antibiotic susceptibility in the *Klebsiella* bacillus. *J. Clin Microbiol 2:*425-429, 1975.
46. Lennette, E. H., Spaudling, E. H., and Truant, J. P. *Manual of Clinical Microbiology,* Ed. 2. Washington, D.C.: American Society for Microbiology, 1974.
47. Lerner, P. I. Meningitis caused by *Streptococcus* in adults. *J. Infect Dis 131*(Suppl):S9-S13, 1975.
48. Light, I. J., Atherton, H. D., and Sutherland, J. M. Decreased colonization of newborn

infants with *Staphylococcus aureus* 80/81; Cincinnati General Hospital, 1960–1972. *J Infect Dis 131:*281–285, 1975.

49. Mackel, D. C., Maki, D. G., Anderson, R. L., Rhams, F. S., and Bennett, J. V. Nationwide epidemic of septicemia caused by contaminated intravenous products; mechanisms of intrinsic contamination. J Clin Microbiol 2:486–497, 1975.
50. Maki, D. G., Hennekens, C. G., Phillips, C. W., Shaw, W. V., and Bennett, J. V. Nosocomial urinary tract infection with *Serratia marcescens*; an epidemiologic study. *J Infect Dis 128:*579–587, 1973.
51. Males, B. M., Roger, W. A., and Parisi, J. T. Virulence factors of biotypes of *Staphylococcus epidermidis* from clinical specimens. *J. Clin Microbiol 1:*256–261, 1975.
52. Maliwan, N., Grietele, H. G., and Bird, J. J. Hospital *Pseudomonas aeruginosa*; surveillance of resistance to gentamicin and transfer of aminoglycoside R factor. *Antimicrob Agents Chemother 8:*415–420, 1975.
53. Marier, R. L., Joyce, N., and Androiot, V. T. Synergism of oxacillin and gentamicin against enterococci. *Antimicrob Agents Chemother 8:*571–573, 1975.
54. Marcoux, J. A., and Washington, J. A., II. Pitfalls in Identification of Methicillin-Resistant *Staphylococcus aureus*. *Appl Microbiol 18:*699–700, 1969.
55. Marsik, F. J., and Parisi, J. T. Significance of *Staphylococcus epidermidis* in the clinical laboratory. *Appl Microbiol 25:*11–14, 1973.
56. McGowan, J. E., Jr., Barnes, M. W., and Finland, M. Bacteremia at Boston City Hospital; occurrence and mortality during 12 selected years (1935–1972) with special reference to hospital-acquired cases. *J Infect Dis 132:*316–335, 1975.
57. McGowan, J. E., Jr., Klein, J. O., Bralton, L., Barnes, M. W., and Finland, M. Meningitis and bacteremia due to *Haemophilus influenza*; occurrence and mortality at Boston City Hospital in 12 selected years, 1935–1972. *J Infect Dis 130:*119–124, 1974.
58. McGowan, J. E., Jr., Terry, P. M., and Nahmias, A. J. Susceptibility of *Haemophilus influenzae* isolates from blood and cerebrospinal fluid to ampicillin, chloramphenicol and trimethoprim-sulfamethoxazale. *Antimicrob Agents Chemother 9:*137–139, 1976.
59. McHenry, M. C., and Gaven, T. L. *Selection and Use of Antibacterial Drugs, Progress in Clinical Pathology*. New York: Grune & Stratton, Inc., 1975.
60. Mendo, K. B., and Gorbach, S. L. Favorable experience with bacterial endocarditis in heroin addicts. *Ann Intern Med 78:*25–32, 1973.
61. Michaels, R. H. Increase in influenzal meningitis. *N Engl J Med 285:*666–667, 1971.
62. Moellering, R. C., Watson, B. K., and Kunz, L. J. Endocarditis due to group D streptococci; comparison of disease caused by *Streptococcus bovis* with that produced by enterococci. *Am J Med 57:*239–250, 1974.
63. Nelson, J. D., and Koontz, W. C. Septic arthritis in infants and children; a review of 117 cases. *Pediatrics 38:*966–971, 1966.
64. Nygaard, G. S. Less frequently reported gram-negative fermentative rods. *Am J Med Technol 42.3:*79–91, 1976.
65. Pittman, J. A. Case report—subacute bacterial endocarditis due to *Acinobacillus actinomycetemcomitans*. *Lab Med 7:*29–31, 1976.
66. Porres, J. M. Miscellaneous gram negative bacilli; key to their identification. *Am J Med Technol 39:*402–416, 1973.
67. Rabson, A. R., Hallett, A. F., and Korrnhof, H. J. Generalized *Yersinia enterocolitica* infection. *J Infect Dis 131:*477–451, 1975.
68. Rapkin, R. H., and Bautista, G. *Hemophilus influenzae* cellulitis. *Am J Dis Child 124:*540–542, 1972.
69. Reyes, M. P., Palutke, W. A., Wylin, R. F., and Lerner, A. M. Pseudomonas endocarditis in the Detroit Medical Center, 1969–1972. *Medicine 52:*173–194, 1973.
70. Robbins, J. B., Schneerson, R., Argaman, M., and Hanzel, Z. T. *Haemophilus influenza* type B; disease and immunity in humans. *Ann Intern Med 78:*259–269, 1973.
71. Ross, S., Rodriguez, W., Controni, G., and Khan, W. Staphylococcal susceptibility to penicillin G. *JAMA 229:*1075–1077, 1974.
72. Sabath, L. D., Barrett, F. F., Wilcox, C., Gerstein, D. A., and Finland, M. Methicillin resistance of *Staphylococcus aureus* and *Staphylococcus epidermidis*. *Antimicrob Agents Chemother 1968:*302–306, 1969.
73. Sharp, P. M., Saenz, C. A., and Martin, R. R. Amikacin (BB-K8) treatment of multiple-drug resistant *Proteus* infections. *Antimicrob Agents Chemother 5:*435–438, 1974.
74. Sogaard, H., Zimmermann-Nielson, C., and Siboni, K. Antibiotic resistant gram-negative bacilli in a urological ward for male patients during a nine-year period; relationship to antibiotic consumption. *J Infect Dis 130:*646–650, 1974.

75. Sonnenwirth, A. C., and Weaver, R. E. *Yersinia enterocolotica. N Engl J Med 283:*1468, 1970.
76. Speller, D. C. E., and Mitchell, R. G. Coagulase negative staphylococci causing endocarditis after cardiac surgery. *J Clin Pathol 26:*517–522, 1973.
77. Sutter, V. L., Vargo, V. L., and Finegold, S. M. *Wadsworth Bacteriology Manual.* Los Angeles: Department of Continuing Education in Health Sciences University Extension, UCLA, 1975.
78. Thorne, G. M., and Farrar, W. E., Jr. Transfer of ampicillin resistance between strains of *Haemophilus influenzae* type B. *J Infect Dis 132:*276–281, 1975.
79. Thornsberry, C., and Kirven, L. A. Ampicillin resistance in *Haemophilus influenzae* as determined by a rapid test for β-lactamase production. *Antimicrob Agents Chemother 6:*653–654, 1974.
80. Thornsberry, C., and Kirven, L. A. Antimicrobial susceptibility of *Haemophilus influenzae. Antimicrob Agents Chemother 6:*620–624, 1974.
81. Toma, S., and Laflear, L. Survey on the incidence of *Yersinia enterocolitica* in Canada. *Appl Microbiol 28:*469–473, 1974.
82. Tuazon, C. U., Cardella, T. A., and Sheagren, J. N. Staphylococcal endocarditis in drug users. *Arch Intern Med 135:*155–1561, 1975.
83. Tuazon, C. U., Perez, A., Kishaba, T., and Sheagreen, J. N., *Staphylococcus aureus* among insulin-injecting diabetic patients; an increased carrier rate. *JAMA 231:*1273, 1975.
84. Weinstein, L., Type B *Haemophilus influenzae* infections in adults. *N Engl J Med 282:*221–222, 1970.
85. Weinstein, M. J., Drube, C. G., Moss, E. U. Jr., and Waitz, J. A. Microbiologic studies related to bacterial resistance to gentamicin. *J Infect Dis 124*(Suppl):S11–S17, 1971.
86. Wilfirt, J. N., Barrett, F. F., Ewing, W. H., Finland, M., and Kass, E. H. *Serratia marcescens*; biochemical, serological and epidemiological characteristics and antibiotic susceptibility of strains isolated at Boston City Hospital. *Appl Microbiol 19:*345–352, 1970.
87. Wise, R. I. Modern management of severe staphylococcal disease. *Medicine 52:*295–304, 1973.

Index